ADAPTIVE, DYNAMIC, AND RESILIENT SYSTEMS

MOBILE SERVICES AND SYSTEMS SERIES
Series Editor: Paolo Bellavista

Adaptive, Dynamic, and Resilient Systems,
edited by Niranjan Suri and Giacomo Cabri
(2014)

Handbook of Mobile Systems Applications and Services,
edited by Anup Kumar and Bin Xie
(2012)

ADAPTIVE, DYNAMIC, AND RESILIENT SYSTEMS

Edited by
Niranjan Suri
Giacomo Cabri

CRC Press
Taylor & Francis Group
Boca Raton   London   New York

CRC Press is an imprint of the
Taylor & Francis Group, an **informa** business
AN AUERBACH BOOK

CRC Press
Taylor & Francis Group
6000 Broken Sound Parkway NW, Suite 300
Boca Raton, FL 33487-2742

© 2014 by Taylor & Francis Group, LLC
CRC Press is an imprint of Taylor & Francis Group, an Informa business

No claim to original U.S. Government works

Printed on acid-free paper
Version Date: 20140501

International Standard Book Number-13: 978-1-4398-6848-5 (Hardback)

Library of Congress Cataloging-in-Publication Data

Adaptive, dynamic, and resilient systems / editors, Niranjan Suri, Giacomo Cabri.
    pages cm -- (Mobile services and systems)
    Includes bibliographical references and index.
    ISBN 978-1-4398-6848-5 (hardback)
    1. Adaptive computing systems. I. Suri, Niranjan, 1972- II. Cabri, Giacomo.

QA76.9.A3A34 2014
004--dc23   2014016457

Visit the Taylor & Francis Web site at
http://www.taylorandfrancis.com

and the CRC Press Web site at
http://www.crcpress.com

Contents

Preface

The origins of this book lie in our combined interest in adaptive systems over the years. Both of us began our research careers with software agent systems, where attributes such as autonomy, learning, self-awareness, and self-adaptation were important. Research into this area expanded into distributed systems, where many of the same concepts proved to be fundamental to building systems that were resilient. Conversations with several like-minded colleagues and researchers led to the genesis of this book. Distributed computing systems now permeate our life and contribute in many ways to our society functioning smoothly. As these systems become increasingly complex and interconnected, they become less understandable, predictable, and controllable. We worry about minor outages or problems quickly snowballing into much larger cascading failures given all of the interdependencies in the systems, thereby resulting in an overall negative impact on society. Complexity and interconnectedness of distributed systems is likely to continue increasing. The purpose of this book is to explore the fundamental building blocks that will support building these distributed systems to be adaptive and resilient, as opposed to being brittle.

We would like to thank all of the authors of the chapters, without whose contributions this work would not have been possible. And a special thanks to John Wyzalek from Taylor and Francis, who gave us the opportunity to publish this book and helped us during the many phases of this project. We look forward to this book starting a discussion among the academic and industrial community centered around this challenging topic.

Finally, we would also like to thank our respective families, who have been very supportive during this endeavor.

Niranjan Suri
Pensacola, FL, USA

Giacomo Cabri
Modena, Italy

Editors

Niranjan Suri is a research scientist at the Florida Institute for Human & Machine Cognition (IHMC) and also a visiting scientist at the U.S. Army Research Laboratory, Adelphi, Maryland. He received his PhD in computer science from Lancaster University, England, and his MSc and BSc in computer science from the University of West Florida, Pensacola. His current research activity is focused on the notion of agile computing, which supports the opportunistic discovery and exploitation of resources in highly dynamic networked environments. His other research interests include coordination algorithms, distributed systems, networking, communication protocols, virtual machines, and software agents.

Giacomo Cabri is an associate professor at the Università degli Studi di Modena e Reggio Emilia, in Italy. He received his PhD in Information Engineering from the Università di Modena e Reggio Emilia and his Laurea degree in Computer Engineering from the Università di Bologna. His current research activities are mainly related to the areas of software agents, both base models and coordination protocols, and autonomic computing; other research interests include Web applications, mobile computing, and programming languages. In these areas, he has published over 140 papers in journals and conferences and has received 6 best paper awards.

Contributors

Tarek Abdelzaher
University of Illinois at Urbana
    Champaign
Urbana, Illinois, USA

Marco Aiello
University of Groningen
Groningen, The Netherlands

Mauro Andreolini
University of Modena and
    Reggio Emilia
Modena, Italy

Rami Bahsoon
University of Birmingham
Birmingham, United Kingdom

Peter Bartalos
Astek Sud Est
Sophia Antipolis, France

Walter Binder
University of Lugano
Lugano, Switzerland

M. Brian Blake
University of Miami
Coral Gables, Florida, USA

Jeffrey M. Bradshaw
Florida Institute for Human and
    Machine Cognition
Pensacola, Florida, USA

Dídac Busquets
Imperial College London
London, United Kingdom

Giacomo Cabri
University of Modena and
    Reggio Emilia
Modena, Italy

Sara Casolari
University of Modena and
    Reggio Emilia
Modena, Italy

Giovanna Di Marzo Serugendo
University of Geneva
Carouge, Switzerland

Ando Emerencia
University of Groningen
Groningen, The Netherlands

Funmilade Faniyi
University of Birmingham
Birmingham, United Kingdom

Jose Luis Fernandez-Marquez
University of Geneva
Carouge, Switzerland

Marc-Philippe Huget
University of Savoy
Annecy le Vieux, France

Alexander Kott
U.S. Army Research Laboratory
Adelphi, Maryland, USA

Roberto Lazzarini
Carpigiani Group
Anzola dell'Emilia, Italy

Alexander H. Levis
George Mason University
Fairfax, Virginia, USA

Peter R. Lewis
Aston University
Birmingham, United Kingdom

Rebecca Montanari
University of Bologna
Bologna, Italy

Mark Pflanz
Booz Allen Hamilton
Falls Church, Virginia, USA

Marcello Pietri
University of Modena and Reggio Emilia
Modena, Italy

Jeremy Pitt
Imperial College London
London, United Kingdom

Matteo Risoldi
University of Luxembourg
Luxembourg City, Luxembourg

David Sanderson
Imperial College London
Birmingham, United Kingdom

Andrew C. Scott
Lancaster University
Lancaster, United Kingdom

Paul L. Snyder
Drexel University
Philadelphia, Pennsylvania, USA

Henk G. Sol
University of Groningen
Groningen, The Netherlands

Cesare Stefanelli
University of Ferrara
Ferrara, Italy

Niranjan Suri
Florida Institute for Human and
Machine Cognition
Pensacola, Florida, USA

and

U.S. Army Research Laboratory
Adelphi, Maryland, USA

Austin Tate
University of Edinburgh
Edinburgh, United Kingdom

Mauro Tortonesi
University of Ferrara
Ferrara, Italy

Stefania Tosi
University of Modena and Reggio Emilia
Modena, Italy

Andrzej Uszok
Florida Institute for Human and
  Machine Cognition
Pensacola, Florida, USA

Giuseppe Valetto
Drexel University
Philadelphia, Pennsylvania, USA

and

Fondazione Bruno Kessler
Trento, Italy

Alex Villazón
Bolivian Private University (UPB)
Cochabamba, Bolivia

Xin Yao
University of Birmingham
Birmingham, United Kingdom

Franco Zambonelli
University of Modena and Reggio
  Emilia
Modena, Italy

Chapter 1

Introduction

Giacomo Cabri and Niranjan Suri

Contents

A True Story

Los Angeles, California, December 10, 2012, 9:00 AM. John has just arrived at his office and is trying to access his email but experiences many problems. The performance is slow, timeouts expire, and he receives a number of "Server Error 502" messages. He is wondering whether the problem is with his client, with the network connection, or with the server. He cannot know that thousands of other users are experiencing similar problems—not only with emails but also with other services from the same provider.

These problems were caused by a bug in what was supposed to be an innocuous update of the load-balancing software from the provider, which was deployed to the systems at 8:45 AM. Neither the developers nor the administrative team could have foreseen the cascade of failures that were caused by the software update. Fortunately, there was a monitoring system in place that detected the problem at 9:06 AM, and the engineers were able to roll back and revert the update in only seven minutes: At 9:13 AM, the system was working well again.

We live in an automated and connected world, one in which almost every system has a computational component attached or embedded. These real-world systems depend on computers and networks to control, coordinate, and communicate with their counterparts as well as with us, humans. As we have increasingly automated these systems, we have realized significant gains in *capability, efficiency,* and *productivity,* causing us to go down the path of further *automation.* This path has led us to our current state, in which real-world systems have evolved into a *complex* and *interconnected* combination of networked computer systems and the human operators who use them. Increasingly, national critical infrastructures depend on these complex systems for their continued successful operation. The vision for the future is driving systems to be highly mobile, dynamic, interdependent, recomposable, and reusable. As the complexity of these systems continues to increase, they become less understandable, predictable, and controllable. Addressing these challenges requires adoption of fundamentally new approaches to build systems that are able to dynamically adapt in order to be resilient.

This book introduces these key issues and their interrelationships and presents new research in support of these areas.

We begin by introducing some abstract relationship graphs between the key attributes of distributed systems. Note that we are not claiming that the relationships depicted by these graphs are *quantitatively* accurate. We do not claim that the relationship is quadratic, linear, exponential, or some other form. Rather, our intention in introducing these relationships is to motivate some discussion about these factors and their dependencies.

Our first assumption, which can be widely observed, is that *automation* helps *productivity.* This relationship is depicted in Figure 1.1. One of the primary motivations for automation at multiple levels is to reduce the need for human operators to manually perform various tedious, repetitive tasks. The manufacturing domain is a classic example in which this can be observed. Automation is used to speed up processes from food preparation and packaging to automobile manufacturing. Historically, automation has had significant impact on societal development (for example, with the industrial revolution).

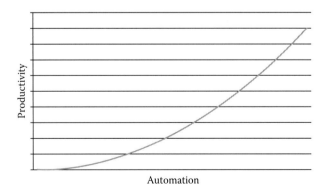

Figure 1.1

In the computing domain, code compilation is a classic example of automation that significantly increases productivity. Automation in this case applies to the generation of executable code from a higher-level language. A second example is garbage collection in the context of memory management. Automation in this case relieves the programmer from worrying about allocation and de-allocation of memory and the errors that often arise from incorrect memory management. Additional examples can be found with techniques such as optical character recognition (OCR) and speech recognition. OCR, in particular, significantly increases productivity by not requiring humans to manually translate and create digitized versions of paper documents.

From a networking perspective, automated routing algorithms and automated network configuration (using DHCP, the dynamic host configuration protocol) are examples in which tasks that were previously manually performed have been automated in order to save time. In the case of DHCP, the savings are primarily in the form of administration time. In the case of automated routing, the advantage is also in terms of reaction time when adapting to failures (such as link failures).

Yet another example is that of automated patch installation, which is now very common for desktop operating systems, such as Windows, Linux, Mac OSX, Android, and iOS.

Our second assumption, depicted in Figure 1.2, is that *interconnectedness* greatly improves *productivity* as well as *efficiency* through resource sharing. The best example that demonstrates this relationship is computer networks and the Internet. Computer networks have enabled resource sharing in multiple ways from sharing hardware (printers, storage devices, etc.) to sharing entire computing resources, such as in the cases of grid computing and cloud computing. Networking (with the Internet being the ultimate example) has significantly improved productivity through email communication, file exchange, document sharing, online repositories of information, search and retrieval, and numerous other capabilities.

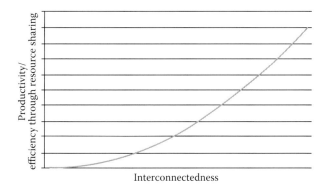

Figure 1.2

Automation and interconnectedness, while boosting productivity and efficiency, also have some negative consequences. In the following graphs, we describe the relationship between these attributes. Again, we emphasize that the graphs are not quantitatively accurate but are meant to convey a general sense of the relationship.

The first consequence of automation is that, as automation increases, so does the *complexity* of the underlying system. This relationship is depicted in Figure 1.3. For example, consider the complexity of a compiler that automates code generation or a virtual machine that automates memory management. Note that we are not arguing against adoption of automation, especially given the productivity gains that result. Compilers and virtual machines are very complex, but once that complexity is addressed, their benefits to productivity are broad and significant.

The second consequence of automation (and the associated increase in complexity) is the decrease in the *understandability* of the system and *predictability* of the system's behavior. These are depicted in Figure 1.4. Consider again the example of compilers—programmers often do not understand (or care) about the internal

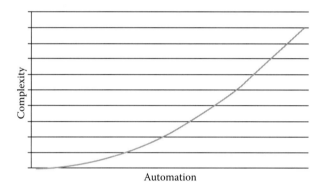

Figure 1.3

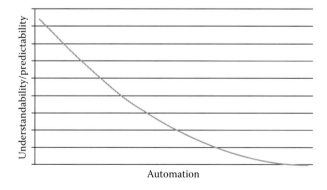

Figure 1.4

operation of a compiler. They do not understand how compilers optimize code, perform register allocation, schedule code for pipelining, or any of the complex operations typically performed by a compiler. As a consequence, hardware requirements for systems are often overestimated in order to ensure performance guarantees. Given that the savings in productivity often easily compensate for the increased cost due to hardware over-provisioning, this has not been an issue except in the most time-critical of systems. Likewise, most programmers do not understand the complexities of automated memory management as provided by modern virtual machines. Consequently, they do not have control over when a virtual machine might stop the program execution and perform a garbage collection cycle. This is just one example of unpredictability in the system introduced by automation.

Note that we would agree with the argument that automation increases the level of abstraction and that this is a beneficial consequence of automation. In computer science, an increase in the level of abstraction often results in an increase in the overall productivity of engineers developing new solutions. We are not suggesting that abstraction is bad but merely that abstraction introduces challenges that need to be addressed.

The second major shift to occur, perhaps more recently than automation, is the rise in interconnectedness of systems. Interconnectedness was enabled by the creation and widespread deployment of computer networks. Like automation, interconnectedness has both positive and negative consequences. On the positive side, interconnectedness has enabled previously unimaginable sharing of information and resources, thereby significantly increasing productivity. On the other hand, interconnectedness creates *interdependencies* between systems. This relationship is depicted in Figure 1.5. As a simple example, consider sharing a printer via a local area network in an office. Interconnectedness saves resources by not requiring that every user have a printer. Instead, users can print to a common, shared printer. On the other hand, all of those users are now dependent on the network and that printer being available. As a more recent example, consider cloud computing and

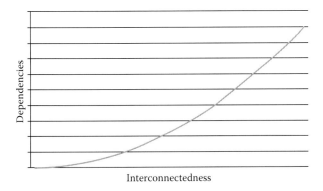

Figure 1.5

cloud storage, which enables users anywhere in the world to access and share data and resources. The negative consequence of this approach is the dependency on the network link to the cloud and the dependency on the cloud always working. As in the motivating example at the beginning of the chapter, this cannot always be assured. Furthermore, unlike with locally located and managed resources, the user is often not in control of resources in the cloud and therefore is dependent on a number of other providers for the connectivity and the services.

The negative consequences described so far (namely, an increase in complexity, a decrease in understandability or predictability, and an increase in interdependencies) may be classified as first-order consequences. Of more concern are the second-order consequences of these negative first-order effects. The first consequence is an increase in the *fragility* of the overall system. This is depicted in Figure 1.6. The increase in fragility is a direct consequence of being dependent on multiple complex components that are not easily understood. Leslie Lamport famously described a distributed system as one in which a user cannot get any work done because some remote system that the user has never heard about is currently down. While this is clearly an exaggeration of the state of distributed systems, there is a kernel of truth to the description.

The second consequence is an increase in *opacity* as illustrated in Figure 1.7. One example of the problem of opacity is mishaps that have occurred with safety-critical systems. For example, automation in some safety-critical systems has led to unfortunate mishaps as described by David Woods and his colleagues in their analysis of safety in nuclear power plants as well as in the aviation industry. In particular, they highlight the problem of "automation surprise," which occurs when the underlying system behaves in an unexpected manner because the human operators do not have a complete understanding of the automation of the system.

The third consequence is a *decrease in predictability* as illustrated in Figure 1.8. Predictability may be defined as the ability to predetermine the behavior of a system when presented with a new environment or situation. As the complexity of

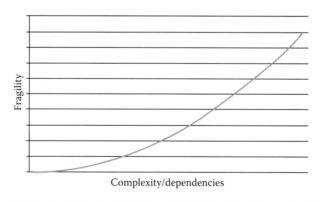

Figure 1.6

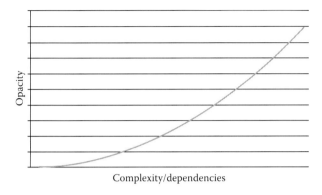

Figure 1.7

the system and the number of dependencies increase, it does become increasingly difficult to predict the actions of the overall system in the face of novel situations.

Finally, the last negative consequence is a *decrease in defensibility* as illustrated in Figure 1.9. We define defensibility as the ability to protect the system against malicious attacks. These attacks may come from inside the organization (typically termed "insider threat") or from outside, over a network link. An increase in inter-connectedness, especially in distributed systems, increases vulnerability to network attacks. An increase in complexity makes it more difficult to detect these vulner-abilities. Finally, an increase in dependencies increases the potential for cascading failures, when a failure in one part of the system, either due to an internal error or an external attack, causes other parts of the system to be affected.

The goal of engineering adaptive and resilient systems is to reduce the rigidity and fragility of the overall system. Our fundamental thesis is that designing sys-tems to dynamically adapt to a continuously changing environment enables them to adapt to a variety of failures caused by the negative consequences of current

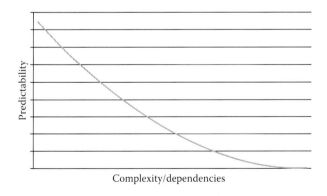

Figure 1.8

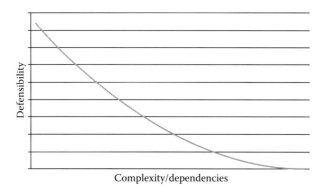

Figure 1.9

systems-engineering approaches. Adaptability is a counter to increased fragility and reduced defensibility. Furthermore, adaptability also increases performance by allowing the system to detect and switch to runtime states that are more efficient, based on resource availability and environmental circumstances.

We do note that increasing the adaptive behavior of systems will make them potentially more opaque and more unpredictable. These drawbacks would have to be countered with architectural mechanisms that can help manage these attributes. For example, opacity can be reduced by tools that allow human introspection into the overall system behavior. Likewise, predictability can be improved by developing policy-based mechanisms that allow adaptation within well-specified guidelines.

1.1 Organization

The book is a collection of 16 chapters, this included, organized into three sections. The first section is titled "Exploring Adaptation and Resilience in Different Domains" and contains six chapters (two through seven). This section provides varied perspectives on the adaptation and resilience aspects in different application domains. Chapters in this section examine service-oriented software engineering, networked systems, e-Health, command and control, and complex systems.

The second section is titled "Different Systems-Level Approaches" and contains five chapters on different architectures and systems that realize, to varying extents, adaptive and resilient systems. The notion of self-organizing systems is a focus here as are developing systems that exhibit agility in their adaptation.

The third section is titled "Orthogonally Enabling Capabilities" and contains four chapters on capabilities that enable the engineering of adaptive and resilient systems. These chapters focus on planning, policy architectures, remote monitoring, resource allocation, and code instrumentation.

One final note is that in this book, we have consciously chosen to not include the cyber-security domain and engineering systems to be able to defend against malicious attacks of various sorts. While this is an extremely important topic, we could only have given it a cursory treatment that would not have been adequate. Therefore, we have deferred any chapters in this topic area.

EXPLORING ADAPTATION AND RESILIENCE IN DIFFERENT DOMAINS

Chapter 2

Modeling Adaptive Software Systems

Marc-Philippe Huget

Contents

The primary difference between adaptive software systems and enterprise software systems is the ability of the former to adapt its behavior (and, as a consequence, its architecture) to changes in the environment. Despite the importance of this topic, existing literature covers only limited research on modeling this adaptation. In this chapter, Marc-Philippe Huget proposes his vision of how adaptation can be designed into computing systems. Three approaches are depicted: (1) goal modeling

13

for defining component objectives, (2) multi-agent systems for reorganizing the system, and (3) model-driven development for transforming the architecture.

2.1 Introduction

A self-adaptive software system, also called an adaptive system or autonomic computing (Kephart and Chess, 2003), is a system capable of modifying its behavior and architecture at runtime to answer unexpected events in the environment. Such unexpected events are, for instance, the reduction of bandwidth; loss of the Internet connection; or the reduction of screen size, memory, or CPU capacity to name a few.

A frequently occurring example of an adaptive system is multimedia players who modify the speed when playing a streaming video to ensure the audio is always synched with the video. A second example is when an application moves from a desktop PC with a large screen to a smartphone or a tablet, and the system needs to adapt the graphical user interface to fit the small screen size.

Self-adaptive systems, such as other software systems, need to be modeled and designed, but we have to admit literature lacks research papers presenting how to model such systems. In this chapter, we describe these few papers and we complement the discussion with references coming from modeling.

To better understand what we are expecting in terms of modeling, we first describe the features of self-* systems. Self-* characteristics are the intrinsic characteristics of self-adaptive systems. These self-* characteristics are described in Section 2.2.

Then, we present the different modeling notations we consider for self-adaptive systems: that is, goal modeling with i* (Yu et al., 2010) in Section 2.3, multi-agent systems (Wooldridge, 2002) in Section 2.4, and model-driven development (Schmidt, 2006) in Section 2.5.

We conclude the chapter in Section 2.6.

2.2 Self-* Characteristics

Kephart and Chess (2003), in their seminal paper about autonomic computing, describe four aspects of self-management in autonomic computing:

- Self-configuration: The different components of the system are able to configure themselves and their interconnections to answer modifications in the environment.
- Self-optimization: The system is able to tune itself to increase its performance.
- Self-healing: The system detects, diagnoses, and repairs faulty components.
- Self-protection: The system protects itself from inner and outer malicious attacks and predicts future failures.

2.2.1 Self-Configuration

Self-configuration facilitates the addition or removal of new components in the system. In a classical system, the integrator is responsible to shut down (part of) the system, then add the new components and configure the interconnections and existing components with regard to these new ones. This task increases in complexity as the system grows.

In a self-adaptive system, it is not necessary to shut down the system because the new components enter the system and advertise their capabilities. The system reconfigures by itself to take count of these new components. In terms of modeling, it means interconnections between components are loose ones because it is possible to modify them at runtime. Two modeling approaches are possible to answer this need: goal modeling and multi-agent systems.

Actually, for self-configuration, it is crucial not to describe the *how* of the system but the *what*, and this could be done by high-level goals. Finally, reconfiguration of the system could be performed by multi-agent systems as shown in Section 2.4.

2.2.2 Self-Optimization

Software systems tend to be more and more complex, and configuring them with an important set of parameters is more and more difficult. Self-optimization may solve the issue because it proposes to tune the system. Components are responsible to find the correct value for their parameters with respect to other components' parameters. This configuration is continuously performed in the system.

Optimizing a system has no relation to modeling a system because it is a constraint satisfaction problem (Tsang, 1993).

2.2.3 Self-Healing

Self-healing in a certain way has to be related to self-configuration because it proposes to detect software and hardware failures and to repair them. It is then necessary to detect faulty components, then to replace them with other ones. We face, once again, the problem of modifying the interconnections between components and the necessity of having loose connections between components.

Apart from the diagnosis, the repair may also use the notion of high-level goals to describe the *what* and not the *how* of the system. Multi-agent systems facilitate the reconfiguration as proposed in the self-configuration aspect described above.

2.2.4 Self-Protection

Self-protection prevents a system from malicious attacks and diagnoses a system for future failures based on activity reports.

Self-protection is not directly related to modeling because it is a diagnosis problem.

2.3 Goal Modeling

As mentioned in Section 2.2, we need to represent goals because self-configuration should not focus on the detailed architecture but on the rules that govern the components and the intention of the application.

i* (Yu et al., 2010) is a modeling framework used in requirements engineering and focus in goals, dependencies, and actors.

Figure 2.1 shows an example of goal modeling with i*. Goals are depicted as rounded-corner rectangles and give the desire of an actor. It corresponds to the *what* that a system is supposed to provide. The *how* associated with the goal is given by task decomposition, for instance, "Purchase By Naming My Own Price."

Actors are depicted by circles. Actors represent active entities that will perform actions given by tasks so as to fulfill goals.

Tasks are depicted as hexagons and correspond to actions to be performed. A task can be decomposed into subtasks.

In the context of self-adaptive software, goal diagrams can describe the business rules governing the application. Actors represent the components or parts of the system, and they are associated with goals they have to fulfill and, with these goals, to the tasks to perform.

2.4 Multi-Agent Systems

Due to space restrictions and in order to stick with self-adaptive software, we reduce our description of multi-agent systems to concepts that are useful for self-adaptive software. A multi-agent system is a system composed of multiple interacting agents within an environment (Wooldridge, 2002). There exist two kinds of agents: reactive ones and cognitive ones. We only consider here cognitive agents because reactive agents are mainly for simulation. Moreover, in the context of self-configuration, we need agents that are able to express their capabilities and to interact with other agents. These characteristics are part of cognitive agents.

The main concepts used for multi-agent systems with cognitive agents are organization, role, agent, capability, protocol, plan, and task. All these concepts can be used in the context of self-configuration for self-adaptive systems.

Indeed, agents are able to enter or to leave an organization at any moment, and when entering, they advertise the services they propose. There is a parallel with self-configuration because components entering the software system have to advertise their capabilities to let the system reconfigure itself.

Moreover, behavior associated with cognitive agents is depicted by a plan and tasks. This could be related to the need to gain abstraction when describing components and their relationships. If we have plans (or goals) to represent components, it is easier to replace a component with another one.

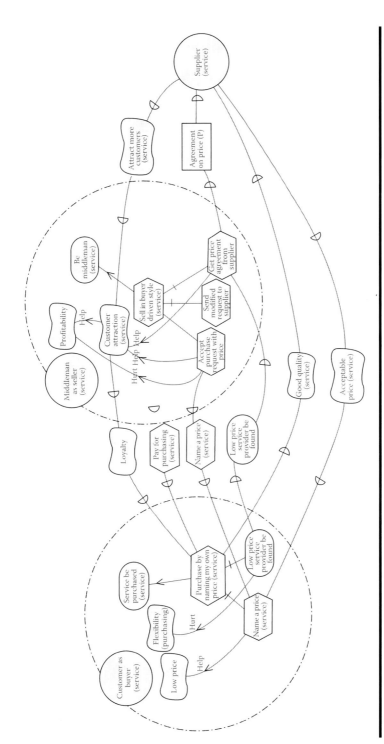

Figure 2.1 Goal modeling: Buyer drive e-commerce example from i* quick guide (http://istar.rwth-aachen.de/tiki-index. php?page=iStarQuickGuide).

Finally, self-organization in multi-agent systems could help in the context of reconfiguring the system. Self-organization is defined as "the mechanism or the process enabling a system to change its organization without explicit external command during its execution time" (Di Marzo Segurendo et al., 2005).

2.5 Model-Driven Development

The last approach we present in this chapter is the notion of models. Model-driven development appears as an answer to software crises and the difficulties in providing, on time and on budget, software (Schmidt, 2006). The objectives of model-driven development are to gain abstraction and focus on models rather on the source code. A model is an abstraction of a physical system focusing on the characteristics required for the application under development. It means, for instance, for an application related to car sales, we will focus on color, number of seats, or trunk volume to name a few characteristics whereas these features are unfounded for car fluid dynamics simulation.

Model-driven development considers three levels:*

1. The application level at which the different objects used in the system are defined. This is equivalent to the UML object diagram (Object Management Group).
2. The model level at which how to model the data used in the application level is described. This is equivalent to the UML class diagram.
3. The metamodel level at which how to model the concepts from the model level is described.

These three levels are of interest in the context of self-configuration. At the application level, we have the different components in use and their interconnections. When a new component enters the system, it is necessary to adapt the architecture to this new component. In this case, the integrator will work on the application model and soon will have two different views of the architecture: the previous one and the new one with the new component. It is necessary to transform from the first architecture to the second one. Model-driven development offers an answer via transformation rules.

Model-driven development and transformation rules are the approach used in Garlan and Schmerl (2002) and Morin et al. (2009).

---

* Actually four levels based on the OMG four-layer architecture, but we do not consider it here.

2.6 Conclusion

In this chapter, we present elements that could be used to model self-adaptive software, especially the two aspects of self-configuration and self-healing. Self-* characteristics are addressed here via three different approaches:

1. Goal modeling helps in clarifying the objectives of the different components and reducing the difficulty of moving from the current architecture to the new one in the context of self-configuration.
2. Multi-agent systems, with the notion of organizations, roles, capabilities, and agents, offer a solution for self-configuration because it is possible to represent components as agents and the component behavior as a plan and tasks. So, when an agent enters the system, the system can reorganize itself to consider the new agent. This is called self-organization and clearly answers to self-configuration and self-healing.
3. Model-driven development and its use at runtime allows self-configuration and self-healing because it is possible to modify the current model to another one via transformation rules.

The different approaches presented here consider the system architecture to answer modifications in the environment.

Reading this chapter, it could be understandable to think these approaches are ideal solutions for modeling software adaptive systems. However, using these approaches has a cost as explained below.

Contrary to formal methods that could be used to model entity behavior in the context of self-configuration and self-healing, these three different approaches present a longer learning curve, especially multi-agent systems and model-driven engineering. These approaches require a good knowledge.

Apart from the difficulty of understanding the notion of organizations, roles, and agents, the main difficulty with multi-agent systems is to obtain a system that converges to a solution. This is particularly true for self-organizing systems.

In model-driven engineering, it is necessary to develop several models corresponding to different situations during runtime and to prepare transformation rules to pass from the current model to another one. This approach requires important work while designing systems.

References

Garlan, D., and B. Schmerl. "Model-based Adaptation for Self-Healing Systems." In *Proceedings of the First Workshop on Self-Healing Systems* (WOSS 2002), D. Garlan, J. Kramer, and A. Wolf (eds.), pp. 27–32, Charleston, 2002.

Kephart, J.M., and D.O. Chess, "The Vision of Autonomic Computing." In *IEEE Computer*, 36(1), pp. 41–50, 2003.

Morin B., O. Barais, G. Nain, and J.-M. Jézéquel. "Taming Dynamic Adaptive Systems with Models and Aspects." In *Proceedings of the 31st International Conference on Software Engineering* (ICSE 2009), pp. 122–132, Vancouver, Canada, 2009.

Object Management Group. Available at http://www.omg.org/spec/UML/2.4.1/Superstructure/PDF.

Schmidt, D.C. "Model-Driven Engineering." In *IEEE Computer*, 39(2), 2006.

Segurendo, G.M., M.-P. Gleizes, and A. Karageorgos, "Self-Organization in Multi-Agent Systems." In *Journal of the Knowledge Engineering Review*, 20(2), pp. 165–189, 2005.

Tsang, E. *Foundations of Constraint Satisfaction*. Academic Press, 1993.

Wooldridge, M.J. *Introduction to Multiagent Systems*. Wiley, 2002.

Yu, E., P. Giorgini, N. Maiden, and J. Mylopoulos. *Social Modelling for Requirements Engineering*. The MIT Press, 2010.

Chapter 3

Service-Oriented Software Engineering Lifecycles: Methodologies and Operations for Adaptability in Enterprise Settings

M. Brian Blake and Peter Bartalos

Contents

Service-oriented computing is a widely recognized approach to engineering modern computing systems. Service-oriented architectures are becoming the base of the software services provided by any enterprise, thanks to their simplicity and robustness. However, they still lack flexibility both in terms of adaptation and resilience. In this chapter, M. Brian Blake and Peter Bartalos propose a perspective on adaptive and resilient computing systems in connection with the service-oriented computing approach and from the point of view of the enterprise world. They examine and emphasize the service features that promote adaptation and resilience in computing systems.

3.1 Introduction

In the 1960s, *software engineering* [17] was introduced as the discipline for decomposing large, domain-specific problems into smaller sub-problems that can be solved through the development of modular, interconnected software components. Software engineering focuses on the software system–design approaches in addition to the management of a group of software developers that can collaboratively develop such systems. The inception of software engineering has been the impetus for modular, self-contained software. From the inception of software engineering to the present, software modularity has been realized incrementally through the introduction of functional programming; object-oriented programming [18]; component-based programming; and most recently, web-scale workflow or *service-oriented computing* (SOC) [4]. In each case, software has become more autonomous, distributed, and adaptive while also more effectively encapsulating functional concerns. As a result, systems are distributed and shared today more efficiently than ever before.

Currently, SOC is a software-engineering paradigm that defines the design, development, and operations of adaptive, dynamic, and resilient systems. Web services are operational units of software that underlie SOC. Web services are accessed via uniform resource locators (URLs), and their distributed invocation is mandated by open, but standard, protocols, such as SOAP and REST. Web services assimilate the general notion of a system of systems also known as *service-oriented architecture* (SOA). SOA makes possible the existence of services both inside and outside of an enterprise with the promise of on-demand inter-organizational systems integration. Within such an environment, independent tasks can be referred to as organizational capabilities. Present information technology professionals have leveraged and extended this model by heavily adopting a practice whereby both software and computing cycles can be outsourced as commodities to other organizations (i.e., *software as a service* and *cloud computing* [25]. Consequently, for systems dynamism and resilience, SOAs in cloud environments require properly developed web services with concise, openly accessible interfaces that embody well-defined organizational capabilities. *A difficult challenge here is for enterprise organizations to modernize their current operations toward (or in some cases create completely new) service-oriented systems.* It is apparent that the first step to addressing this question is understanding how service-oriented environments are engineered; in other words, what are the enterprise approaches to *service-oriented software engineering.*

3.2 Methodologies for Designing Resilient Software Services

Because web services are the backbone of SOA, software engineers must be skilled in developing modular software services that decompose business processes into manageable sub-capabilities. In this way, organizations can share their offerings at multiple levels of granularity while also creating unique access points for their peer organizations. Software engineers must incorporate leading methodologies for design and development of such systems to ensure their adaptability and resilience [8,16].

3.2.1 Service-Oriented Modeling and Architecture (SOMA)

SOMA [2,21] is an approach for organizations to understand their current IT environment and characterize their infrastructure with respect to the SOA paradigm. The SOMA approach suggests that target SOA enterprise environments can be decomposed into layers. These layers are (1) the *operational systems layer*, consisting of existing custom-made, legacy applications with core operational responsibilities; (2) the *enterprise components layer*, comprised of nonfunctional enterprise-wide components that ensure load-balancing and systems availability; (3) the *service layer*, consisting of the atomic capabilities that the organization exposes as services that potentially *wrap* the operational layer; (4) the *business process composition layer*, the area in which services are bundled as higher-level workflows; (5) the *access or presentation layer*, the layer in which workflows are presented to customers as organizational-level capabilities; (6) the *integration layer (enterprise service bus)*, this layer, through routing and protocol mediation, enables service interoperability; and (7) the *quality of service (QoS layer),* comprises modules that guarantee security, performance, and availability of domain-specific capabilities.

SOMA introduces a method to help organizations visualize their IT systems as the previously described multilayer SOA. The method consists of steps, *business modeling and transformation, service identification, specification and realizations, implementation,* and *deployment/monitoring/management* (i.e., choreography of services). These steps are illustrated in Figure 3.1. The business modeling and service identification steps are composed of top-down, bottom-up, and middle-out processes. The top-down process decomposes services from the capabilities offered

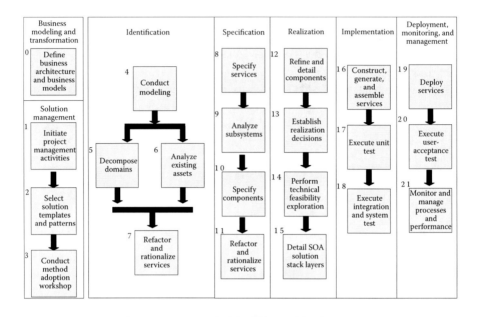

Figure 3.1 SOMA lifecycle high level flow. (Reconstructed from Arsanjani, A. et al., *IBM Systems Journal*, 47, 3, 377–396, 2008.)

to the end user. Bottom-up processes attempt to describe and characterize the modular functions of the existing systems and extract services from them. Finally, middle-out processes extract unanticipated services by examining the goals of the organization. The specification and realization steps comprise the specification—a design of the inner-workings of the system based on models of various subsystems and various components. Subsystems consist of interconnected components. At these steps, components are specified by their actions and methods. The realization and implementation phases consist of the partitioning of specific components and the actual development of services while respecting the variations in implementation techniques. During deployment/monitoring/management, services are placed into operation and iteratively monitored and managed.

3.2.2 Service Development Lifecycle and Service-Centric System Management

In prior work of the authors [7], two key activities were introduced for software engineers that develop service-centic software systems (SCSS), *service development* and *service-centric system management*. The general phases of an incremental service development lifecycle consist of *conceptualization, analysis, design, development, testing, deployment,* and *retirement*. This lifecycle is shown in Figure 3.2. In the conceptualization and analysis phases, the software engineer manages the elicitation

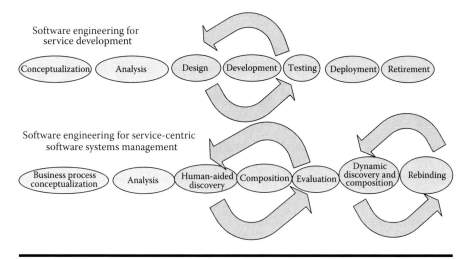

Figure 3.2 Integrated service development and service operations lifecycles.

of requirements that clarify the business needs. Design, development, and testing comprise an iterative set of phases in which services are designed, created, and evaluated through testing. Results of the testing phase are used as feedback to the design phase. Services are then deployed for universal access and, at some time in the future when the business offering changes, the services are removed and retired.

Once web services are developed and deployed, software engineers must engage in service-centric system management lifecycles to allow services to be discovered, analyzed, and composed/consumed on demand. Because service-based capabilities are managed in distributed locations outside of the consumers' control, this environment requires more specialized development lifecycles for maintaining predictable operations. This operational lifecycle resembles the previously mentioned development lifecycle; however, the focus is on managing available services in real time. In addition, the seven phases are slightly different: *business process conceptualization, domain analysis, discovery, composition, evaluation, on-demand composition,* and *rebinding.* Although the business process conceptualization and domain analysis are phases that attempt to capture the business needs, these phases must take into account the fact that the solution services already exist. Discovery, composition, and evaluation are phases that represent the core of the service-oriented computing paradigm of finding, blueprinting, and analyzing candidate services at design time. Finally at run time, on-demand composition and rebinding allow composite service-oriented business processes to be constructed and to evolve over time. These parallel lifecycles are imminently woven. Service-oriented software engineers, in the first lifecycle, must understand the implications of establishing long-standing business processes that must be capable of automation and service integration in the second operational lifecycle.

3.2.3 SOA Design Patterns

Considering a more agile, ad-hoc lifecycle, software engineers might consider reusing best practices from other organizations. *SOA Design Patterns* is a repository of best practices when modernizing systems toward the SOA paradigm [9,23]. SOA design patterns derive from success in object-oriented design patterns. Via the approach, system developers recognize occurrences in legacy software and convert them into service-centric approaches based on guidance from existing patterns. There are more than 50 design patterns to date, and the repository and body of knowledge continues to grow and evolve.

3.3 Adaptive Service-Based Operations That Promote Resiliency

With the variation in corporations and the fluctuations in international economy, enterprises must be agile with respect to partnering with organizations that were not considered a priori. Organizations must subsume the services as a result of new partnerships on a continual basis. The aim of service-centric environments is for predictable interoperability of inter-organizational capabilities in response to and sometimes irrespective of changes in business needs.

3.3.1 Information Disambiguation for Inter-Enterprise Integration

A challenge in integrating services across organizational boundaries is the automatic interpretation of how the input and output messages relate. Several documents that describe web service messages, such as WSDL (web service description language), the OWL-S (web ontology language for services), and WSDL-S (web service description language-semantics), all facilitate messages composed of sub-messages. These messages describe data corresponding to the services' input/output parameters serialized in markup languages, many times the eXtensible Markup Language (XML). The message consists of one or more *parts*, each corresponding to one parameter.

The challenge in interconnecting web services is disambiguating messages to find equivalence. While languages, such as WSDL, heavily rely on syntactic interpretation of the service messages, other languages, such as OWL-S, WSDL-S, and WSML, allow systems to reason about messages semantically. Semantic web services are web services enhanced by semantic description. Both the requestor and the consumer must understand the semantics of services in a consistent way. As such, semantic annotation binds the elements of the web service to domain terms described by shared ontologies. When a shared ontology is not available, concepts from different ontologies must be disambiguated. There are approaches that estimate the semantic relationship (distance) of two concepts associated with web service elements. Several works deal with this problem explicitly in the context of service discovery and composition [26].

If one assumes that equivalence can be derived from web service elements and messages, either by semantic or syntactic processing [14], then there are fundamental methods that can be applied to web services integration or composition. Discovering workflows of web services within operational systems, sometimes referred to as *web service composition* [6,13,22], has been addressed with numerous techniques. *Uninformed search* [20], such as bread-first and depth-first [27]; *heuristic search*, such as greedy search or A* search [1]; *artificial planning* [3], such as order planning or metric planning; and *metaheuristic search and dynamic programming* [10,15]. In industrial settings, the discovery and composition of services must occur across multiple distributed service repositories. As such, ontology creation and maintenance and service annotation are important issues that can only be handled manually in some cases. However, in the majority of cases, automation is required for practical usability, and this automation must be performed in concert with the general approaches for service composition (e.g., search, planning, and dynamic programming).

3.3.2 Predictable Quality of Service

In these dynamic environments, organizations must have QoS guarantees when services are provided by their collaborators. Service-level agreements (SLAs) have been used to ensure the QoS standards. In an ad-hoc workflow scenario, a business may need to perform real-time composition of existing services in response to consumer requests. Interpretation of SLAs and cost estimation must occur in parallel with service composition.

SLAs are technical contracts between two types of businesses, producers and consumers. In service-centric environments, these SLAs are captured with web-service agreements (WS-agreements) [24]. A SLA captures the agreed-upon terms between organizations. In simple cases, one consumer forms a SLA with a producer. In more complex cases, a consumer may form a SLA that defines a set of producer businesses. Considering a SOC environment, capabilities are shared via the implementation of web services exposed by a producer organization. The ultimate goal of SOC is for consumers to access these shared capabilities on demand.

These SLA technologies and specifications present new challenges. When a consumer organization creates a new business capability that requires the workflow composition of multiple web services, then that organization will also need to understand if the product of all the SLAs is within the required threshold of feasibility as defined by the end users [5]. Information disambiguation is also a problem in the context of managing SLAs. Service-level objectives (SLOs) can be articulated differently across organizations. During real-time enterprise integration, these SLOs must be correlated in order to understand the expectations of each organization. Semantics have been leveraged to address this open problem.

3.3.3 Real-Word Example: Infrastructure Support for End-to-End SOA Lifecycles

There are several infrastructures that support end-to-end service-oriented software engineering lifecycles [11,12,19]. To provide a concrete case, we detail the lifecycle support offered by IBM in their Business Process Management Suite [12]. This lifecycle, shown in Figure 3.3, derives from the basic traditional software engineering lifecycles that informed the three methodologies described in the previous sections. Explicitly, the phases in this suite are *model, assemble, deploy*, and *manage*. For each phase, there are specific tools, and for each tool, there are views for the business subject matter expert, system architect, software engineer, software developer, and production engineer. In the modeling phase, the business SME develops the capability into a high-level workflow using the WebSphere Business Modeler. In the assemble phase, the system architect and software engineer work with the workflows and associate the steps to web services. A key to overall process is the WebSphere Integration Developer that manages this automated process. In the deployment and monitor phases, the software engineer and production engineer leverage the WebSphere Process Server to deploy the web service workflows, and in production, the system is monitored with the WebSphere Business Monitor Server. The overall process is illustrated in Figure 3.3.

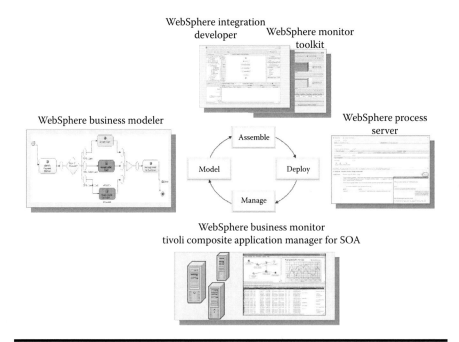

Figure 3.3 A concrete, real-world lifecycle: IBM Business Process Management Suite.

This software infrastructure is favorable because the tools are relatively homogeneous and can be integrated together easily. As such, when web-service elements are developed in one part of an organization, then these elements can be reused or easily integrated with elements in other parts of the organization. Multiple organizations sharing this infrastructure also can easily couple their respective enterprises.

3.4 Conclusion

Conceptualizing, modeling, designing, developing, and deploying adaptive, dynamic, and resilient systems requires modularity at its core. Principled design processes and lifecycles coupled with next-generation modular software will ensure systems that are easily modernized for new paradigms, such as Web 3.0, and social networking and beyond.

References

1. Aiello, M., Platzer, C., Rosenberg, F., Tran, H., Vasko, M., and Dustdar, S. Web Service Indexing for Efficient Retrieval and Composition, *Proceedings of 2006 IEEE Joint Conference on E-Commerce Technology and Enterprise Computing*, E-Commerce and E-Services (CEC/EEE '06), 2006, pp. 424–426.
2. Arsanjani, A., Ghosh, S., Allam, A., Abdollah, T., Ganapathy, S., and Holley, K. SOMA: A Method for Developing Service-Oriented Solutions. *IBM Systems Journal*, 47(3):377–396, 2008.
3. Akkiraju, R., Srivastava, B., Ivan, A., Goodwin, R., and Syeda-Mahmood, T. Semantic Matching to Achieve Web Service Discovery and Composition, *Proceedings of 2006 IEEE Joint Conference on E-Commerce Technology and Enterprise Computing*, E-Commerce and E-Services (CEC/EEE '06), pp. 445–447.
4. Blake, M.B., and Huhns, M. Web-Scale Workflow: Integrating Distributed Services, *IEEE Internet Computing*, 12(1):55–59, 2008.
5. Blake, M.B., and Cummings, D.J. Workflow Composition of Service Level Agreements, *International Conference on Services Computing* (SCC 2007), July 2007.
6. Blake, M.B., and Gomaa, H. Agent-Oriented Compositional Approaches to Services-Based Cross-Organizational Workflow, Special Issue on Web Services and Process Management, *Decision Support Systems*, 40(1):31–50, 2005.
7. Blake, M.B. Decomposing Composition: Service-Oriented Software Engineers, *IEEE Software*, 24(6):68–77, 2007.
8. Cox, D.E., and Kreger, H. Management of the Service-Oriented Architecture Lifecycle, *IBM Systems Journal*, 44(4), 2005, http://www.research.ibm.com/journal/sj/444/cox.html.
9. Erl, T. *SOA Design Patterns*, Prentice Hall, NJ, 2009.
10. Glover, F., and Kochenberger, G.A., Eds., *Handbook of Metaheuristics*, ser. International Series in Operations Research & Management Science. Kluwer Academic Publishers: Norwell, MA, USA and Springer Netherlands: Dordrecht, Netherlands, 2003, vol. 57.
11. HP SOA Systinet (2011) http://www8.hp.com/us/en/software/software-product.html?compURI=tcm:245-936884.

12. IBM Business Process Management Suite (2011) http://publib.boulder.ibm.com/infocenter/ieduasst/v1r1m0/topic/com.ibm.iea.wpi_v6/wpswid/6.1.2/Overview/WPIv612_BPMOverview.pdf.

13. Koehler, J., and Srivastava, B. Web Service Composition: Current Solutions and Open Problems, *Proceedings of the Workshop on Planning for Web Services in conjunction with ICAPS03*, 2003.

14. McIlraith, S., Son, T., and Zeng, H. Semantic Web Services. *IEEE Intelligent Systems*, 16(2):46–53, 2001.

15. Michalewicz, Z., and Fogel, D.B. *How to Solve It: Modern Heuristics*, 2nd ed. Springer-Verlag: Berlin/Heidelberg, 2004.

16. Mittal, K. (2007) SOA Unified Process, http://www.kunalmittal.com/html/soup.shtml.

17. Naur, P., and Randell, B. Software Engineering. Report on a conference sponsored by the NATO Science Committee, Garmisch, Germany, 7–11 October 1968, Scientific Affairs Division, NATO, Brussels, 1969, pp. 138–155.

18. Nygaard, K. Basic Concepts in Object Oriented Programming, *Proceedings of the 1986 SIGPLAN Workshop on Object-Oriented Programming*, Yorktown Heights, NY, 1986, pp. 128–132.

19. Oracle WebLogic (2011) http://www.oracle.com/technetwork/middleware/weblogic/overview/index.html.

20. Russell, S.J., and Norvig, P. *Artificial Intelligence: A Modern Approach* (AIMA), 2nd ed. Prentice Hall International Inc.: Upper Saddle River, NJ, USA and Pearson Education: Upper Saddle River, NJ, USA, 2002.

21. Service-oriented modeling and architecture (SOMA) (2011) http://www.ibm.com/developerworks/library/ws-soa-design1/.

22. Sirin, E., Hendler, J., and Parsia, B. Semi-Automatic Composition of Web Services Using Semantic Descriptions, In *Proceedings of Web Services: Modeling, Architecture and Infrastructure,* Workshop in conjunction with ICEIS2003, 2002.

23. SOA Patterns (2011) http://www.soapatterns.org/.

24. Web Service Agreements (WSAG), http://www.gridforum.org/Public_Comment_Docs/Documents/Oct-2005/WS-AgreementSpecificationDraft050920.pdf

25. Wei, Y., and Blake, M.B. Service-Oriented Computing and Cloud Computing: Challenges and Opportunities, *IEEE Internet Computing*, 14(6), 2010.

26. Williams, A.B., Padmanabhan, A., and Blake, M.B. Experimentation with Local Consensus Ontologies with Implications to Automated Service Composition, *IEEE Transactions on Knowledge and Data Engineering*, 17(7):1–13, 2005.

27. Xu, B., Li, T., Gu, Z., and Wu, G. SWSDS: Quick Web Service Discovery and Composition in SEWSIP, *Proceedings of 2006 IEEE Joint Conference on E-Commerce Technology and Enterprise Computing*, E-Commerce and E-Services (CEC/EEE '06), 2006, pp. 449–451.

Chapter 4

On Measuring Resilience in Command and Control Architectures

Mark Pflanz and Alexander H. Levis

Contents

Adaptation and resilience have always been important from a military systems perspective. In addition to naturally occurring causes, most military scenarios involve an enemy (or enemies) that is actively trying to disrupt the organization and operation of the military and their missions. Therefore, engineering military systems to be resilient through adaptation is a key requirement and ranges from the challenges at the infrastructure level (e.g., communications, logistics) to the organizational level for command and control. This chapter by Mark Pflanz and Alexander Levis addresses the topic of resilience in military command and control organizations. In particular, the focus of the chapter is on developing metrics to measure the resiliency of organizational structures, to model the structures using petri nets, and to then experimentally measure the resiliency through simulations. While the metrics and approach are generic, the chapter focuses on a Maritime Operations Center scenario.

4.1 Introduction

The word "resilience" is derived from the Latin word "resilire," which means "the ability to rebound or jump back." The International Council on Systems Engineering (INCOSE) defines resilience as "the ability of organizational, hardware and software systems to mitigate the severity and likelihood of failures or losses, to adapt to changing conditions, and to respond appropriately after the fact" [1]. Many other highly related definitions for resilience have been developed; however, all involve the following common themes: avoidance, survival, recovery, disruption. Of these, the notion of disruption requires further definition. INCOSE defines disruption as "the initiating event of a reduction in performance. A disruption may be either a sudden or a sustained event …" [1]. Jackson [2] defines disruptions as events that jeopardize a system's ability to perform its intended capabilities. Madni and Jackson [3] define disruptions as "conditions or events that interrupt or impede normal operations by creating discontinuity, confusion, disorder or displacement." Disruptions can be naturally precipitated, unintended, maliciously intended, or non-maliciously intended. They are typically difficult to predict and often unavoidable. Systems able to avoid, survive, and recover from disruptions are referred to as *resilient systems* [2].

The objective of this chapter is to describe a quantitative approach for measuring the expected resilience of a command and control organization based on its architecture. The main idea is that resilience can be measured through its attributes, and these measures may be combined into a holistic evaluation of resilience. The approach can be used to support selection among alternative architectures and to identify areas for improvement in the proposed architecture. To illustrate this approach, we consider the resilience of a Maritime Operations Center's (MOC) command and control system to exercise or implement a capability when a disruption occurs. Section 4.2 highlights key aspects in resilience that must be considered in any evaluation. Section 4.3 describes the attributes of resilience along with their measures and introduces a holistic means of combining the measures. Section 4.4 contains the MOC case study. Section 4.5 presents observations and suggestions for future work.

4.2 On Resilience

The concept of "resilience" is meaningful only if one considers the resilience "of what, to what, and under what conditions." Carpenter et al. [4] introduce a notion "of what, to what" when describing the magnitude of disturbance an ecosystem can withstand prior to changing states. The evaluation of resilience should also consider "under what conditions." Here we are concerned with resilience in what situations or in which scenarios. For example, the resilience of a command and control system may be different if the command and control system experiences a disruption during peak combat operations rather than a disruption that occurs during normal peacetime activities. Our approach extends that notion into the systems engineering and development field to include application of threshold requirements and the resilience of capabilities to disruptions. A capability is defined as "the ability to achieve a desired effect under specified performance standards and conditions, thru combinations of ways and means [activities and resources] to perform a set of activities" [5]. Capabilities are a key consideration in this research because disruptions inherently affect the capabilities of a system or an organization to perform its tasks and mission.

An evaluation of resilience must also include temporal aspects. Time scales may vary based upon the system under consideration. However, the time scale can be normalized to allow for fairer comparisons. Figure 4.1 illustrates the significance of time when examining resilience and is originally described in [6]. Phases of resilience identified by Jackson [2] are overlaid on the time axis. The evaluation begins at some initial time, defined as time t_0. A disruption occurs at time t_d. The system reaches some minimum operating performance level at time t_{min} and returns to an acceptable operating state (possibly the predisruption state) at time t_{ret}. The *avoidance* phase of resilience runs from time t_0 to time t_d, the *survival* phase runs from time t_d to t_{min}, and the *recovery* phase runs from t_{min} to t_{ret}.

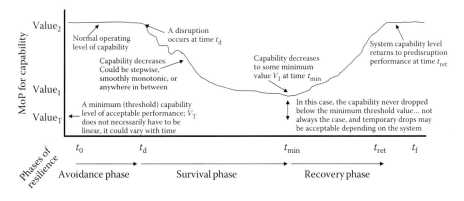

Figure 4.1   Temporal aspects in evaluating resilience.

The performance of the system under consideration is evaluated using some measure of performance (MoP) for a single capability of the system as described by the architecture. During the avoidance phase, a system is operating at some normal operating level of capability, defined above as $Value_2$ (V_2). When a disruption occurs at time t_d, the level of capability decreases to some minimum value $Value_1$ (V_1) at time t_{min}. The system has a minimum threshold level of capability denoted as $Value_T$ (V_T) below which performance is deemed unacceptable or below which a catastrophic failure could result. The same approach is used for other MoPs appropriate to other capabilities.

This approach uses the architecture of a system to evaluate its resilience. Architecture is defined as "the fundamental organization of a system, embodied in its components, their relationships to each other and the environment, and the principles governing its design and evolution" [7]. In simple terms, we can view the architecture as the high-level design of the system. Architects develop the overall design, and engineers design and deliver systems that conform to that architecture. By representing the architecture of a system in a rigorous way, one can analyze the architecture design for key properties and use the executable model of the design to examine for desired performance and behavior aspects. In this manner, one can make decisions and improvements far earlier in the process, saving time and money and ultimately delivering better results.

The executable model of the architecture is expressed as a Colored Petri Net [8]. Petri Net–based architecture models are used for a number of reasons. They are rigorous (meaning that defined mathematical models underlie all aspects of Petri Net theory), visualizable because of the graph theoretic underpinnings, and executable. These properties of Petri Nets support analyzing structural, behavioral, and performance characteristics of the architecture via simulation as well as through the algorithmic evaluation of the properties of the Petri Net model. Finally, established and traceable means exist for translating other architectural approaches (for example, business process model and notation or BPMN [9]) into Petri Net format.

4.3 Attributes of Resilience and Their Measures

On the basis of the existing body of resilience knowledge, Jackson [2] defines four primary attributes that characterize resilience: *tolerance, flexibility, capacity,* and *inter-element collaboration*. In the *Guide to the Systems Engineering Body of Knowledge* [10] inter-element collaboration is re-described as cohesion. Our approach retains the term "inter-element collaboration" because cohesion has already been used in [6] and [11] in a distinctly different manner. The approach described in this work partially redefines the attributes in [2] and extends them to better support the overall evaluation of resilience.

Tolerance is the ability to degrade gracefully after a disruption or attack. Flexibility is the ability of a system to reorganize its elements to maintain its capabilities at degraded or even predisruption levels. Capacity is the ability to operate at a certain level as defined by a given measure. We further define capacity as the available capability margin between current operating levels and minimum threshold operating levels. Inter-element collaboration describes unplanned cooperation within a system (typically an organization) to share resources or work together in new ways. Inter-element collaboration involves the emergent properties, often human-related, of many systems and is not considered in this evaluation approach.

To measure graceful degradation as the key attribute of tolerance, we consider the rate of departure (Tol_{RD}) from normal operating conditions, and Tol_{RD} is defined as the rate of change over time in a system's effectiveness in meeting its requirements. This encapsulates both the temporal aspects of resilience (t_d and t_{min}) as well as the effectiveness aspects of how the system performs with respect to its requirements and how effectiveness changes during the survival phase (postdisruption). Effectiveness can be measured by comparing the system performance with respect to defined MoPs against the corresponding requirements.

Bouthonnier and Levis [12] and Cothier and Levis [13] describe a methodology of comparing system performance to system requirements as the intersection of the locus of performance (L_p) and the locus of requirements (L_r). System performance is characterized by the applicable MoPs selected by the stakeholders and the system development team. The L_p describes the range of system performance in the defined MoP space as the parameters of various situations are varied according to expected conditions. The L_r defines the required system performance levels over the same MoP space. To examine the intersection of the L_p and L_r, a scenario is required. Parameters of interest (e.g., response time or inter-arrival time) are varied to form a parameter locus. Consequently, the mapping of the parameter locus through the architecture generates the L_p. The executable architecture is simulated at each point in the parameter locus to determine the L_p. The two loci, L_p and L_r, are then depicted in a common reference frame. System effectiveness at meeting the established requirements is determined by measuring the (weighted) intersection of the two loci in the common reference frame. Where the approach in [12] and [13] is static, our approach adds time. Specifically, the intersection of L_p and L_r is measured at predisruption (prior to t_d) and postdisruption (at t_{min}) time periods and computed using Equation 4.1, yielding in a change of effectiveness per unit of time. Figure 4.2 shows an abstract visualization of rate of departure.

$$Tol_{RD} = \frac{\left[\dfrac{L_p \cap L_r}{L_p}, t_d\right] - \left[\dfrac{L_p \cap L_r}{L_p}, t_{min}\right]}{t_{min} - t_d} \tag{4.1}$$

Other means of measuring tolerance exist and are discussed in Pflanz [6]. For example, resilient systems also typically exhibit high fault tolerance: They continue providing their main functionality despite the occurrence of one or more

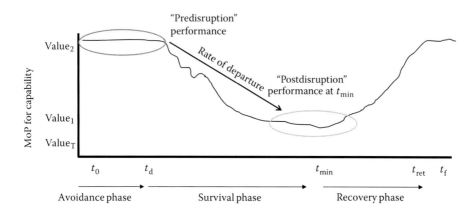

Figure 4.2 Abstract visualization of rate of departure.

element-level failures. A second measure of tolerance, fault tolerance, examines the elements that can fail prior to a loss of capability using cut vertexes. A third measure of tolerance, point-of-failure tolerance, examines the relatedness of individual failures to a loss of overall capability. When considering faults, it is important to understand the relatedness of failures at the element level to a loss of functionality or a loss of capability whether single element-level failures tend to induce a failure of the entire system or large portions of the system.

In contrast to tolerance, *flexibility* is the ability of a system to reorganize and adapt itself to changing conditions. Flexibility is an enabler of adjustment used by many systems to maintain their functionality during the changing conditions that follow a disruption. The graph theoretic interpretation of Petri Nets can be used to examine flexibility. Valraud and Levis [14] demonstrated the use of Petri Net place-invariants to describe information flow paths and functionalities in an architecture. In their approach, a simple information flow path (obtained from a place-invariant) corresponds to a simple functionality of the system described by the architecture. A complete information flow path is obtained by coalescing all of the simple information flow paths terminating in a common sink. A complete information flow path corresponds to a complete functionality described by the architecture and defined as the partially ordered set of functions that generate a specific output. A capability is then the instantiation of one or more related complete functionalities. A well-known technique to solve for the place-invariants of Petri Nets is provided in [15].

The flexibility of an architecture proposed for a certain capability can be measured by proportion of use (PoU), which reflects the fraction of the total elements used by any given simple functionality to deliver the overall capability. For example, does the average functionality use 10% of the elements or 80% of the elements supporting that capability? Systems with low PoU are more resilient to a disruption because each element is involved in comparatively fewer simple functionalities and easier to reorganize because elements are less extensively used in the capability.

Systems with high PoUs are less resilient to disruption because elements tend to be involved in comparatively more simple functionalities for a given capability and more difficult to reorganize because each element is extensively involved in multiple simple functionalities needed to deliver the overall capability. PoU implies substitutable elements. A separate measure further described in [6], fault tolerance, uses cut vertices to indirectly examine element criticality and loss of functionality. PoU is defined in Equation 4.2. A second means of measuring flexibility using the graph-theoretic properties of Petri Nets is defined in Liles [11].

$$
\text{PoU} = \frac{\displaystyle\sum_{i=1}^{r} B_i}{\frac{E}{r}} = \frac{\displaystyle\sum_{i=1}^{r} B_i}{rE} \tag{4.2}
$$

where

r = total number of information flow paths ℓ
B_i = number of elements E contained by path ℓ_i
E = total number of elements

There are three primary means of addressing *capacity* when time is also considered. Buffering capacity is the capability margin available immediately at the time of disruption or attack. Reactive capacity accounts for the fact that certain systems are able to bring additional capacity on line after a given reaction time, defined as t_{rc}. This allows for the system to increase capacity to some maximum value, V_{max}. Given that a system survives a disruption, residual capacity describes the remaining capacity above the threshold requirements and captures system vulnerability to a follow-on disruption that might occur in quick succession to the original disruption. Figure 4.3 describes how to compute each aspect of capacity when considering time.

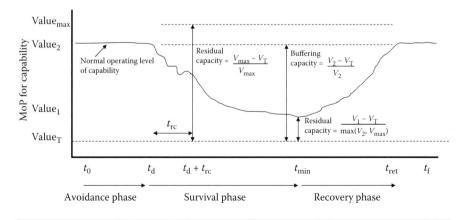

Figure 4.3 Measuring capacity.

Assuming that measures for each attribute of resilience (capacity, tolerance, flexibility) can be computed, a holistic evaluation of system resilience can be produced. This is accomplished by following these steps:

1. Select the appropriate metric for each attribute.
2. Measure the architecture's performance against each metric.
3. Compare the architecture's performance against a required performance level for each attribute.
4. Apply the system effectiveness analysis approach [12,13] to obtain an overall resilience measure.

Resilience-related improvements to the design can now be quantified, and alternative architectures can be compared. The idea is to evaluate the resilience performance of the baseline architecture against the resilience requirements established by the system developers. Then either compare the baseline against alternative architectures or make improvements to the baseline to move its performance into a desired range. This is illustrated in Figure 4.4. The three attributes of resilience, tolerance, flexibility, and capacity, are the axes of the performance space. The evaluation of Architecture A leads to the point on the left in Figure 4.4. The L_r is depicted

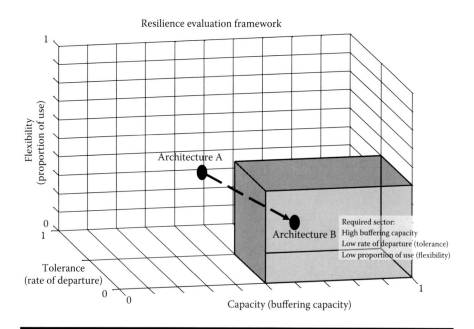

Figure 4.4   Resilience evaluation.

as the shaded box in the figure; any combination of values of the three attributes that leads to a point inside the box satisfies the resilience requirements. Clearly, Architecture A does not meet the resiliency requirements. However, redesign of the architecture leads to Architecture B, which results in a set of values for the attributes that meet all resiliency requirements. How well they meet them depends on the selected measure of effectiveness.

4.4 Decision-Making Organization Case Study

This case study involves a new organization called the Maritime Operations Center (MOC). As the United States' presence and engagement continues on a global scale, the US Navy is transitioning portions of its command and control organizations to a MOC structure. A MOC is a large, distributed organization at the fleet level with command and control responsibilities to "manage [routine] operations and be able to smoothly transition from peacetime operations to disaster relief operations and major combat operations, while still handling fleet management functions" [16]. The MOC is organized to support a Joint Force Maritime Component Commander (JFMCC). The MOC receives orders from the JFMCC, conducts planning operations, and generates operations orders (OPORD) for execution by the units assigned to the MOC. Figure 4.5 shows a picture of ships from the US Navy Fourth Fleet undergoing training exercises during MOC certification accreditation [16]. This case study used a baseline and augmented MOC models constructed by the GMU System Architectures Laboratory staff as a foundation, made several modifications, and then applied the approach described in this chapter.

Figure 4.5   US Fourth Fleet MOC, International Exercise PANAMAX 2008.

4.4.1 Process Description

The MOC model used in this case study* involves six major decision-making (DM) organizations: assessment, operational intelligence, future plans, command, current plans, and current operations. These organizations work in concert to conduct command and control of Naval and Joint forces on the surface, below the surface, in the airspace, and ashore.

Augmentation is a typical strategy used by many human organizations for dealing with crises and uncertainty in workload. This case study compares two different candidate architectures for the MOC: a baseline MOC and an augmented MOC, and the augmented MOC includes additional nodes for operational intelligence and future plans, such that cross talk exists between nodes. These additional nodes, once invoked, require time to establish and are available after a given reaction time.

A primary capability of the MOC is to generate mission orders for subordinate unit execution based on incoming JFMCC orders (higher HQ). The appropriate MoPs in this case are the mission order generation rate, stated as the number of mission orders generated per 24 h, and the average system time from when an order from higher headquarters is received to the time at which it is disseminated to subordinate units as an OPORD at a rate per 24 h. For example, if an order takes 4 h to process, the mission order generation rate is 6 orders per 24 h.

Orders arrive at the MOC from the JFMCC approximately every 3.5 to 4 h with an execution time of 24 h later. If the MOC spends more than 8 h to generate mission orders for its subordinate units, then the subordinate units do not have sufficient time to conduct their own planning, move into position, and execute the mission. This is essentially an extension of the traditional one third : two thirds planning rule that the military uses, according to which higher units do not take more than one third of available time to ensure lower units can successfully execute the mission. Therefore, if the MOC takes longer than approximately 8 h to generate mission orders (i.e., falls below a mission order generate rate of 3 per 24 h), the mission is put in jeopardy because subordinate units may not be able to execute in time.

Like most operations centers, the MOC is dependent upon software to automate and improve its functioning. In this case study, we are examining the resilience of the MOC's capability to generate mission orders to the disruption "loss of situational awareness software." When a new release of the software was received, the update caused both versions to crash, and attempts to restart were unsuccessful.

Loss of this software affects the information-fusion stage of each DM organization, extending the process time associated with that step. Each DM organization can still complete the "generate mission order" process, but the process transitions to a manual backup and requires a longer time to complete. In this case, the manual

---

* Information regarding the MOC used in this case study is based on a George Mason University Systems Architecture Laboratory (SAL) research project for the Office of Naval Research (ONR) under contract number N00014-08-1-0319.

process takes three to five times as long as the software-supported information-fusion process. The software failure occurs at t_d = 48 h. Twenty-four hours are required to bring additional (augmented) capacity online; therefore, t_{rc} = 24 h.

4.4.2 Architecture

An organizational architecture, a potential design for the MOC, is depicted in Figures 4.6 (base MOC) and 4.7 (augmented MOC). Note that in the augmented MOC, an additional operational intelligence and additional future plans cells are

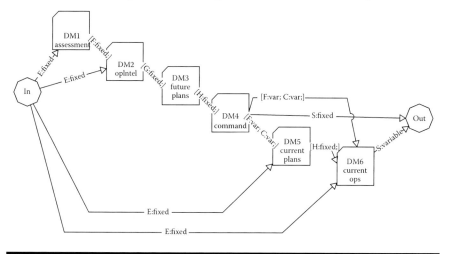

Figure 4.6   Base MOC organizational design.

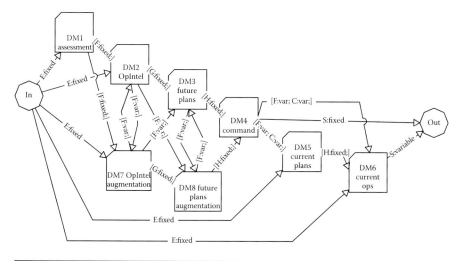

Figure 4.7   Augmented MOC organizational design.

added with cross talk to the original cells. Each DM organization is shown as a modified rectangle; the arcs represent fixed and variable connections (interactions) between DM organizations by which information (or signals) is passed. Fixed connections between decision nodes indicate interactions that do not vary whereas variable connections may change between situations.

Remy et al. [17] introduced a four-stage (later expanded to a five-stage [18]) interacting DM model based upon Petri Net theory and the lattice algorithm. Each DM organization is modeled using the five-stage DM model; therefore, each DM organization shown as a rectangle in Figures 4.6 and 4.7 can be mathematically described using a Petri Net with interactions defined in Remy et al. [17]. See Figure 4.8 from Kansal et al. [19].

The individual DM nodes receive either a signal from the external environment or from another DM node. The situation-assessment (SA) stage represents the processing of the incoming signal to obtain the assessed situation that may be shared with other DMs. The DM can also receive situation-assessment signals from other DMs within the organization; these signals are then fused together in the information-fusion (IF) stage to produce the fused situation assessment. The fused information is then processed at the task-processing (TP) stage to produce a signal that contains the task information necessary to select a response. Command input from superiors is also received. The command-interpretation (CI) stage then combines internal and external guidance to produce the input to the response-selection (RS) stage. The RS stage then produces the output to the environment or to other organization members [19]. Using the theory outlined in [17], the software implementation of the Lattice algorithm is used to generate feasible solutions that represent all possible architectures that meet a set of defined constraints. These solutions are represented as ordinary Petri Nets. Figure 4.9 is a Petri Net representation of the DM organization shown in Figure 4.7.

In this case, the primary output of the "generate mission orders" capability is a mission order. The places P53 and P55 shown in Figure 4.9 are the primary components of that mission order, and T5 is the transmission of that mission order to

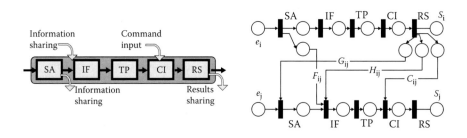

Figure 4.8 Five-stage model of each DM node. (From S. K. Kansal et al., Computationally Derived Models of Adversary Organizations, *Proc. IEEE Symp. on Computational Intelligence for Security and Defense Applications*, Honolulu, HI, April 2007.)

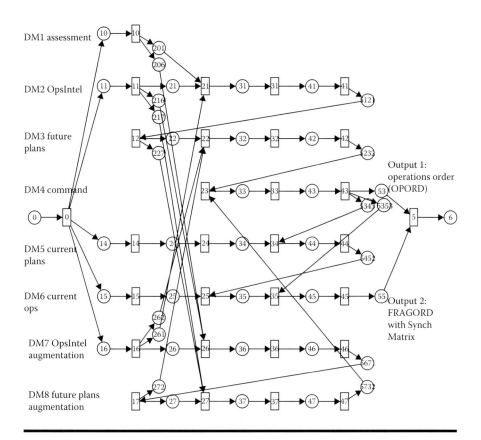

Figure 4.9 Augmented MOC universal net in Petri Net form.

subordinate units. For example, the primary output of the future operations cell is an OPORD, corresponding to P53. The primary output of the current operations cell is a fragmentary order (FRAGORD) situation report and a synchronization matrix, corresponding to P55. A subordinate unit will execute the mission when either or both components are present; however, it will not execute if neither is present.

Using the generated Petri Net, we can next add further necessary logic to the net and instrument it to support simulation. Care is taken to ensure that changes do not affect the overall structural properties of the original net (for example, to change the nature of the information flow paths). Time was added to the original Petri Net and appropriate stochastic delays estimated for each step to represent the amount of time required for each task. The arcs were inscribed to ensure a single incoming mission order from a higher headquarters is matched up correctly as different organizations within the MOC perform their roles (i.e., when the OPORD is approved in the command cell, it matches the Synch Matrix and FRAGORD from the current operations cell.) The resulting Petri Net is shown in Figure 4.10.

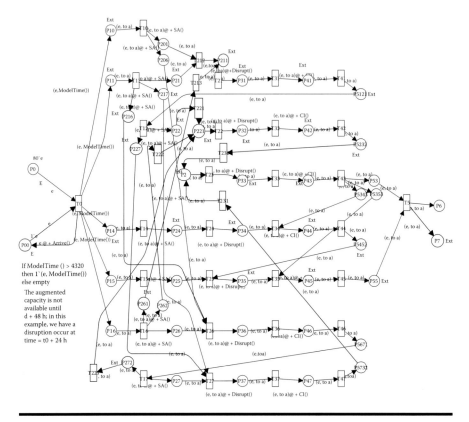

Figure 4.10 Petri Net for the augmented MOC used in simulation.

4.4.3 MOC Case Study Results

Once the architecture is developed, sufficiently verified, and any errors/revisions are addressed, it may be used to support the analyses described in Section 4.2. To demonstrate fully the approach, all measures of capacity, tolerance, and flexibility are calculated. However, an architect with the overall development team could, in principle, investigate only those measures that are of special interest and have been endorsed by the stakeholders.

4.4.3.1 Capacity

The results of simulations using the Petri Net model of the MOC are used to calculate measures for buffering, reactive, and residual capacity. The MOC operates under normal conditions between times t_0 and t_{48}. At t_{48} ($t_d = t_{48}$), a disruption occurs; in this case, it is the failure of the information-fusion software. The time to execute the information-fusion step in the MOC mission order generation capability increases as the MOC staff switches from the automated software-based approach to a manual approach. At time t_d, augmented capacity is requested; however, it takes 24 h to stand up this augmented capacity and integrate it into the existing MOC command and control structure. Therefore, $t_{rc} = 24$. The augmented capacity comes in the form of an additional intelligence cell and an additional future plans cell. Essentially, the MOC is augmenting with additional manpower to retain its capability to generate mission orders in a timely manner.

Figure 4.11 reflects the architecture's modeled simulation results during the course of the scenario. The MoP for the generate mission orders MoE is shown on the vertical axis as "mission order generate rate" (orders/24 h). The time is shown on the horizontal axis. Starting at time t_0, the MOC performs under normal predisruption performance levels with respect to the capability to generate mission orders. At this point, the MOC is capable of generating mission orders in just over 4 h, approximately. From the model results, this translates into an average of 5.67 mission orders every 24 h. The situational awareness (information fusion) software fails at time t_{48}, and the mission order generation rate falls off dramatically as the MOC switches to manual backup procedures. At the minimum point of performance ($t_{min} = t_{53}$), the MOC is barely at the threshold level of performance of approximately 8 h to generate a mission order or three mission orders per 24 h. By time t_{72}, additional capacity (manpower) has been integrated to stand up an augmented future plans cell and augmented operational intelligence cell. These additional cells are able to restore a part but not all of the original capability in terms of the mission order generation rate MoP.

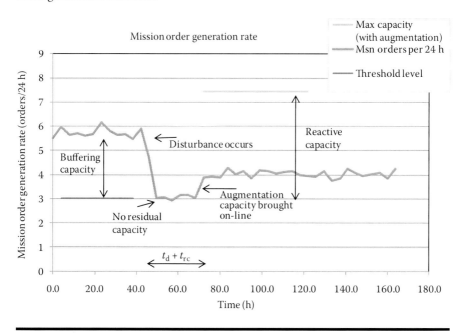

Figure 4.11   Measuring capacity in the augmented MOC.

In Figure 4.11, the lower line denotes the threshold capacity, set in this scenario as three mission orders generated every 24 h as described earlier. The topmost line indicates the maximum performance the MOC could achieve with respect to this capability if the augmented capacity were in place but no disruption had occurred. This was calculated by simulating the architecture with the augmented capacity in place but without the effects of the disruption. Comparing the mission order generation rate to the threshold value establishes the measure of effectiveness for capacity.

The primary difference between the two alternative architectures under examination in this case study is that the augmented MOC is able to generate reactive (spare) capacity, and the base MOC is not. Maximum capacity, when augmentation is available, was determined by running the simulation model without the effects of the disruption and with the spare capacity in place. This was completed by slight modifications to the inscriptions on the arcs in the Petri Net shown in Figure 4.10.

Using the equations for buffering, reactive, and residual capacities (see Figure 4.3), we find the results shown in Table 4.1. When operating under predisrupted conditions, approximately half (47%) of the MOC's capability was above the required threshold of three mission orders per 24 h. During the survival phase (postdisruption), the MOC was operating at the threshold of three orders per 24 h. However, as the calculations indicate, no residual capacity exists, meaning that any further disruption could have resulted in catastrophic failure in terms of mission completion. The MOC was operating close to an edge in performance. Additional manpower assisted in raising MOC performance above the threshold but did not return it to predisruption levels. The simulation results indicate that only restoration of the failed situational awareness software would return the MOC to predisruption performance levels. If the reactive capacity (the augmentation cells) were in place with no disruption, then 60% of the MOC's capacity would be above threshold.

4.4.3.2 Tolerance

As with the capacity-related metrics, the rate of departure metric is determined by employing an executable model of the architecture to assess performance achieved

Table 4.1   Determining Capacity in the MOC

Max Capacity (w/Augmentation)	7.44	V_{max}
Threshold Level	3.00	V_T
Normal Operating Level	5.67	V_2
Buffering Capacity	47%	
Reactive Capacity	60%	
Residual Capacity	0%	V_1

against performance required. The rate of departure (Tol$_{RD}$) was defined as the rate of change over time in system effectiveness in meeting its requirements after a disruption occurs. This encapsulates both the temporal aspects of resilience (t_d and t_{min}) as well as the effectiveness aspects of how the system performs with respect to its requirements and how effectiveness changes during the survival phase (postdisruption).

A parameter locus is generated to account for how key parameters affecting performance may vary during the scenario. The mission order inter-arrival time is an important parameter because it represents how quickly mission orders arrive from the JFMCC. Inter-arrival time of orders from higher HQ (JFMCC) is varied to examine the effect of queuing as the MOC executes the mission order process based on those JFMCC orders. The disruption involved loss of the situational-awareness software supporting the information-fusion stage of the MOC. Because this drives the cells within the MOC to use manual means, the time required for the information-fusion tasks performed is varied to reflect various manual task durations. These two variables are included in the parameter locus shown in Figure 4.12.

The L$_r$ is determined based on the specific variables of interest to the system under study. In the case of the MOC, the average mission order generation rate and the percentage of orders delivered late to subordinate units are important. Figure 4.13 shows the L$_r$.

Average mission order generation rate: Number of mission orders generated per 24 h. Per the one third:two thirds planning rule, higher units do not take more than one third of available time to ensure lower units can successfully execute the mission. If the MOC takes longer than approximately 8 h to generate mission orders (three per 24 h), subordinate units may not be able to execute in time.

% orders delivered late: Percentage of mission orders delivered to subordinates more than 8 h after receipt at MOC out of the total in the first 48 h following the

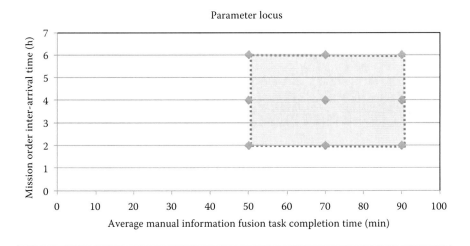

Figure 4.12   Parameter locus for the MOC.

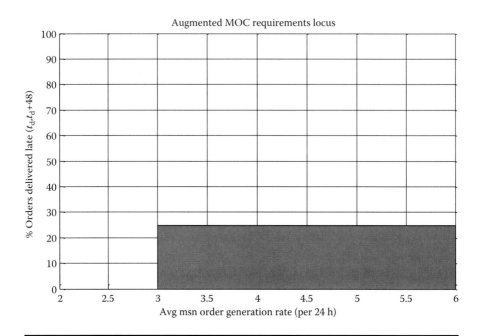

Figure 4.13   Augmented MOC requirements locus.

disruption. This addresses the effect on subordinate units. A threshold of one in four (25%) is established for this requirement.

Executing the architecture at each point in the parameter locus (Figure 4.12) yields the L_p. Figure 4.14 displays predisruption performance during which data is collected before the disruption occurs. In the augmented MOC, mission order generation times are well within requirements, and zero orders are delivered late to subordinate units in any portion of the parameter space.

Prior to the disruption, we can see that the augmented MOC is very effective at the mission order generation process. Zero orders are delivered late to subordinate units within the parameter space, and the order generation rate is well within the required level of effectiveness. Executing the architecture again at each point in the parameter locus but *after* a disruption yields a second L_p. Figure 4.15 displays postdisruption performance during which data is collected during the survival phase after the disruption occurs. After the disruption, the mission order generation rate is slowed as the information-fusion process requires more and more time. For certain cases, up to half of the orders in the 48 h following the disruption were delivered late. While augmented capacity is available in the augmented MOC, it is not available until after the augmentation cells are established, approximately 24 h after being called for. Mission order generation is highly dependent on software to enable tasks. Loss of situational awareness software causes a reversion to manual information-fusion methods with much longer

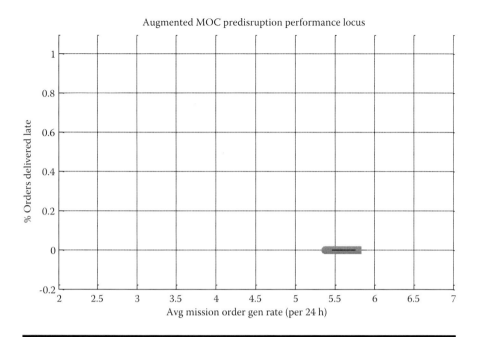

Figure 4.14   Augmented MOC predisruption performance locus.

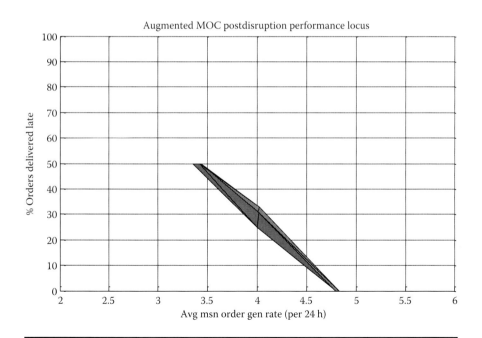

Figure 4.15   Augmented MOC postdisruption performance locus.

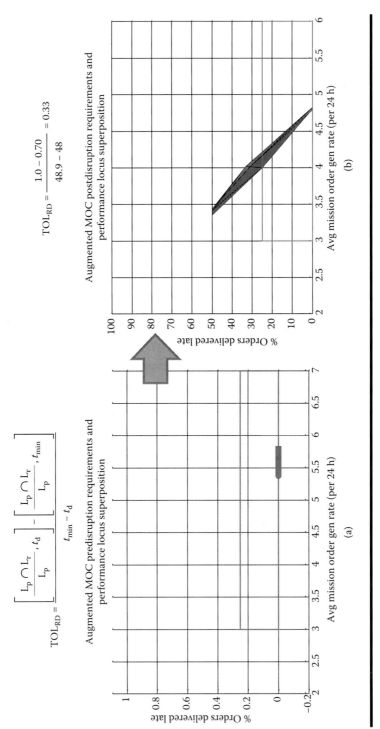

Figure 4.16 Computing rate of departure in the MOC case study.

processing times. These problems are reflected in the degraded performance seen postdisruption.

Prior to time t_d, the performance of the system meets the requirements over 100% of the parameter space (see Figure 4.16a). After the disruption occurs, the system performance meets the requirements in only 70% of the parameter space, showing a significant loss of capability after the disruption (see Figure 4.16b). The MOC DM architecture degraded from 100% to 70% effectiveness over a course of ~1 h on average (although the event was instantaneous, the effects take time to occur fully). The rate of departure is ~33% per hour loss of effectiveness (Figure 4.16).

Note here that because the augmented capacity is not available in time prior to the disruption reaching its full effect, the rate of departure for the base MOC and augmented MOC are essentially equivalent. Additionally, care should be taken in determining the time at which the minimum performance is assessed using Equation 4.2 as shown in Figure 4.16. Because we are typically considering stochastic systems, the absolute point of minimum performance could skew the calculation of Equation 4.2. It is recommended to use the point at which the system enters this new state of degraded performance rather than the numerically absolute minimum performance that could significantly change the denominator of Equation 4.2. In the MOC case study, we used the time at which the system enters the area of minimum (i.e., disrupted) performance versus the absolute time of minimum performance. See Figure 4.17.

In addition to being executable (supporting simulation), Petri Nets have a graph theoretic interpretation that enables the analysis of properties. The identical model

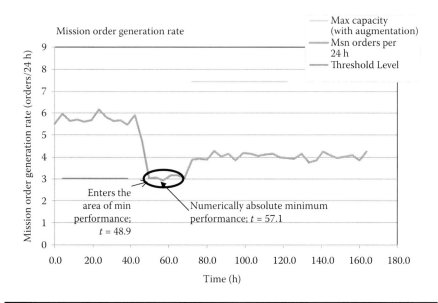

Figure 4.17 Area of minimum performance versus numerically absolute time.

used in the simulations above (see Figures 4.10 through 4.17) was also analyzed in static form to assess other aspects of tolerance as well as flexibility. As described in Section 4.2, examining these other aspects of tolerance and flexibility require an ability to determine the information flow paths that form the simple functionalities describing the overall capability under study. The information flow paths are derived from the place invariants in the architecture. The software application CAESAR III generates the simple information flow paths associated with this net. Figure 4.18 shows an example simple information flow path of the augmented MOC.

Fault tolerance (Tol_{FT}) is a measure that uses the graph-theoretic properties of Petri Nets. Recall that fault tolerance (Tol_{FT}) is the ratio of simple information flow paths that may be disrupted prior to the loss of the capability to the total number of simple information flow paths. Those elements of the sub-graph (vertices) that can be removed without disconnecting the sub-graph or eliminating the complete functionality (capability) are those that may be disrupted.

For the Petri Net shown in Figure 4.10, there are 44 simple information flow paths, containing as many as 37 elements or as few as 13 elements out of a total of

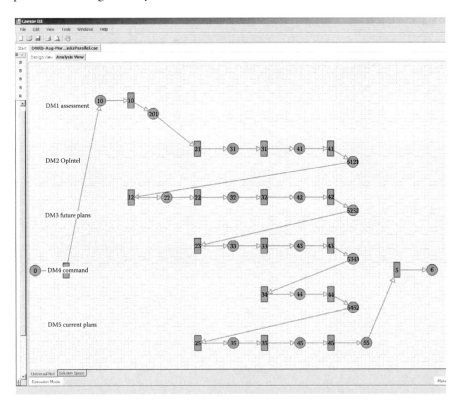

Figure 4.18   Example simple information flow path of the augmented MOC.

74 elements contained in the Petri Net of the augmented MOC. This large number of information flow paths results from the high level of interconnectivity and redundancy within the augmented MOC organizational design. The base MOC contains only eight information flow paths; all eight are contained by the base MOC.

The methodology described in Section 4.2 is used to determine the cut vertices needed to compute the fault tolerance. Any vertex that is on every path from sources S_i to sink U_j is a cut vertex. The augmented MOC includes one source ($p0$) and two sinks ($p53$, $p55$). (With sinks defined as $p53$ and $p55$, this eliminates the need to consider elements $t5$ and $p6$, which are elements after the sinks as designated above.) We can partition the above set of information flow paths into the 14 that have $p53$ as a sink and the 30 that have $p55$ as a sink. Solving for the elements common to every path from source $p0$ to sink $p53$ and from source $p0$ to sink $p55$, we find the following cut vertices:

> sink p53 cut vertices: {p33,p43,t0,t23,t33,t43}
> sink p55 cut vertices: {p45,t0,t35,t45}

There are no cut vertices common to both sinks except for $t0$. The set of non-cut vertices (V_{nc}) includes every other element. In this example, every information flow path contains many non-cut vertices, meaning multiple flow paths exist to connect each source to its corresponding sink. In the augmented MOC, there are 44 information flow paths so that $r = 44$. Because every flow path contains multiple non-cut vertices, meaning multiple flow paths exist to connect each source to its corresponding sink,

$$\ell_1 \dots \ell_{44} \text{ each contain members of } V_{nc} \text{ therefore } x_{1..44} = 1$$

$$\text{Tol}_{FT} = \frac{x}{r} = \frac{\sum_{i=1}^{r} x_i}{r} = \frac{44}{44}$$

The MOC with augmented capacity displays maximum fault tolerance. Every information flow path can be disrupted in some way without a loss of capability. This result is not surprising, given the extensive redundancy and interconnectivity built into the augmented MOC organizational design. Additionally, the parallel dissemination of information from $t0$ fosters an embedded redundancy in the transmission of information. This parallel transmission structure is typical in military organizations, in which a "warning order" is often broadly disseminated to initiate early planning activities. What this means is that multiple paths exist connecting the source to each sink, such that the elimination of a single element does not result in elimination of the overall capability.

The point of failure tolerance (Tol_{PF}) examines a situation different from fault tolerance. Tol_{PF} captures the relatedness of a local failure of any given element to the broader loss of functionality or loss of capability. More generally, Tol_{PF} addresses whether an element-level failure induces a system-level failure. This is accomplished by examining the localization of failures during a disruption. Elements that are a member of only one simple information flow path are said to have localized failure effects. The equation for tolerance yields

For the augmented MOC,

$$\text{Tol}_{\text{PF}} = \frac{\sum_{j=1}^{E} q_j}{E} = \frac{q}{E} = \frac{8}{71} = 0.11$$

For the base MOC,

$$\text{Tol}_{\text{PF}} = \frac{\sum_{j=1}^{E} q_j}{E} = \frac{q}{E} = \frac{8}{50} = 0.16$$

These results imply highly non-localized failures in both the augmented and base MOC. Only about 11% of the elements are associated with one information flow path in the augmented MOC and 16% in the base MOC. In the case of point of failure tolerance, higher is better because this indicates a higher proportion of elements with localized failure effects.

Point of failure tolerance is also intended to draw an architect's attention to areas in the design where greater attention may be required. In the augmented MOC, of particular interest is that 42 of the 44, or about 95%, of the information flow paths use elements *t23, p33, t33, p43,* and *t43,* and 30 of the 44, or almost 70%, of the information flow paths use elements *p45* and *t45.* From Figure 4.9, we can see this represents the command and current operations cells, respectively. While it is natural for a military organization to rely heavily on the commander to make decisions, a disruption affecting this portion of the organizational design would have broad-ranging consequences. The architect should direct attention at these portions of the architecture to determine if changes are required.

4.4.3.3 Flexibility

The final set of metrics to examine deal with flexibility, and flexibility refers to the ability of a system to reorganize and adapt itself to changing conditions. One measure of flexibility is cohesion as defined by Liles [11]. The cohesion measure examines the average cohesion of the individual nodes. A set of nodes with higher cohesion implies that the individual nodes are less flexible and less resilient.

We will examine flexibility in which each DM organization within the MOC identified in Figures 4.6 and 4.7 is treated as a node (i.e., assessment, operational intelligence, future plans, command, current plans, and current operations). Equation 4.3 measures cohesion within a node, and Equation 4.4 yields the average cohesion across all nodes.

$$\text{Coh}(nx_{ki}) = \frac{z_{ki}}{x_{ki}} \tag{4.3}$$

where k is the specific system, z_{ki} = number of paths in node n_{ki} and $x_{ki} = I_{ki} * Q_{ki}$ where I_{ki} = number of inputs for the node and Q_{ki} = number of outputs for the node. The overall cohesion is measured as

$$\text{Coh}(f_k) = \sum_{i=1}^{m} \text{Coh}(n_{ki}) \frac{1}{m} \tag{4.4}$$

where m is the number of nodes in the system—in this case, the organization's cells.

Executing Equations 4.3 and 4.4 yields the results shown in Figure 4.19. These results show that the augmented MOC is less cohesive (cohesion is 81%) than the

Cohesion (Mult nodes) augmented MOC

Node	Inputs	Outputs	Paths	$\text{Coh}(n_{ki})$
DM1	1	2	2	1.00
DM2	3	3	5	0.56
DM3	3	2	4	0.67
DM4	2	3	6	1.00
DM5	2	1	2	1.00
DM6	3	1	3	1.00
DM7	3	3	5	0.56
DM8	3	2	4	0.67

$m = 8$ $\qquad\qquad$ $\text{Coh}(f_k) = 0.81$

Cohesion (Mult nodes) base MOC (nonaugmented)

Node	Inputs	Outputs	Paths	$\text{Coh}(n_{ki})$
DM1	1	1	1	1.00
DM2	2	1	2	1.00
DM3	1	1	1	1.00
DM4	1	3	3	1.00
DM5	2	1	2	1.00
DM6	3	1	3	1.00

$m = 6$ $\qquad\qquad$ $\text{Coh}(f_k) = 1.00$

Figure 4.19   Calculating cohesion in the MOC.

base MOC (cohesion is 100%; the base MOC is very inflexible) and therefore more flexible.

These results are somewhat intuitive. In this case study, we are essentially adding capacity for the intelligence and future planning functionality through augmentation cells, which provide a redundant capability in those areas. This should naturally increase the flexibility of the MOC as an organization. This approach quantifies that increase.

Liles [11] also introduces a second measure of flexibility, which we have renamed as common use. It measures the extent of reuse of the elements to support multiple simple functionalities that comprise the overall capability. Executing the equation for the common use yields the following:

Augmented MOC:

$$\text{Common Use (CU)} = \frac{\sum_{j=1}^{E} A}{E} = \frac{1308}{74} = 17.7$$

Base MOC:

$$\text{Common Use (CU)} = \frac{\sum_{j=1}^{E} A}{E} = \frac{222}{50} = 4.4$$

From common use alone, it is difficult to determine whether 4.4 versus 17.7 is an improvement or not. This is because there are 44 information flow paths in the augmented MOC but only eight in the base MOC. Therefore, the numbers for common use will be inherently different. The next section helps explain these metrics in a more comparable fashion to support evaluation.

We defined PoU as the relative proportion of elements used by any given simple functionality to deliver the overall capability. We note two principal advantages to this metric. First, it describes what proportion of the elements is contained within the average simple functionality of a capability. For example, does the average functionality use a relatively small (say 10%) or a relatively large (say 80%) number of the elements? As PoU increases, a disruption to a given element within a capability is more likely to have broad-ranging effects. Systems with a high proportion of use are more difficult to reorganize (less flexible) because each element is more related to each functionality. Second, proportion of use normalizes the common use such that one can compare different architectures from a common perspective. This allows us to determine whether a particular value for common use is comparatively high or low. Figures 4.20 and 4.21 show the results of computing PoU for the augmented and base MOC alternatives.

Inf flow path	# Elements contained by ℓ_i	Inf flow path	# Elements contained by ℓ_i
$\ell = 1$	27	$\ell = 24$	30
$\ell = 2$	27	$\ell = 25$	37
$\ell = 3$	37	$\ell = 26$	31
$\ell = 4$	31	$\ell = 27$	37
$\ell = 5$	37	$\ell = 28$	31
$\ell = 6$	31	$\ell = 29$	36
$\ell = 7$	19	$\ell = 30$	30
$\ell = 8$	13	$\ell = 31$	37
$\ell = 9$	27	$\ell = 32$	31
$\ell = 10$	26	$\ell = 33$	29
$\ell = 11$	27	$\ell = 34$	23
$\ell = 12$	27	$\ell = 35$	37
$\ell = 13$	26	$\ell = 36$	31
$\ell = 14$	27	$\ell = 37$	37
$\ell = 15$	19	$\ell = 38$	31
$\ell = 16$	27	$\ell = 39$	37
$\ell = 17$	27	$\ell = 40$	31
$\ell = 18$	27	$\ell = 41$	37
$\ell = 19$	27	$\ell = 42$	31
$\ell = 20$	19	$\ell = 43$	29
$\ell = 21$	37	$\ell = 44$	23
$\ell = 22$	31	$\Sigma B = 1308$	
$\ell = 23$	36	$E = 74$	
		$r = 44$	

$$\text{PoU} = \frac{\dfrac{\sum_{i=1}^{r} B_i}{E}}{r} = \frac{\sum_{i=1}^{r} B_i}{rE}$$

$$= \frac{1308}{74 \cdot 44} = \frac{1308}{3256} = 40.4\%$$

Figure 4.20 Proportion of use in the augmented MOC.

For the MOC with augmentation, PoU is 0.4, meaning that each simple functionality involves about 40% of the elements required to deliver the capability. In the base MOC without augmentation, each simple functionality involves approximately 56% of the total elements. From this perspective, we can say that the augmented MOC is more flexible. In the augmented MOC, a disruption to a given

Inf flow path	# Elements contained by ℓ_i
$\ell = 1$	27
$\ell = 2$	27
$\ell = 3$	37
$\ell = 4$	31
$\ell = 5$	37
$\ell = 6$	31
$\ell = 7$	19
$\ell = 8$	13
$\Sigma B = 222$	
$E = 50$	
$r = 8$	

$$\text{PoU} = \frac{\dfrac{\sum_{i=1}^{r} B_i}{E}}{r} = \frac{\sum_{i=1}^{r} B_i}{rE}$$

$$= \frac{222}{50 \cdot 8} = \frac{222}{400} = 55.5\%$$

Figure 4.21 Proportion of use in the base MOC.

element can be expected to affect about 40% of the overall functionality of the system under evaluation. In the base MOC, a disruption to a given element can be expected to affect about 56%.

4.4.3.4 MOC Resilience Results

In this case study, we have applied the individual components of the resilience evaluation approach for both the base MOC and the augmented MOC alternatives. The MOC is designed as a series of DM nodes with each node as a five-stage DM structure with interactions between nodes. This architecture was transformed into an ordinary Petri Net using the theory outlined in [17]. Necessary logic and instrumentation were added to the Petri Net such that it became an executable form of the MOC architecture suitable for behavioral and performance analyses.

It is critical to consider the resilience "of what, to what," focusing on the resilience of a capability to a disruption in a particular environment (under what conditions). In this case study, we examined the resilience of the MOC's capability to "generate mission orders" to the disruption "loss of situational awareness software." When a new release was received, the update caused both versions to crash, and attempts to restart were unsuccessful. The MOC transitioned from automated procedures based on the software to manual procedures and called upon augmented capabilities in the form of additional operations intelligence and future plans cells. These required additional time to establish. However, this augmented capability could not be established until well after the full effect of the disruption was felt.

For both cases of the MOC, the disruption brought the MOC's capability to the brink of not meeting the threshold. If another disruption occurred before the augmenting capacity could be brought online, the MOC would have been incapable of completing one of its key tasks, the generation of mission orders. The augmented capability did enable a portion of the MOC's mission order generation capability but not back to predisruption levels. Table 4.2 reports the results for each metric in the base MOC and the augmented MOC.

The architect, along with the overall development team, must consider which aspects of resilience are most important to the system under consideration. Table 4.2 assists the architect to determine which aspects of resilience are most applicable to the architecture definition and resilience issues at hand. In the case of the MOC and capacity, the most appropriate metric is buffering capacity. As was shown, reactive capacity is not available in time to play a role in the survival phase although it does distinguish the two candidate architectures, and, once in place, the augmented capacity offers a number of benefits. Further, because this was a non-malicious type of disruption, the residual capacity is of less concern because a follow-up disruption is not necessarily likely. For tolerance, the most appropriate metric is again rate of departure. In this case, rate of departure directly measured the time-sensitive nature of the MOC's capability to generate mission orders by assessing the rate of generation and the number of "late" orders delivered to subordinate units. In

Table 4.2 Resilience Metrics for the Base and Augmented MOC

Attribute	Metric	Measures	Question Answered	Augmented MOC	Base MOC
Capacity: "the ability to operate at a certain level as defined by a given measure"	Buffering capacity	Available capability margin between current operating levels and a defined minimum threshold operating level at the time preceding a disruption.	Can a disruption be absorbed with immediately available (on-hand) resources?	47%	47%
	Reactive capacity	Available capability margin between maximum operating levels (i.e., including any spare capacity) and a defined minimum threshold operating level.	Can a disruption be absorbed with the addition of spare capacity?	60%	0%
	Residual capacity	Available capability margin between operating levels at the end of the survival phase and a defined minimum threshold operating level.	Given survival, how vulnerable is the system to a follow-on disruption that occurs before the system can recover?	~0%	~0%

(continued)

Table 4.2 Resilience Metrics for the Base and Augmented MOC (Continued)

Attribute	Metric	Measures	Question Answered	Augmented MOC	Base MOC
Tolerance: "the ability to degrade gracefully after a disruption"	Rate of departure	Rate of change in system performance with respect to its requirements (i.e., rate of loss of effectiveness) after a disruption.	What level of capability is lost per unit of time during the survival phase?	0.33	0.33
	Fault tolerance	The ratio of simple functionalities that may be disrupted without a loss of capability to the total number of simple functionalities.	How many simple functionalities can be disrupted prior to losing the capability? Primarily a tool to draw the architect's attention to key areas in the design.	1.0	1.0
	Point of failure tolerance	Relatedness of failures at the element level to an overall loss of capability.	Are element level failures relatively localized, or do failures incur broad system-level effects? Primarily a tool to draw the architect's attention to key design areas.	0.11	0.32

Flexibility: "the ability of a system to reorganize its elements to maintain its capabilities"				
Cohesion	Relatedness of the elements within a node or module, which support a given capability.	How difficult is it to reorganize the system at the node/module level?	0.81	1
Common use	Extent of common use of the elements among the simple functionalities, which support the overall capability.	Can a system execute multiple functionalities concurrently, or is it limited by competition for resources?	17.6	4.4
Proportion of use	The ratio of the total elements used by any given simple functionality to deliver the overall capability.	Are most of the elements needed for a given functionality, making it more difficult to reorganize?	0.40	0.56

terms of flexibility, PoU was also selected because it directly addresses the ability of the system to be reorganized based on the average use of the elements across the simple functionalities supporting that capability. The cohesion metric is not as useful because the disruption is likely to take effect much more quickly than any reorganization could occur. This makes cohesion a metric potentially more useful in the "recovery" phase of resiliency. Additionally, the overlap in the common use and PoU was discussed, leading to a selection of PoU for this assessment.

Having determined which metrics are most important, resilience can be considered from the perspective of the extent to which performance meets requirements, that is, the extent to which the L_p intersects the L_r. The two alternative architectures can be compared in this manner. Determining the resilience L_r requires value judgment. The example is shown with a L_r:

> Buffering capacity ≥ 33%
> Rate of departure (tolerance) < 50%
> Proportion of use (flexibility) < 50%

Figure 4.22 shows the results of the overall evaluation. The augmented MOC meets the resilience attribute requirements of capacity, tolerance, and flexibility, making it the preferred candidate architecture from the point of view of resiliency. The base MOC lacks required flexibility but meets the other requirements.

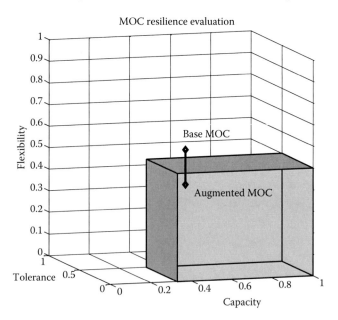

Figure 4.22 Resilience evaluation of the base and augmented MOC.

As we consider the performance of the two designs, base MOC and augmented MOC, the evaluation framework allows us to make useful quantitative comparisons. In the case of capacity, the two designs have equivalent buffering capacity and residual capacity. While the augmented MOC has greater reactive capacity, the augmented MOC cannot bring that reactive capacity on line fast enough to make a difference in the survival phase. In the case of point of failure tolerance, the base MOC actually performs better. This is because its failures are the most localized. More specifically, about one third (0.32) of the base MOC element failures are localized as compared to ~1/10 (0.11) of the augmented MOC. The greater interconnectivity of the augmented MOC accounts for this difference. In the case of flexibility, the augmented MOC performs best in terms of PoU. A smaller proportion of its elements, on average, are needed for a given functionality as compared to the base MOC (40% vs. 56%).

The primary advantage of the augmented MOC (additional reactive capacity) is not relevant during the survival phase; the augmented and base MOC perform equivalently in terms of buffering capacity. However, the reactive capacity does allow the augmented MOC to restore performance, making a future disruption less likely to have a catastrophic effect when compared to the base MOC. In terms of tolerance, the augmented MOC performs worse because its failures are less localized and therefore more likely to have broader, system-level effects. In terms of flexibility, the augmented MOC performs better; the fact that fewer elements are needed for the average functionality will allow the augmented MOC to be more easily reorganized.

4.5 Comments

In this chapter, the concept of resilience was described from a systems engineering perspective and a set of attributes and measures for calculating the resilience of a system were presented, that is, a quantitative means for evaluating the expected resilience of a system was described. The definitions, the evaluation process, and the calculations were presented through an example of an organizational architecture performing command and control. A key idea was to use measures for each of the attributes of resilience and then to combine these measures into a holistic assessment. By representing the architecture in a rigorous way using Petri Nets, the approach supports simulation of the architecture and the analysis of behavioral properties based on structure. This allowed us to examine the expected performance (by executing the Petri Net–based architecture) and the structural characteristics (e.g., the analysis of the simple information flow paths of the architecture using Petri Net s-invariants). The work focused on the survival phase of resilience and did not address the recovery or avoidance phases. While flexibility does, in part, address characteristics beneficial during a recovery phase, such as the ability to reorganize, further research is needed to identify a complete end-to-end assessment of resilience that would include the avoidance, survival, and recovery phases.

References

1. INCOSE, Resilient Systems Working Group homepage, http://www.incose.org/practice/techactivities/wg/rswg/.
2. S. Jackson, *Architecting Resilient Systems: Accident Avoidance and Survival and Recovery from Disruptions.* Hoboken, NJ: Wiley Series in Systems Engineering and Management, 2010.
3. A. Madni and S. Jackson, Towards a Conceptual Framework for Resilience Engineering, *IEEE Systems Journal*, Special issue on Resilience Engineering, No. 2, pp. 181–191, 2009.
4. S. R. Carpenter, B. H. Walker, J. M. Anderies, and N. Abel, From Metaphor to Measurement: Resilience of What to What? *Ecosystems*, Vol. 4, pp. 765–781, 2001.
5. Department of Defense (DoD), *DoD Architecture Framework (DoDAF) version 2.02*, August 2010. [Online] Available at http://cio-nii.defense.gov/sites/dodaf20/.
6. M. A. Pflanz, On the Resilience of Command and Control Architectures, PhD Dissertation, Volgenau School of Engineering, George Mason University, Fairfax, VA, November 2011.
7. ISO/IEC Systems and Software Engineering, *Recommended Practice for Architectural Description of Software-Intensive Systems*, ISO/IEC 42010, 2007.
8. K. Jensen and L. M. Kristensen, *Coloured Petri Nets: Modelling and Validation of Concurrent Systems*, Berlin, Germany: Springer 2007.
9. M. Weske, *Business Process Management: Concepts, Languages, Architectures*, Berlin, Germany: Springer-Verlag, 2010.
10. Guide to the Systems Engineering Body of Knowledge (SEBoK), *Resilience Engineering*, Available at http://www.sebokwiki.org/index.php/Resilience_Engineering.
11. S. W. Liles, On the Characterization and Analysis of System of Systems Architectures, PhD Dissertation, Department of Information Technology and Engineering, George Mason University, Fairfax, VA, August 2008.
12. V. Bouthonnier and A. H. Levis, Effectiveness of C³ Systems, *IEEE Transactions on Systems, Man, and Cybernetics*, Vol. SMC-16, No. 14, January/February 1984.
13. P. H. Cothier and A. H. Levis, Timeliness and Measures of Effectiveness in Command and Control, *IEEE Transactions on Systems, Man, and Cybernetics*, Vol. SMC-16, No. 6, November/December 1986.
14. F. R. H. Valraud and A. H. Levis, On the Quantitative Evaluation of Functionality in C³ Systems, in *Information and Technology for Command and Control*, S. J. Andriole and S. M. Halpin, Eds. IEEE Press, pp. 558–564, 1991.
15. J. Martinez and M. Silva, A Simple and Fast Algorithm to Obtain All Invariants of Generalized Petri Nets, *Informatik-Fachbrichte*, Vol. 52, pp. 301–303, Springer-Verlag, 1982.
16. US Navy Public Affairs, 2009. US 4th Fleet, Specialist 2nd Class Alan Gragg, *US. 4th Fleet Stands Up Maritime Operations Center*, 26 March 2009, [Online] Available at http://www.navy.mil/search/display.asp?story_id=43775 [Accessed August 2011].

17. P. A. Remy, V. Y. Jin, and A. H. Levis, On the Design of Distributed Organization Structures, *Automatica*, Vol. 24, No. 1, 1988.
18. A. H. Levis, Executable Models of Decision Making Organizations, in *Organizational Simulation*, W. B. Rouse and Ken Boff, Eds., Wiley, NY, 2005.
19. S. K. Kansal, A. M. AbuSharekh, and A. H. Levis, Computationally Derived Models of Adversary Organizations, *Proc. IEEE Symp. on Computational Intelligence for Security and Defense Applications,* Honolulu, HI, April 2007.

Chapter 5

Resiliency and Robustness of Complex Systems and Networks

Alexander Kott and Tarek Abdelzaher

Contents

The burgeoning field of network science has started to examine the fundamentals of multi-genre networks, which include traditional communication networks but also encompass information flow networks, social networks, and potentially other networks as well. The dependencies between these different genres (or types) of networks introduce additional complexities that must be addressed. However, the existence of these different genres can also be a benefit to resilience. In this chapter, Alexander Kott and Tarek Abdelzaher first introduce the notion of complex networks or multi-genre networks, followed by definitions of robustness and resiliency

67

in complex networks. The core part of the chapter examines several types of failures and approaches to make complex networks robust and/or resilient to these failures.

5.1 Introduction to Complex Networks

In this chapter, we explore the resiliency and robustness of systems while viewing them as complex, multi-genre networks. We show that this perspective is fruitful and adds to our understanding of fundamental challenges and tradeoffs in robustness and resiliency as well as potential solutions to the challenges.

When using the term *complex, multi-genre networks*, we refer to networks that combine several distinct genres: networks of physical resources, communication networks, information networks, and social and cognitive networks. As illustrated in Figure 5.1, a complex network may utilize a physical resource network (e.g., a sensor network to measure and report on phenomena of interest) to store information (e.g., a data cloud) and to communicate it to the stakeholders (e.g., the Internet or a courier network) and a network of human decision makers (and possibly artificial agents) that assess and comprehend the information and produce decisions.

One of the key commodities flowing through such a network are data elements that themselves are connected by links, either implicit (e.g., semantic relations) or explicit (e.g., URLs), forming a web, called an information network, from which reliable information is to be distilled. Other elements and links in the complex network could be of a physical nature, such as actuators that execute the decisions made by human or artificial agents or a power grid, a road network, etc. The combination of these heterogeneous networks forms the network ecosystem whose resiliency we wish to understand. A resilient ecosystem will continue to support sound decisions and actions even as physical infrastructure is damaged, sources are infiltrated, data is corrupted, or access to critical resources is denied.

Study of systems such as multi-genre networks is relatively uncommon; instead, it is customary in research and engineering literature to focus on a view of a network comprised of homogeneous elements, (e.g., a network of communication

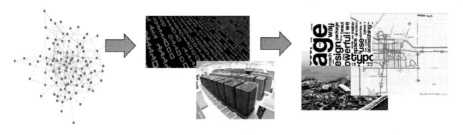

Figure 5.1   A complex network often includes networks of several distinct genres: networks of physical resources, communication networks, information networks, and social and cognitive networks.

devices or a network of social beings). Yet most, if not all, real-world networks are multi-genre; it is hard to find any real system of a significant complexity that does not include a combination of interconnected physical elements, communication devices and channels, data collections, and human users forming an integrated, interdependent whole.

The multi-genre perspective becomes more important as our society and sociotechnical systems are permeated with a growing number of communication links between humans and physical systems (e.g., the Internet, cell phones) and corresponding social links (e.g., the broad popularity and impact of social networking applications). Note that, when we consider multiple genres of networks, the total number of links that connect any two elements grows significantly. For example, two physical devices may not be connected by any physical or communication links, but if their human owners are part of the same social network, the devices are, in fact, connected through their human owners in ways that may be significant for understanding the devices' behavior. Additional links, and especially additional links of a heterogeneous nature, are likely to affect the network's robustness and resiliency.

For this reason, our ability to design, manage, and operate many industrial, financial, and military systems has become increasingly dependent on our understanding of a complex web of interconnected physical infrastructure, distributed computation, and social elements. It is the combined, interactive behavior of multiple genres of networks that determines the ultimate properties of the overall system, including robustness and resiliency. Further complexity arises because such systems often exist in adversarial environments and in the presence of resource loss, uncertainty, and data corruption (Figure 5.2).

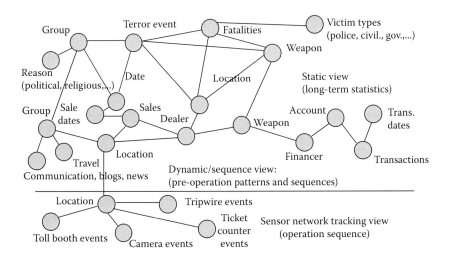

Figure 5.2 A counter-terrorism network: This complex network includes nodes and links that belong to multiple genres; information, communications, social.

Making the right decisions and, more generally, exerting the desired influence upon such systems, requires proper modeling of networked physical, informational, and social elements to understand their combined behavior, assess their vulnerabilities, and design proper recovery strategies. In recent years, research programs have been initiated to understand the interactions of dissimilar genres of networks (e.g., [32]).

5.2 Resiliency and Robustness

As discussed earlier in this book, robustness and resiliency are related yet distinct properties of systems. Robustness denotes the degree to which a system is able to withstand an unexpected internal or external event or change without degradation in system's performance. To put it differently, assuming two systems—A and B—of equal performance, the robustness of system A is greater than that of system B if the same unexpected impact on both systems leaves system A with greater performance than system B.

We stress the word *unexpected* because the concept of robustness focuses specifically on performance not only under ordinary, anticipated conditions (which a well-designed system should be prepared to withstand), but also under unusual conditions that stress its designers' assumptions. For example, in IEEE Standard 610.12.1990, "Robustness is defined as the degree to which a system operates correctly in the presence of exceptional inputs or stressful environmental conditions." Similarly, "robust control refers to the control of unknown plants with unknown dynamics subject to unknown disturbances" [25].

The resiliency, on the other hand, refers to the system's ability to recover or regenerate its performance after an unexpected impact produces a degradation of its performance. For our purposes, assuming two equally performing systems, A and B, subjected to an unexpected impact that left both systems with an equal performance degradation, the resiliency of system A is greater if after a given period, T, it recovers to a higher level of performance than that of system B.

For example, ecology researchers describe resiliency as "the capacity of a system to absorb disturbance and reorganize while undergoing change so as to still retain essentially the same function, structure, identity, and feedbacks" [26]. Alternatively, one may focus on the temporal aspect of the definition, in which resiliency is "the time required for an ecosystem to return to an equilibrium or steady-state following a perturbation" [27]. Note that, in complex nonlinear systems, the length of recovery typically depends on the extent of damage. There may also be a point beyond which recovery is impossible. Hence, there is a relationship between robustness (which determines how much damage is incurred in response to an unexpected disturbance) and resiliency (which determines how quickly the system can recover from such damage). In particular, a system that lacks robustness will often fail beyond recovery, hence offering little resiliency. Both robustness and resiliency, therefore, must be understood together.

Does the multi-genre network perspective offer anything to our understanding of robustness and resiliency? It does because it changes our perspective on the complexity of the links within the system. In his pioneering work, Perrow [28] explains that catastrophic failures of systems emerge from high complexity of links, which lead to interactions that the system's designer cannot anticipate and guard against. A system's safety precautions can be defeated by hidden paths, incomprehensible to the designer because the links are so numerous, heterogeneous, and often implicit. The greater connectivity we recognize in a multi-genre network helps us see more of the overall network's complexity and, hence, the potential influences on its robustness and resiliency.

For example, greater complexity of connections between two elements of the systems, such as the number of paths that connect them through links of various natures, may lead to increased robustness of the system by increasing the redundancy of its functions. The same increase in complexity, however, may also lead to lower robustness by increasing—and hiding from the designer—the number of ways in which one failed component may cause the failure of another. When we consider an entire multi-genre network—and not merely one of the heterogeneous, single-genre sub-networks that comprise the whole—we see far more complexity of the paths connecting the network's elements.

A system of military command—a multi-genre network that comprises human decision-makers, sensing, communication and computing devices, and large collections of complex interlinked information—exhibits complex modes of failures that become more likely as the degree of collaboration between the system's elements increases. Experimental studies of such systems [29] show how increased availability of information delivered through an extensive communications network may paralyze decision makers and induce them into a confusion or endless search for additional information. They also show how increased collaboration—made possible by improved networking—may mislead collaborators into accepting erroneous interpretations of the available information.

Of particular importance are those paths within the system that are not recognized or comprehended by the designer. Indeed, the designer can usually devise a mechanism to prevent a propagation of failure through the links that are obvious. Many, however are not obvious, either because there are simply too many paths to consider—and the numbers rapidly increase once we realize that the paths between elements of a communication system, for example, may also pass through a social or an information network—or because the links are implicit and subtle. Kott [30] offers examples of subtle feedback links leading to a failure in organizational decision making. It is even more difficult to comprehend an enormous number of complex, multi-genre links in large-scale societal effects, such as insurgency dynamics [31] (Figure 5.3).

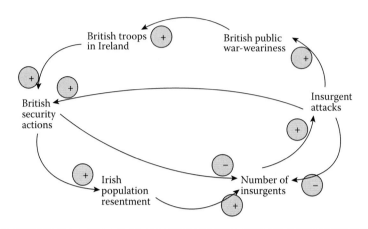

Figure 5.3 Even in this highly simplified network of insurgency dynamics in Northern Ireland, there are multiple, often nonobvious, paths between phenomena in the system.

Similarly, greater complexity of a network may have a range of influences on resiliency. As the number and heterogeneity of links grow, they offer the agents (or other active mechanisms) within the network more opportunities to regenerate the network's performance. These agents may be able to use additional links to more elements in order to reconnect the network, to find replacement resources, and ultimately to restore its functions. This does however require that the network include agents capable of taking advantage of the opportunities offered by increased complexity. On the other hand, greater complexity of the network may also reduce the resiliency of the network. For example, agents may be more likely to be confused by the complexity of the network or to be defeated in their restoration work by unanticipated side effects induced by hidden paths within the network. In either case, by considering the holistic, multi-genre nature of the network, the designers or operators of the system have a better chance to assess its robustness and resiliency.

5.3 Fundamental Tradeoffs

Most approaches to improving resiliency and robustness involve compromises, and the key challenge is to find a favorable compromise [33]. Such compromises involve reducing or managing the complexity of the network: coupling, rigidity, and dependency. We discuss several of these approaches in the ensuing sections of this chapter.

Resources versus resiliency: Additional resources in a network can help improve resiliency. For example, adding capacity to nodes in a power-generation and distribution network may reduce the likelihood of cascading failures and speed up the service restoration (as discussed in Sections 5.4 and 5.5). Adding local storage and influencing the distribution of nodes of different functions in a supply network also leads to improved resilience at the expense of additional resources (Section 5.6).

Performance versus resiliency: A network optimized for higher performance may experience greater degradation under an impact, and it may be more difficult and time-consuming to restore its operation (Section 5.4). For example, it was shown that "suboptimal" design that is risk-averse will generally avoid disastrous failure with a higher probability than the optimal one [2]. A power-generation and distribution network controlled by a sophisticated computer network will yield better performance yet may experience greater and harder-to-restore cascading collapse (Section 5.5).

Resiliency to one type of disruption versus resiliency to another disruption type: In order to improve a network's resiliency against a certain type of impact, designers may have to sacrifice its resiliency against another type of impact. For example, in a scale-free network, a targeted attack may produce much greater damage (and be more difficult to rectify) than a random attack or random failure (Sections 5.4 and 5.5).

Complexity versus resiliency: Improved resiliency can be obtained by incorporating more sophisticated mechanisms into a network. Complexity goes beyond merely adding resources to the system. Complexity implies potentially a more costly system and, most likely, greater investment into development, operation, and maintenance of the system. For example, resiliency may be improved by adding multiple functional capacities to each node (Section 5.6) by processing more input sources (Section 5.7) or by combination of multiple inference algorithms (Section 5.8). Yet the same complex measures might cause greater difficulties in restoring the network's capability if degraded by an unexpected—and probably harder to understand—failure.

Several tradeoffs often coexist and amplify the designer's challenge.

Finally, it is important to notice that, in multi-genre networks, there is an inseparable relationship between resiliency and robustness in that the latter is a precondition of the former. This relationship comes from the interactions that such systems have with their physical and social environments. To explain, observe that when we have a system with low robustness (e.g., small failures cascade to catastrophes) that is partly "physical," it is hard to build resiliency on top. This is because physical failures and human losses are often irreversible or expensive to remedy (e.g., once there is a reactor meltdown, it is hard to "roll back"). This characteristic is unlike purely logical systems (e.g., databases), in which roll back from failure is cheaper, making robustness and resiliency somewhat orthogonal concerns.

5.4 Robustness to Topology Modifications

The first precondition of resiliency in multi-genre networks is robustness to topology modifications. This robustness allows the network to survive local damage (that results in topological change). Most biological beings, for example, are extremely resilient in that they are able to recover remarkably well from local injury or damage. However, resiliency is not limitless. There are bounds on injury or illness within which resiliency mechanisms work well. If damage exceeds those bounds, recovery, in general, cannot be attained. Hence, in order to engineer resilient systems, it is important to understand their underlying robustness properties first because these properties help understand the bounds within which recovery (and hence resiliency) is possible. Systems with low robustness will have a very limited range in which recovery is efficient. The infrastructure on which multi-genre networks rest consists of multiple interdependent physical networks, such as a road network, a data network, or the power grid. The first key characteristic that determines response to failure of such networks is their topology. It is therefore interesting to ask the following question: In a system described by an arbitrary graph or interconnected network, what topologies are inherently more robust?

Much prior research addressed the fundamental vulnerabilities of different networks as a function of their topological properties. Of particular interest has been the classification of properties of networks according to their node degree distribution. While some networks (such as wireless and mesh networks) are fairly homogeneous and follow an exponential node degree distribution, others, called scale-free networks (such as the web or the power grid), offer significant skew in node degrees, described by a power law. In a key result by Albert, Jeong, and Barabási [3], it was shown that scale-free graphs are much more robust to random node failures (errors) than graphs with an exponential degree distribution but that these scale-free graphs are increasingly more vulnerable to targeted attacks (namely, removal of high-degree nodes). The above observations are intuitive and can be explained by the difference in topology between scale-free and exponential networks.

An exponential node degree distribution means that nodes with a larger degree are exponentially less probable. This makes the network more homogeneous. In contrast, in a scale-free network, node degree is distributed according to a power law. The power law distribution has a heavy tail, which means that some nodes are extremely well connected whereas most nodes are not. The aforementioned difference in the degree of homogeneity versus skew in the two types of networks explains the difference in their robustness properties. Specifically, homogeneous networks, by the very nature of their homogeneity, feature uniform degradation as nodes are removed and offer no significant difference in behavior when the most connected nodes are removed first. This is because all nodes contribute roughly equally to network connectivity. In contrast, power-law networks, due to their skewed distribution, are less sensitive to random removal, which tends to affect the less connected nodes more (because there is more of them) and hence tends to have a less significant effect on connectivity compared to inhomogeneous networks. Targeted removal, on the other hand, causes much more damage by eliminating the relatively few highly connected nodes.

For the same reason, as one might expect, when nodes are randomly removed, the size of the largest remaining connected component in the network is bigger in the case of power-law networks than in the case of exponential networks. However, if highest-degree nodes are removed first, the opposite is true. This has implications on the extent of effort needed, for example, to partition the network, both in the case of an attack and in the case of random failures. Callaway, Newman, Strogatz, and Watts [4] describe an analytic technique for quantifying the above trends by relating them to appropriate percolation problems in random graphs with arbitrary degree distribution. The underlying basic theory for deriving properties of networks of arbitrary degree distribution is described in [1].

The above results are consistent with the intuition that performance optimizations exacerbate worst-case behavior. Scale-free networks perform better in the sense of remaining connected longer despite random errors, but the very mechanism that allows them to perform better (namely, the existence of a few very well-connected nodes) also causes increased vulnerability in the worst-case scenario. This property is often called *robust yet fragile* and is a characteristic of many large-scale complex networks [23].

While we cannot change the topology of a large network, such as the Internet, in order to improve robustness as a foundation for resiliency, the network should have a mechanism to reduce dependencies. Dependencies, in this context, exist because performance depends on the availability, capacity, and latency of the few well-connected nodes. In a later section, we discuss how such dependencies are relaxed. For example, on the world wide web, one of the most common power-law networks in use, content is simply replicated (cached). Retrieving the most popular content from a cache reduces dependence on the well-connected sources (e.g., data centers) and backbone routers at the expense of some degradation in content freshness.

5.5 Robustness to Cascaded Resource Failures

The next step in understanding robustness as a precondition of resiliency is to understand robustness of a network to cascaded failures. Such failures are non-independent in that one triggers another. A network that is prone to large "domino effects" will likely sustain severe damage in response to even modest disturbances, which significantly limits the scope within which efficient recovery (and hence resilient operation) remains possible. Hence, below, we extend the results presented in the previous section to non-independent failures.

Multi-genre networks exhibit pathways by which component failures may cascade from one node to another—an issue not discussed above. It is of great interest to analyze the susceptibility of networks to cascaded failures, such as large blackouts or congestion collapse. Early mathematical analysis of failure cascades as a function of different network topologies was published by the National Academy of Sciences [8]. A model was assumed, in which a node fails if a given *fraction* of its neighbors have failed. The model, although applicable to a class of failure propagation scenarios, was intended to be general enough to describe other cascades as well, such as rumor propagation and propagation of new trends. Consistently with results analyzing independent failures, it was shown that, under this propagation model, scale-free networks are less susceptible to propagation cascades emanating from random failures. This may have been expected given their higher robustness to *independent* random failures as well. Note also that, under this model, because the failure threshold is set on the *fraction* (not the absolute number) of failed neighbors, nodes with a higher degree generally need more neighbors to fail first before they collapse as well, which mitigates propagation of the cascade in scale-free networks.

A failure-propagation model that is more relevant to flow networks was described in [12]. In this model, each node is assumed to send a unit of flow to each other node via the shortest path. Hence, the expected load on a node and thus its capacity should be proportional to its centrality or how many paths it falls on. The higher the proportionality factor between expected load and maximum capacity, the more extra leeway exists to accommodate overload. A node is assumed to fail if its actual load exceeds capacity. This may occur if other nodes fail, causing flow rerouting. Hence, this model aims at understanding overload-related failure cascades. Unsurprisingly, it was shown that scale-free networks, according to this model, are more vulnerable to failure cascades due to attacks on *selected* (i.e., high-degree) nodes than homogeneous networks but are less susceptible to cascades that result from *random* failures.

Recent work [13] further analyzed the tradeoff between network throughput (or conductance) and robustness to failure in scale-free networks with *weighted* links. It was shown that while these networks remain vulnerable to cascades due to targeted destruction of highly connected nodes, their vulnerability is minimized when link conductance (weight) is set inversely proportional to the product of the linked node degrees (Figure 5.4).

While the above results appear to be mutually consistent, later evidence shows that the exact nature of failure propagation dramatically affects the robustness properties of networks, often contradicting established intuitions. For example, an interesting analysis of cascaded failures is found in recent work on interdependent networks [6]. It is inspired by a blackout example, in which failure of nodes in a power grid causes failure of nodes in a powered computing and communication network, but because some of these computing nodes are used to control the grid, their failure, in turn, causes further failures in the power grid, resulting in a vicious cycle. If the dependencies between the networks are modeled by links, it was shown that the set of interdependent networks was more susceptible to cascaded failures than the individual networks in isolation. Moreover, in an apparent contradiction to prior intuition, an interdependent set of scale-free networks was shown to be *more susceptible to random failures* than an interdependent set of exponential networks. This may appear to be puzzling, considering that scale-free networks are known to be *more robust to random failures* than their homogeneous counterparts. The authors explained that the phenomenon was because, in interdependent scale-free networks, low-degree nodes in one network are more likely to propagate

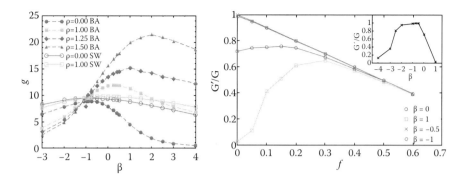

Figure 5.4 Network vulnerability can be minimized by managing the relationship between properties of links and node degrees. Left shows network conductance, *g*, as a function of edge weight parameter, β, for different source/target distributions, ρ. Right shows the relative size of the surviving component, *G'/G* as a function of the fraction of removed nodes, *f*, for different β. (With kind permission from Springer Science+Business Media: *Handbook of Optimization in Complex Networks*, edited by My T. Thai and P. Pardalos, Optimizing synchronization, flow, and robustness in weighted complex networks, G. Korniss, R. Huang, S. Sreenivasan, and B.K. Szymanski, 2011.)

failures to high-degree nodes in another [6]. Indeed, contrary to prior work, in which failure of a node was conditioned on failure of a *fraction* of its neighbors [8], the interdependent-network study assumed that a *single* failed neighbor in the other network brings down a node. Hence, nodes with a higher-degree in one network would tend to die quickly when random failures occur in the other, making scale-free networks (in which connectivity depends on such high-degree nodes) increasingly susceptible.

Indeed, much of the results on network robustness in the presence of failure cascades depend on the model of the failure cascade. This dependence on failure propagation models was articulated well in a recent study [7], in which it was shown that radically different graphs, such as cliques versus trees, can become more robust under different models of failure propagation. For example, if a node's failure probability *increases with the number of its failed neighbors* (as in the web of interdependent financial establishments), having diversity among neighbors (i.e., lack of "dependency" edges) is preferable in the sense of minimizing failure risk. This is so regardless of the size of the overall connected component to which the nodes belong. For a given number of edges per node, a tree topology minimizes edges among a node's neighbors and hence minimizes risk in this scenario. In contrast, if the failure of *any one neighbor* is sufficient to bring down a node (as in the propagation of infectious diseases) then the failure probability is a function of the size of the connected component to which the node belongs because a domino effect that starts anywhere in the component will eventually bring down all nodes. Given the same number of edges per node, a clique (not a tree) would minimize the size of the connected component and hence minimize risk. By a similar argument, scale-free versus homogeneous networks would also be better for different failure propagation models. The example suggests the importance of modeling failure propagation correctly when analyzing network robustness. Proper modeling of failure propagation is especially challenging in networks in which node interactions are mitigated by computing or other intelligent engineered components (e.g., computer-controlled switches in a power grid or on the Internet). The algorithms implemented by such components will likely significantly affect the failure propagation model and hence the resiliency properties of the network at hand. An interesting challenge therefore becomes to engineer these components in ways that maximize the resiliency of the current topology.

5.6 Buffering and Resiliency to Function Loss

The vulnerability of the underlying resource network to connectivity breaches (e.g., as measured by the size of the surviving connected component) is only one factor affecting the overall vulnerability of the entire multi-genre network. In general, the overall performance of the process depends on other factors besides connectivity, such as the degree to which it experiences *function loss* as a result of various failures and perturbations.

In data and commodity flow networks, the function of the network is to offer its clients access to a set of delivered items. In such networks, buffers (e.g., local distributors) constitute a resiliency mechanism that obviates the need for continued access to the original source. Should the original source become unreachable, one can switch to a local supplier. Hence, local access can be ensured despite interruption of the global supply network as long as access to a local distributor (buffer) is available. Local access is an especially valuable solution in the case of a data flow network, in which the commodity (namely, the data content) is not consumed by user access in the sense that a local distributer can continue to serve a content item to new users irrespective of its use by others.

Much work on network buffering has been done to increase the resiliency of data access to fluctuations in resource availability. For example, buffering (or caching) has been used to restore connectivity and performance upon topology changes in ad hoc networks [11] as well as to reduce access latency in disruption-tolerant networks [14].

The concept of buffering and its relationship to resiliency, however, transcends data and commodity flow networks in general. In recent work on biological modeling [9], the concept of *buffering* was generalized to model the extent to which degeneracy in multi-functional agents offers "spare capabilities" for adaptation. Degeneracy, a term borrowed from biology, refers to a condition in which (i) agents (network nodes) can perform one of multiple functions depending on context and (ii) the same function can be performed by one of several agents. For example, storage agents (buffers) in a commodity flow network can use their space to store any of a set of possible items. Also, the same item can be stored by any of multiple agents, hence fulfilling the conditions of degeneracy. The degeneracy model applies in other contexts as well. For example, individuals in an organization can allocate their time to any of a set of possible projects. Similarly, the same project can be performed by any of multiple individuals. It is shown that functional degeneracy (i.e., the combination of versatility and redundancy of agents) significantly improves resiliency of network functions [9] by facilitating reconfiguration to adapt to perturbations. Intuitively, the higher the versatility of the individual agents and the higher the degree to which they are interchangeable, the more resilient the system is to perturbation because it can reallocate functions to agents more flexibly to restore its performance upon resource loss. Degeneracy and networked buffering also lead to a better adaptive behavior to changing conditions, as argued, for example, in the context of agile manufacturing [10].

A dimension not typically explored in biology (in which agents, such as cells, can perform one main function at a time) is the dimension of *capacity*; that is, the number of different functions that an agent can simultaneously perform. Capacity quantifies, for example, the number of items a storage node can simultaneously hold or the number of projects on which a given individual can simultaneously work. Quantifying the resiliency of multi-genre networks as a function of both their degeneracy and their capacity remains an interesting open problem. The problem is especially interesting in the context of non-independent failures; that is to say in situations in which failures of some nodes or functions can propagate to others, according to some appropriate propagation model.

In the context of data-fusion processes, one example of functional loss in networks with degeneracy and non-independent failures comes from distributed real-time computing. Processors (nodes) in such computing systems run several computational workflows with end-to-end latency constraints (to ensure timeliness of results) and exhibit a great amount of degeneracy. A given processor can typically work on any of several computational workflows. Similarly, a workflow can be performed by (a sequence of) any of multiple processors. Moreover, failures that delay one workflow (such as anomalous conditions that lead the workflow to consume an unnaturally large amount of time and resources) can also delay other workflows that compete on the same resources. Recent work addressed greedy heuristics for allocating resources to real-time flows in such a way that resiliency of end-to-end flow latency is maximized with respect to local failures [15]. It was shown that the end-to-end latency could be made significantly more resilient to perturbations by a proper resource reallocation that minimizes dependencies between flows that arise from resource sharing. A broader look at resiliency to function loss under a more general failure model and time constraints remains a matter of future investigation (Figure 5.5).

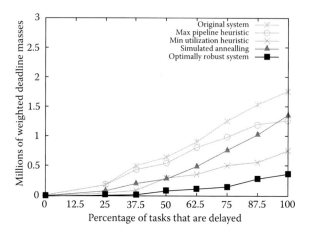

Figure 5.5 Resiliency of end-to-end flow latency can be maximized with respect to local delays by properly allocating resources to flows. (Adapted from P. Jayachandran and T. Abdelzaher, On Structural Robustness of Distributed Real-Time Systems Towards Uncertainties in Service Times, in Proc. IEEE Real-Time Systems Symposium (RTSS), December 2010.)

5.7 Resiliency to Input Corruption

In multi-genre networks that span physical, information, and social spaces, correctness of network functions and timing, addressed above, is not, in itself, sufficient to guarantee resiliency. Network functions only transform inputs to outputs. The correctness of outputs depends not only on the correctness of the transformation, but also on the integrity of inputs. Hence, a functionally resilient network may remain vulnerable to input corruption. The problem is especially severe in networked information and social systems. Such networks operate in the presence of high uncertainty (as in a battlefield or a disaster-response scenario). Techniques are needed to ensure resiliency of the quality of information delivered.

An "optimal" way of ensuring quality of information might trivially be to use highly reliable sources. This approach, even when feasible, is vulnerable to failure or corruption of those sources. Therefore, of much interest is the development of mechanisms that maintain high quality of information in a manner resilient to individual source failure or corruption. Hence, although significant literature exists on various data fusion and outlier elimination methods [16], we focus on techniques with which we do not rely on any one source for correctness of information but rather infer both the quality of information and the credibility of sources directly from the data itself.

The core of these resiliency mechanisms (to input corruption) lies in exploiting the topology of linkages that exist between data inputs to rank information dynamically in a way that allows the system to adapt to increasing amounts of noise and bad data by identifying and removing such data (and its sources) from the input pool. A static version of this idea takes root from Google's PageRank [17]. PageRank infers the most credible sources of information for a given query with no *a priori* knowledge of the authority of each source. It does so by recursively observing a network of links between pages, and more links to a page (from more credible sources) imply more credibility to the page, leading to an iterative assessment (or ranking) of credibility of all pages. The technique can be generalized to the joint estimation of credibility of arbitrary *sources* and *claims*. Given a collection of sources and a collection of claims, the credibility of claims is a function of the number and credibility of their sources whereas the credibility of sources is a function of the credibility of their claims. Several prior heuristics exist for expressing this recursive relationship with the purpose of simultaneously arriving at a ranking of sources and a ranking of claims by credibility, given the reported data itself whose credibility is being determined (e.g., [18–20]). Most recently, an iterative function was derived that was shown to be *optimal* in that it converges to the maximum likelihood estimate for the probability of correctness of sources and claims. Furthermore, a bound was derived on the confidence interval of this maximum likelihood estimate [21]. Applying these results in real-time to an incoming data stream allows for continual assessment of input quality. Should quality of some inputs be suddenly degraded, the mechanism soon catches up and downgrades the credibility ranking of the inputs in question. This adaptation offers resiliency to input contamination (Figure 5.6).

The approach can easily accommodate additional prior knowledge, when available, but does not have to depend on such knowledge. The main advantage of this recent result is that it offers not only the best (i.e., maximum likelihood) hypothesis regarding the probability of correctness of data and sources, but also a confidence value in this hypothesis. Quantifying the degree to which the confidence interval (and hence quality of information) is susceptible to failures in individual sources (as a function of the topology of the source-claim network) remains an interesting open problem.

5.8 Resiliency of Inference

The last element common to multi-genre networks lies in the algorithms that extract information from data. The ultimate success of these networks often lies in delivering high-quality, high-confidence support for decision making. A related question becomes how much resiliency can be built into the decision support algorithms themselves, such that they may gracefully adapt to failures of the underlying networks to deliver sufficient data, failures to deliver it at sufficient quality, or failures to deliver it at an affordable cost?

The goal is to make sure that high-quality outputs can be restored even as the amount and quality of available resources and data decrease. This is achieved

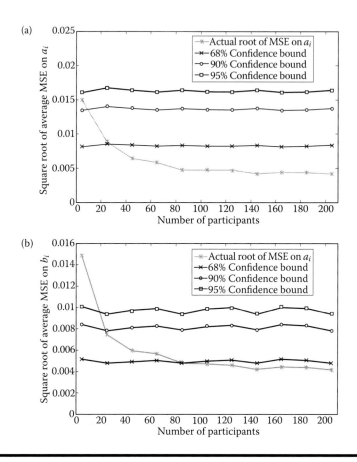

Figure 5.6 Confidence in correctness of sources (described by parameters a_i and b_i) can be managed with resiliency within a desired bound. (a) Confidence bound on a_i and (b) confidence bound on b_i. (Adapted from D. Wang et al., On Quantifying the Accuracy of Maximum Likelihood Estimation of Participant Reliability in Social Sensing, In Proc. 8th International Workshop on Data Management for Sensor Networks, Seattle, WA, August 2011.)

by finding "adaptation knobs" that restore acceptable performance by reallocating scarce resources to where they are needed most for the situation at hand. Applying this wisdom to the design space of algorithms (e.g., algorithms that make predictions from current observations), one may observe that by breaking up the data space appropriately into the right sub-cases, one is able to construct specialized algorithms for each sub-case. Being specialized, these algorithms are likely to contain much fewer parameters, which decreases their cost. Moreover, being specialized, they can also do a better job at estimating or predicting system behavior correctly in the special case to which they pertain. The combination of decreased cost and increased accuracy offers the desired improvement in quality/cost ratio. The underlying adaptation capability lies in recognizing which algorithm to use in a particular scenario. Consistent with the above ideas, recent work presented sparse regression cubes [22], an approach borrowed from data mining that jointly (i) partitions sparse, high-dimensional data into subspaces within which reliable prediction models apply and (ii) determines the best such model for each partition. It was shown that by adapting the model to the current situation, as indicated by the structure of the sparse regression cube, significantly more resilient predictions can be made in the face of uncertainty, errors, and resource constraints.

With different resiliency mechanisms explored at different layers of the multi-genre network, a significant research question that remains is to understand the combined behavior of such a heterogeneous network ecosystem, given our understanding of the behavior of its individual components. This question has been a key motivation and grand challenge of network science: the science of prediction, measurement, and adaptation of complex multi-genre networks. While this chapter touched on some of its key results, fundamental discoveries are yet to be made in understanding performance, resiliency, and adaptation properties of complex systems operating in physical, information, and social spaces.

References

1. M. E. J. Newman, S. H. Strogatz, and D. J. Watts, "Random graphs with arbitrary degree distributions and their applications," *Physical Review E*, Vol. 64, No. 2, 2001.
2. M. E. J. Newman, M. Girvan, and J. Doyne Farmer, "Optimal design, robustness, and risk aversion," *Physical Review Letters*, Vol. 89, No. 2, July 2002.
3. R. Albert, H. Jeong, and A.-L. Barabási, "Error and attack tolerance of complex networks," *Nature*, pp. 378–482, 2000.
4. D. S. Callaway, M. E. J. Newman, S. H. Strogatz, and D. J. Watts, "Network Robustness and Fragility: Percolation on Random Graphs," *Physical Review Letters*, Vol. 85, No. 25, December 2000.
5. L. Sha, "Using Simplicity to Control Complexity," *IEEE Software*, July–August, 2001.
6. S. V. Buldyrev, R. Parshani, G. Paul, H. Eugene Stanley, and S. Havlin, "Catastrophic Cascade of Failures in Interdependent Networks," *Nature Letters*, April 2010.
7. L. Blume, D. Easley, J. Kleinberg, R. Kleinberg, and E. Tardos, "Which Networks Are Least Susceptible to Cascading Failures?" In *Proc. 52nd Annual IEEE Symposium on Foundations of Computer Science*, Palm Springs, California, October 2011.
8. D. Watts, "A Simple Model of Global Cascades on Random Networks," *Proceedings of the National Academy of Sciences*, Vol. 99, No. 9, April 2002.
9. J. M. Whitacre and A. Bender, "Networked Buffering: A Basic Mechanism for Distributed Robustness in Complex Adaptive Systems," *Theoretical Biology and Medical Modeling*, Vol. 7, No. 20, June 2010.
10. R. Frei, and J. M. Whitacre, "Degeneracy and Networked Buffering: Principles for Supporting Emergent Evolvability in Agile Manufacturing Systems," *Journal of Natural Computing*, Special Issue on Engineering Emergence, Vol. 11, No. 3, 2012.

11. A. C. Valera, W. K. G. Seah, and S. V. Rao, "Improving Protocol Robustness in Ad Hoc Networks through Cooperative Packet Caching and Shortest Multipath Routing," *IEEE Transactions on Mobile Computing*, Vol. 4, No. 5, September 2005.

12. A. E. Motter and Y.-C. Lai, "Cascade-Based Attacks on Complex Networks," *Physical Review E*, Vol. 66, 2002.

13. G. Korniss, R. Huang, S. Sreenivasan, and B. K. Szymanski, "Optimizing Synchronization, Flow, and Robustness in Weighted Complex Networks," in *Handbook of Optimization in Complex Networks*, edited by My T. Thai and P. Pardalos, Springer, 2011.

14. W. Gao, G. Cao, A. Iyengar, and M. Srivatsa, "Supporting Cooperative Caching in Disruption Tolerant Networks," *IEEE International Conference on Distributed Computing Systems (ICDCS)*, 2011.

15. P. Jayachandran and T. Abdelzaher, "On Structural Robustness of Distributed Real-Time Systems Towards Uncertainties in Service Times," in *Proc. IEEE Real-Time Systems Symposium (RTSS)*, December 2010.

16. M. E. Liggins, D. L. Hall, and J. Llinas, *Handbook of Multisensor Data Fusion*, CRC Press, 2009.

17. S. Brin and L. Page, "The Anatomy of a Large-Scale Hypertextual Web Search Engine," In 7th International Conference on World Wide Web (WWW '07), pp. 107–117, 1998.

18. J. M. Kleinberg, "Authoritative Sources in a Hyperlinked Environment," *Journal of the ACM*, Vol. 46, No. 5, pp. 604–632, 1999.

19. J. Pasternack and D. Roth, "Knowing What to Believe (When You Already Know Something)," In *Proc. International Conference on Computational Linguistics (COLING)*, 2010.

20. X. Yin, J. Han, and P. S. Yu, "Truth Discovery with Multiple Conflicting Information Providers on the Web," *IEEE Transactions on Knowledge and Data Engineering*, Vol. 20, pp. 796–808, June 2008.

21. D. Wang, T. Abdelzaher, L. Kaplan, and C. Aggarwal, "On Quantifying the Accuracy of Maximum Likelihood Estimation of Participant Reliability in Social Sensing," In *Proc. 8th International Workshop on Data Management for Sensor Networks*, Seattle, WA, August 2011.

22. H. Ahmadi, T. Abdelzaher, J. Han, R. Ganti and N. Pham, "On Reliable Modeling of Open Cyber-Physical Systems and Its Application to Green Transportation," ICCPS, Chicago, IL, April 2011.

23. J. C. Doyle, D. L. Alderson, L. Li, S. Low, M. Roughan, S. Shalunov, R. Tanaka and W. Willinger, "The Robust Yet Fragile Nature of the Internet," *Proceedings of the National Academy of Sciences of the United States of America*, 2005.

24. Z. Bai, P. M. Dewilde, and R. W. Freund, "Reduced-Order Modeling," *Numerical Methods in Electromagnetics, of Numerical Analysis*, Vol. XIII, Amsterdam: Elsevier 2005, pp. 825–895.

25. P. C. Chandrasekharan, *Robust Control of Linear Dynamical Systems*, Academic Press, 1996.

26. B. Walker, C. S. Holling, S. R. Carpenter, A. Kinzig, "Resiliency, Adaptability and Transformability in Social–Ecological Systems," *Ecology and Society*, Vol. 9, No. 2, 2004, http://www.ecologyandsociety.org/vol9/iss2/art5.

27. L. H. Gunderson, "Ecological Resiliency—In Theory and Application," *Annual Review of Ecology & Systematics*, Vol. 31, pp. 425, 2000.

28. C. Perrow, *Normal Accidents: Living with High Risk Technologies*, Princeton University Press, 1984.

29. A. Kott, *Battle of Cognition: The Future Information-Rich Warfare and the Mind of the Commander*, Greenwood Publishing Group, 2008, pp. 205–211.
30. A. Kott, *Information Warfare and Organizational Decision-Making*, Artech House, 2006.
31. A. Kott and G. Citrenbaum, *Estimating Impact: A Handbook of Computational Methods and Models for Anticipating Economic, Social, Political and Security Effects in International Interventions*, Springer, 2010, pp. 2–14.
32. Network Science Collaborative Technology Alliance, www.ns-cta.org.
33. F. Chandra, D. F. Gayme, L. Chen, and J. C. Doyle, "Robustness, Optimization, and Architectures," *European Journal of Control*, pp. 472–482, 2011.

Chapter 6

Resilient and Adaptive Networked Systems

Mauro Andreolini, Sara Casolari,
Marcello Pietri, and Stefania Tosi

Contents

Nowadays, networks form the backbone of most of the computing systems, and modern system infrastructures must accommodate continuously changing demands for different types of workloads and time constraints. In a similar context, adaptive management of virtualized application environments among networked systems is becoming one of the most important strategies to guarantee resilience and performance of available computing resources. In this chapter, Mauro Andreolini et al. analyze the management algorithms that decide, in an adaptive manner, the transparent reallocation of live sessions of virtual machines in large numbers of networked hosts. They discuss the main challenges and solutions related to the adaptive activation of the migration process.

6.1 Introduction

Most data centers were characterized by high operating costs, inefficiencies, and a multitude of distributed, heterogeneous servers that added complexity in terms of resilience and management. In the last years, enterprises consolidated their systems through virtualization solutions in order to improve data center efficiency and offer the benefit of performance and fault isolation, flexible migration, resource consolidation, and easy creation of specialized environments (Ibrahim et al. 2011). Consolidating system resources in as few nodes as possible and centralizing resource management allows the increment of the overall node utilization while lowering management costs.

In this scenario, *live migration* of virtual machines is one of the main building blocks to guarantee resilience and high utilization of networked resources. Live migration is defined as the transfer of running virtual machine instances from one physical server to another with little or zero downtime and without interrupting virtualized services (Chen et al. 2011). There are multiple benefits of live migration among which are fault tolerance, load balancing, and consolidation. For instance, to avoid failover of the virtual machines, it is necessary to live migrate one or more guests running on one physical server to another physical server that guarantees continued and uninterrupted service. Live migration has been supported by the vast majority of hypervisors, such as VMware, Xen, KVM, and VirtualBox (Shetty et al. 2012). Moreover, some recent versions (e.g., VMware, VMotion) support adaptive migration although it is supported by static threshold-based activation mechanisms. This is a step ahead, but we state that full resilience and high resource utilization require the research for *adaptive* management algorithms and supports that are able to continuously tune their behavior as a function of changing operating conditions. Adaptive capacity management requires continuous monitoring services and runtime decision algorithms for deciding when a physical host should migrate a portion of its load, which portion of the load should be moved, and where it should be moved to. These problems and solutions represent the focus of this chapter.

The chapter is organized as follows. Section 6.2 discusses the virtual machine migration process. Section 6.3 presents a taxonomy of live migration strategies. Section 6.4 analyzes a case study showing the differences between a *static* and an *adaptive* selection strategy. Conclusions are drawn in Section 6.5.

6.2 Migration of Virtual Machines

A typical networked architecture consists of a huge set of physical machines (*hosts*), each of them equipped with some virtualization mechanisms from hardware virtualization up to micro-partitioning, operating system virtualization, and software virtualization. These mechanisms allow each machine to host a concurrent execution of several virtual machines (*guests*) each with its own operating system and applications.

To accommodate varying demands for different types of processing, the most modern infrastructures, such as clouds, include adaptive management capabilities and virtual machine mobility, that is, the ability to move transparently virtual machines from one host to another. In this scenario, the decision algorithm orchestrating the live migrations has to select one or more *sender* hosts from which some virtual machines are moved to other destination hosts, namely *receivers*. By migrating a guest from an overloaded host to an unloaded one, it is possible to improve resource utilization and resilience.

The migration algorithms define three sets: sender hosts, receiver hosts, and migrating guests, and their cardinalities are denoted as S, R, and G, respectively. Also let N be the total number of hosts. The algorithm has to guarantee that $N \geq S + R$ and that the intersection between the set of sender hosts and of receiver hosts is null (Andreolini et al. 2009).

Every virtual machine migration approach shares a common management model made up of four distinct phases, outlined in Figure 6.1.

Selection of sender hosts. The first action requires the selection of the set of sender hosts that require the migration of some of their guests. The idea is to have a migration algorithm so that the cardinality S of the set of senders is much smaller than the total number of hosts, that is, $S \ll N$.

Selection of guests. Once the senders are selected, we have to evaluate how many and which guests it is convenient to migrate. Even for this phase, the goal is to limit the number of guests for each host that should migrate, so that $G < (N - S)$. If this does not occur after the first evaluation, the guest selection can proceed

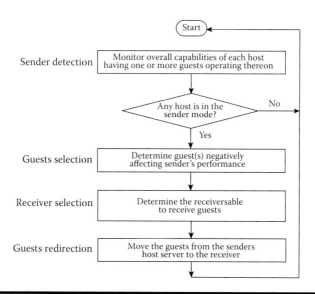

Figure 6.1   Flow diagram of the virtual machine migration process.

iteratively until the constraint is satisfied. (It is worthwhile to observe that, in our experiments, no instance required an iteration.)

Selection of receiver hosts. Once the guests that have to migrate are selected, we have to define the set of receiver hosts. In these networked infrastructures, the major risk we want to avoid is a dynamic migration that tends to overload some receiver hosts so that, at the successive checkpoint, a receiver may become a sender and so on. Similar fluctuations devastate performance and resilience.

Assignment of guests. The guests selected for migration are assigned to the receivers through a management algorithm aiming to satisfy some architectural and/or application constraints. For example, a greedy algorithm may begin to assign the most onerous guests to the lowest loaded hosts and so on until the sender list is completed. Many other possibilities exist.

6.3 Taxonomy of Live Migration Strategies

Existing approaches to live migration of virtual machines can be broadly classified according to different decisions intervening during the four phases of the migration process. These decisions, shown in Figure 6.2, include the *activation strategy*, *monitoring*, *component selection*, *destination*, and *migration mechanism*.

Monitoring. The monitoring system used throughout all the operations is one of the most important components of the live migration mechanism. The first decision regards the performance indicators (*measures*) used to evaluate the load

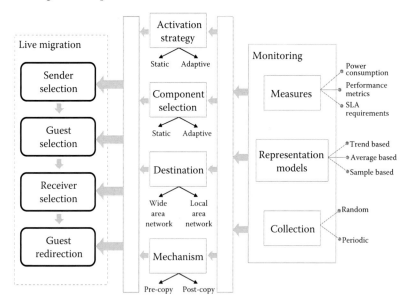

Figure 6.2 Taxonomy of live migration strategies.

conditions of the hosts and virtual machines. Different choices are possible, according to the goal of the migration.

- *Performance*: throughput, utilization of the most relevant hardware resource components (CPU, disk, memory, network) as in Stage et al. (2009) and Gerofi et al. (2010)
- *Power consumption*: CPU voltage at the different power states (Jung et al. 2010)
- *Reliability*: ping response times, service response probes as HTTP response times (Stage et al. 2009)

The next decision regards the *collection* of individual samples through the definition of an appropriate sampling interval. The collection may be periodic (e.g., every second, every minute) or random. This decision may influence the statistical properties of the collected measures, and therefore, the results of the algorithm may be applied in the virtual machine migration process, especially at low sampling frequencies.

Once a sampling strategy is defined, we define *representation models* that read the sampled time series and produce "reduced views" filtered from outliers and reflecting the behavioral load trend of hardware and software resources. These representation models may be linear (e.g., moving average, ARMA) or nonlinear (e.g., polynomial, spline).

Activation strategy. The activation strategy decides when the live migration mechanism has to start. Current solutions operate in two ways: *statically* or *adaptively*.

Static approaches, such as in Khanna et al. (2006) and Wood et al. (2007), set some threshold value, and live migration is triggered as soon as the host load overcomes that predefined value. For example, the scheme proposed by Khanna et al. (2006) classifies a host as a sender or as a receiver depending on whether its load is beyond or below some fixed thresholds. When the load of a host overcomes the threshold, this solution moves all the selected guests to the physical host having the least available resources sufficient to run them without violating the SLA. If there is no available host, it activates a new physical machine. Static approaches cannot work in a networked system consisting of thousands of hosts in which, at a checkpoint, a threshold may signal hundreds of senders and, at the successive checkpoint, the number of senders can become a few dozen, or even worse, most servers differ from those of the previous set. Also, the decision about which guests it is useful to migrate from one server to another is affected by similar problems if we adopt some threshold-based method. Adaptive approaches in which no static thresholds are used to activate the live migration process are preferable. In the adaptive scenario, the activation is triggered by adaptive factors, such as *significant changes* in the host load conditions. A significant change is any modification in the statistical behavior of the load that lasts for a considerable number of consecutive measurements

(Casolari et al. 2012). Considering different statistical behaviors brings the detection of different significant changes, like *trend changes* or *state changes*. A significant trend change happens when the trend patterns of the load vary over time in their direction (e.g., upward or downward) or in their distribution (e.g., linear or exponential). For example, Figure 6.3a shows a host load profile (concerning host CPU utilizations) presenting two trend changes. An exponential increasing trend (until sample 156) is followed by an exponential decreasing trend (from sample 157 to sample 336) and then by a horizontal linear trend (from sample 337 to sample 600). A significant state change is a considerable variation of the mean load value that occurs either instantaneously or rapidly with respect to the period of sampling and that lasts for a significant number of consecutive measurements (Casolari et al. 2012). The load profile in Figure 6.3b presents two significant state changes, one upward and one downward. At sample 200, the mean load value passes from ≈0.3 to ≈0.6, and then it returns to ≈0.3 at sample 420.

The ability to capture such changing conditions in host and guest load profiles is crucial for an activation strategy to reduce the number of migrations to just the most significant ones.

In the case study described in Section 6.4, we present the implementation of an approach that adaptively selects as the senders only the hosts subject to significant state changes of their load. On the other hand (Stage et al. 2009), consider the bandwidth consumption during migration and propose a system that classifies the various loads in order to consolidate more guests on each host based on typical periodic trends if they exist.

Similar considerations are also valid for the *component selection*, which is responsible for choosing hosts and virtual machines that will participate in the live migration process.

Destination. The destination host can be in the same local subnet or, in the most modern networked systems, such as the clouds, even located in another geographical network. Common solutions for the selection of receiver hosts are limited to guests that can migrate only among hosts that are within the same subnet and share common storage (Kamna & Sugandha 2012). This limit prevents adaptive migration of guests among federated data centers that will be more and more important in the future for performance and resilience reasons.

Some works (Ramakrishnan et al. 2007; Travostino et al. 2006; Clark et al. 2005) address the problems of guest migration across WANs and aim to reduce downtime during migration. For example, the solution proposed in Clark et al. (2005) is very efficient because it is able to transfer an entire machine causing a downtime of a few hundreds of milliseconds. Travostino et al. (2006) migrated virtual machines on a WAN area with just 1–2 s of application downtime through light path (DeFanti et al. 2003). Ramakrishnan et al. (2007) propose cooperative, context-aware migration mechanisms through existing server virtualization technologies and by proposing dynamic storage replication technologies to facilitate migration across geographically interconnected machines.

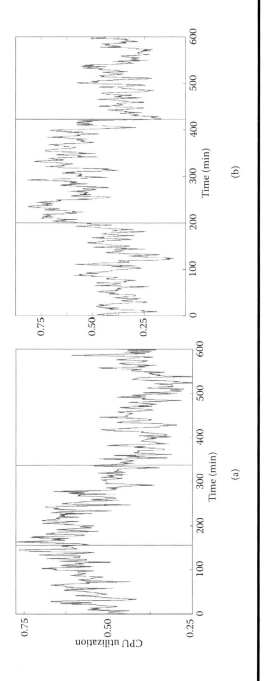

Figure 6.3   Load profiles of hosts. (a) Trend change profile. (b) State change profile.

Migration mechanism. Current live migration mechanisms are mainly based on the *pre-copy* or *post-copy* schemes.

In the pre-copy scheme, the bulk of the guest's memory state is migrated to a target node even as the guest continues to execute at a source host (Hines & Gopalan 2009). If a transmitted page is dirtied, it is resent to the target in the next round. This iterative copying of dirtied pages continues until either a small, writable working set has been identified or a preset number of iterations is reached, whichever comes first. This represents the end of the memory transfer phase and the beginning of service downtime. The guest is then suspended, and its processor state plus any remaining dirty pages are sent to a target node. Finally, the guest is restarted, and the copy at source is destroyed.

On a high-level, post-copy migration defers the memory transfer phase until after the guest's CPU state has already been transferred to the target and resumed there. Post-copy first transmits all processor states to the target, starts the guest at the target, and then actively pushes memory pages from source to target. Concurrently, any memory pages that are faulted on by the guest at target and not yet pushed are demand-paged over the network.

6.4 Case Study

In this chapter, we present a case study application of adaptive migration algorithms in an experimental test bed of 30 physical machines hosting 140 virtual machines. The adopted virtualization solution is VMware ESXi 4.0. We collect periodic samples of the CPU utilization through the VMware monitor with a frequency of one every 20 s. We show an application of a guest selection algorithm that is able to adaptively select the most critical guests for each server on the basis of a load trend–based model instead of traditional approaches based on instantaneous or average load measures.

The focus is on the first two phases (selection of the sender host and selection of guests) that concern open research issues and that represent the core of Sections 6.4.1 and 6.4.2, respectively. The last two phases (selection of receiver hosts and assignment of guests) are hinted at in Section 6.4.3.

6.4.1 Selection of Sender Hosts: CUSUM versus Static Models

The identification of the set of sender hosts represents the most critical problem for the adaptive management of a networked architecture characterized by thousands of machines in which an abuse of guest migrations would devastate performance and resilience.

We present an adaptive algorithm for sender host selection that guarantees high performance and low overheads because it is able to limit the number of migrations to a few really necessary instances. The algorithm considers the load profile evaluated through the CUSUM-based stochastic model (Page 1957). The goal is to signal only the hosts subject to *significant state changes* of their load, and we define a state change as *significant* if it is intensive and persistent. This is not an easy task when the application context consists of large numbers of hosts subject to many instantaneous spikes, non-stationary effects, and an unpredictable and rapidly changing load. As examples, Figure 6.4a and b show two typical profiles of the CPU utilization of two hosts in a cloud architecture. The former profile is characterized by a stable load with some spikes, but there is no *significant state change* in terms of the previous definition. On the other hand, the latter profile is characterized by some spikes and by two significant state changes around sample 180 and sample 220. A robust detection model should raise no alarm in the former case and just two alarms in the latter instance. In a similar scenario, it is clear that any detection algorithm that takes into consideration an absolute or average load value as an alarm mechanism tends to cause many false alarms. This is the case of threshold-based algorithms (Khanna et al. 2006; Wood et al. 2007) that are widely adopted in several management contexts. Just to give an example, let us set the load threshold to define a sender host to 0.8 of its CPU utilization (done for example in VMware). In Figure 6.4, the small triangles on the top of the two figures denote the checkpoints at which the threshold-based detection algorithm signals the host as a sender. There are 10 signals in the former case and 17 in the latter case instead of the expected zero and two. This initial result denotes a clear problem with a critical consequence on performance: We have an excessive number of guest migrations even when not strictly necessary.

The proposed algorithm adopts a different approach for selecting sender hosts by evaluating the entire load profile of a resource and aiming to detect abrupt and permanent load changes. To this purpose, we consider a stochastic model based on the CUSUM algorithm (Page 1957) that works well even at runtime. We consider both the simpler *baseline CUSUM* implementation and the more sophisticated *selective CUSUM* implementation presented in Andreolini et al. (2009). The baseline CUSUM model statically uses reference values for its parameters, and the selective CUSUM algorithm adaptively adjusts its parameters according to data characteristics.

In Figure 6.5, we report the results obtained by using both baseline and selective CUSUM for sender host selection. If we compare the triangles plotted in Figure 6.5 with those in Figure 6.4 (referring to a threshold-based algorithm), we can appreciate that the total number of detections is significantly reduced because it passes from 27 to 11. In particular, the baseline CUSUM is able to avoid detections due to load oscillations around the threshold value. On the other hand, it is unable to address the issue of unnecessary detections related to short-time spikes, such as those occurring at samples 30, 45, 55, and 90 in Figure 6.5a.

The three small boxes on the top of Figure 6.5b, instead, denote the activations signaled by the selective CUSUM. This algorithm determines robust and selective detections of the sender hosts because it is able to remove any undesired signal

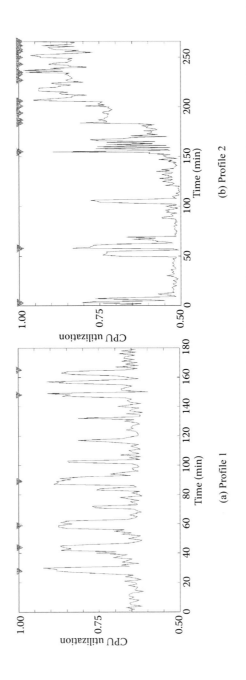

Figure 6.4   CPU load in two hosts.

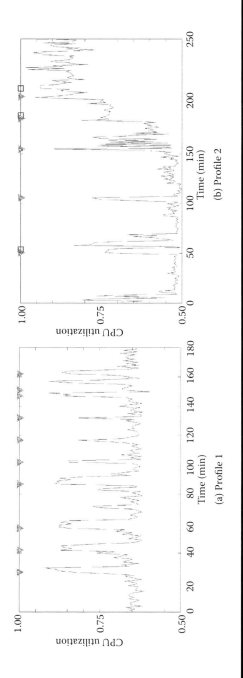

Figure 6.5   Baseline and selective CUSUM models.

caused by instantaneous spikes in Figure 6.5a and to detect only the most significant state changes at samples 55, 185, and 210 in Figure 6.5b—actually, just one more (at sample 55) than the optimal selection of two signals.

6.4.2 Selection of Guests: Trend-Based versus Sample-Based

When a host is selected as a sender, it is important to determine which of its guests should migrate to another host. As migration is expensive, it is important to rely on adaptive solutions that are able to select the few guests that have contributed to the significant load change of their host. For each host, the proposed adaptive solution is based on the following three steps:

1. Evaluation of the load of each guest
2. Sorting of the guests depending on their loads
3. Choice of the subset of guests that are on top of the list

The first step is the most critical because there are several alternatives to denote the load of a guest. Let us consider, for example, the CPU utilization of five virtual machines (A–E) in Figure 6.6 obtained by the VMware monitor.

The typical approach of considering the CPU utilization at a given sample as representative of a guest load (e.g., Khanna et al. 2006; Wood et al. 2007) is not a robust choice here because the load profiles of most guests are subject to spikes. For example, if we consider samples 50, 62, 160, 300, and 351, the highest load is shown by the guest B although these values are outliers of the typical load profile of this guest.

Even considering the average of the past values as a representative value of the guest load may bring us to false conclusions. For example, if we observe the guests at sample 260, the heaviest guest would be A followed by E. This choice is certainly preferable to a representation based on absolute values, but it does not take into

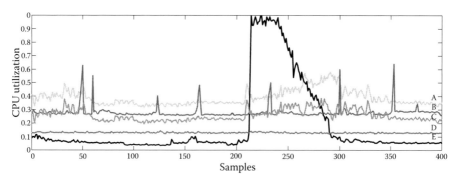

Figure 6.6 Profiles of guest machines.

account an important factor of the load profiles: The load of the guest E is rapidly decreasing while that of the guest A is continuously increasing.

The idea is that a guest-selection model should not consider just absolute or average values, but it should also be able to estimate the *behavioral trend* of the guest profile. The behavioral trend gives a geometric interpretation of the load behavior that adapts itself to the not stationary load and that can be utilized to evaluate whether the load state of a guest is increasing, decreasing, oscillating, or stabilizing. Consequently, it is possible to generate a load representation of each guest based on the computation of a weighted linear regression of *trend coefficients* and the actual load value of a server. After having obtained a load representation for each guest, they sort them from the heaviest to the lightest and then select only the guests that contribute to one third of the total relative load. To give an idea, let us consider two hosts H_1 and H_2 characterized by the following load values: (*0.25, 0.21, 0.14, 0.12, 0.11, 0.10, 0.03, 0.02, 0.01*) and (*0.41, 0.22, 0.20, 0.10, 0.04, 0.02, 0.01*), respectively.

In H_1, we select the first two guests because the sum of their relative loads, 0.46, exceeds one third. On the other hand, in H_2, we select just the first guest that alone contributes to more than one third of the total load.

6.4.3 Selection of Receiver Hosts and Assignment of Guests

Although migration mechanisms are rapidly improving (Kamna & Sugandha 2012; Ibrahim et al. 2011), live migration remains an expensive task that should be applied selectively, especially in a cloud context characterized by thousands of physical machines and about one order more of virtual machines. The receiver-selection process must be carefully designed to avoid migration loops that could occur between sender and receiver hosts. For example, an overloaded resource-consuming guest being moved to a receiver host may trigger a further migration process if the receiver host becomes part of the senders at next activation. The scheme proposed by Khanna et al. (2006) moves all the selected guests to the physical host having the least available resources sufficient to run them without violating the SLA. If there is no available host, it activates a new physical machine. Next to performance and SLA requirements, bandwidth consumption also should be considered in the selection of receiver hosts. The authors in Bobroff et al. (2007) introduce prediction techniques and a bin-packing heuristic to allocate and place virtual machines while minimizing the number of activated physical machines. They also propose an interesting method for characterizing the gain that a virtual machine can achieve from dynamic migration. Stage et al. (2009) propose dynamic scheduling models for the assignment of guests to hosts, taking into account additional migration control parameters, such as bandwidth adaptation behavior, minimum and maximum bandwidth usage, iterated pre-copy migration algorithms, and different termination iteration conditions.

Once the receiver hosts are selected, we have to assign them guests selected for migration. As we want to spread the migrating load to the largest number of receiver hosts, we want that no receiver should get more than one guest, that is, $G = R$. Hence, we have to guarantee that the number of guests we want to migrate is $G < (N - S)$. Typically, this constraint is immediately satisfied because S is a small number, $S << N$, and typically $G \le 2S$.

However, if for certain really critical scenarios it results that $G > (N - S)$, we force the choice of just one guest for each sender host. This should guarantee a suitable solution because otherwise we have that $S > R$, that is, the entire platform tends to be overloaded. Similar instances cannot be addressed by an adaptive migration algorithm, but they should be solved through the activation of standby machines (Khanna et al. 2006) that typically exist in a networked data center. It is also worthwhile to observe that all our experiments were solved through the method based on the one third of the total relative load with no further intervention.

6.5 Conclusions

Adaptive live migrations of virtual machines are an interesting opportunity to allow networked infrastructures to guarantee resilience and accommodate changing demands for different types of processing subject to heterogeneous workloads and time constraints. Nevertheless, there are many open issues about the most convenient choice about when to activate migration, how to select guest machines to be migrated, and the most convenient destinations. These problems are becoming even more severe in modern data centers and cloud architectures characterized by hundreds of thousands of virtual machines. The proposed adaptive algorithms and models are able to identify just the real critical host and guest devices by considering the load profile of hosts and the load trend behavior of the guest instead of thresholds, instantaneous, or average measures that are typically used in the literature.

References

Andreolini, M., Casolari, S., Colajanni, M., Messori, M., Dynamic load management of virtual machines in a cloud architecture, In: *Proc. of First Int. Conference on Cloud Computing*, 2009.

Bobroff, N., Kochut, A., Beaty, K., Dynamic placement of virtual machines for managing SLA violations, In: *Proc. of the 10th IFIP/IEEE International Symp. on Integrated Network Management*, 2007.

Casolari, S., Tosi, S., Lo Presti, F., An adaptive model for online detection of relevant state changes in Internet-based systems, *Performance Evaluation*, vol. 69, no. 5, 2012.

Chen, X., Gao, X., Wan, H., Wang, S., Long, X., Application-transparent live migration for virtual machine on network security enhanced hypervisor. Research paper. *China Communications*, 2011.

Clark, C., Fraser, K., Steven, H., Gorm Hansen, J., Jul, E., Limpach, C., Pratt, I., Warfield, A., Live migration of virtual machines, In: *Proc. of the 2nd ACM/USENIX Symp. on Networked Systems Design and Implementation*, 2005.

DeFanti, T., de Laat, C., Mambretti, J., Neggers, K., St. Arnaud, B., TransLight: A global-scale LambdaGrid for e-science, *Communications of the ACM*, 2003.

Gerofi, B., Fujita, H., Ishikawa, Y., An efficient process live migration mechanism for load balanced distributed virtual environments, In: *Proc. of 2010 IEEE International Conference on Cluster Computing*.

Hines, M. R., Gopalan, K., Post-copy based live virtual machine migration using adaptive pre-paging and dynamic self-ballooning, In: *Proc. of the ACM SIGPLAN/SIGOPS Int. Conf. on Virtual Execution Environments*, 2009.

Ibrahim, K. Z., Hofmeyr, S., Iancu, C., Roman, E., Optimized pre-copy live migration for memory intensive applications, In: *Proc. of 2011 International Conference for High Performance Computing, Networking, Storage and Analysis*, 2011.

Jung, G., Hiltunen, M. A., Joshi, K. R., Schlichting, R. D., Pu, C., Mistral: Dynamically managing power, performance, and adaptation cost in cloud infrastructures, In: *Proc. of 2010 IEEE International Conference on Distributed Computing Systems*.

Kamna, A., Sugandha, S., A survey on infrastructure platform issues in cloud computing, *International Journal of Scientific & Engineering Research*, vol. 3, no. 6, 2012.

Khanna, G., Beaty, K., Kar, G., Kochut, A., Application performance management in virtualized server environments, In: *Proc. of Network Operations and Management Symp.*, 2006.

Page, E. S., Estimating the point of change in a continuous process, *Biometrika*, vol. 44, 1957.

Ramakrishnan, K. K., Shenoy, P., Van der Merwe, J., Live data center migration across WANs: A robust cooperative context aware approach, In: *Proc. of the 2007 SIGCOMM Workshop on Internet Network Management*, 2007.

Shetty, J., Anala, M. R., Shobha, G., A survey on techniques of secure live migration of virtual machine, *International Journal of Computer Applications*, vol. 39, no. 12, Feb. 2012.

Stage, A., Setzer, T., Network-aware migration control and scheduling of differentiated virtual machine workloads, In: *Proc. of 31st Int. Conf. on Software Engineering*, 2009.

Travostino, F., Daspit, P., Gommans, L., Jog, C., de Laat, C., Mambretti, J., Monga, I., Van Oudenaarde, B., Raghunath, S., Wang, P. Y., Seamless live migration of virtual machines over the MAN/WAN, *Future Gener. Computer System*, vol. 22, no. 8, 2006.

Wood, T., Shenoy, P., Venkataramani, A., Yousif, M., Black-box and gray-box strategies for virtual machine migration, In: *Proc. of the 4th USENIX Symp. On Networked Systems Design and Implementation*, 2007.

Chapter 7

Bringing Adaptiveness and Resilience to e-Health

Marco Aiello, Ando Emerencia, and Henk G. Sol

Contents

E-health is a rapidly growing field that is quickly expanding as it is motivated by two driving forces. On the one hand, the information and communication technology (ICT) support to doctors can significantly improve the quality of the entire health care system by improving the quality of care while, at the same time, lowering costs. On the other hand, ICT can also help older people residing at their homes by means of computing systems comprised of sensors, actuators, agents, and other technologies. Given the large variety of situations, adaptation is a very important requirement. At the same time, resilience is an equally important requirement in order to provide guarantees about safe behavior of the systems in spite of unpredictable situations. In this chapter, Marco Aiello, Ando Emerencia, and Henk G. Sol present the challenges in the field by means of three case studies and propose their vision about the future of the field.

7.1 Introduction

Technological and scientific advancement in the informatics are essential to improving the quality of the health care system although it still has to meet the expectations of revolutionizing the field.[1] In fact, there is a large gap between the postulated and empirically demonstrated benefits of e-Health. Many of the clinical claims made about the most commonly used e-Health technologies are yet to be substantiated by empirical evidence, and this will require substantial research resources and effort.[1] Many attempts are taken to increase the efficiency and effectiveness of e-Health, especially because health care and cure are picking up between 10%–15% of GDP in many countries. The advancements of the last decades in information and telecommunication technologies (ICT) are particularly relevant for the sector, especially if we look at health care as a pattern of identifying health data, processing the data based on historical knowledge, and finally acting on an event. The application of ICT to the health sector is often referred to as *e-Health*. Some authors use the term also to refer to the business, legislative, and social issues relating the application of ICT to the health sector.

G. Eysenbach[2] identifies 10 requirements for e-Health to be useful and successful, among which are increased efficiency and effectiveness of the systems and also ease and enjoyment of use. Clearly, there should be an added value in using technology for any given sector, and for e-Health, the most natural targets seem to be large data management of historical patient and disease information, data mining of large collections of data to find cause-effect relationships, delivering care from a distance, and supporting or even automating the process of diagnosis and cure. But navigating the abundance of data to find appropriate treatments or to develop new policies is critical as many data sources are unreliable or heterogeneous in nature. Therefore, the attempt should be to take the patient/client/citizen as point of departure.

Our claim is that the times are mature for a step ahead in the e-Health system as long as we are able to infuse adaptivity and resilience into the system engineering while putting the actual end-user of the system at the center of the design, that is, the patient and not, as it was identified in the past, the caregiver. We identify the challenges and principles for achieving this step and make the description concrete by looking at three examples of state-of-the-art projects: one regarding smart homes to support the needs of people in their homes, one about use of mobile technology in Africa, and one about self-management of a psychiatric disease.

7.2 Addressing the Emerging Challenges

We identify a set of primary challenges that e-Health is facing from the perspective of the engineering of solid ICT systems. These are the following:

   i. Systems are becoming increasingly *complex* in terms of the number of components and their interactions and also in the size and quality of the exchanged data. Therefore, the challenge is to deal with such complexities in e-Health by building adaptive, dynamic, and resilient systems. A local deterministic study of possible behaviors and failures will be not possible in general; therefore, other techniques that look at the interface of the whole complex system and are able to cope with unpredicted faults are necessary.

ii. Current approaches tend to consider very narrow specific health problems and provide a fixed solution. Such approaches will become unfeasible when dealing with the integration of support systems tackling several health aspects at once. Systems will need to interoperate and address cases at *run-time*. For instance, it is possible today to have heart-monitoring, diabetes insulin–dispensing, and fall-monitoring devices, but these do not interoperate. Furthermore, certain health situations are only detectable and addressable when information from the various subsystems is fused.

iii. Putting the *patient at the center of the design.* e-Health systems have been designed traditionally to support the caregivers in providing better diagnosis and cures. But e-Health is facing the challenge of realizing that the patient is really the end-user of the system and has to feel supported, comfortable, and cured by the new technology. This includes addressing challenges regarding acceptability, confidentiality, privacy, and legislative and ethical issues. For the engineering of resilient systems, we additionally consider the use of the contributions of decision-enhancement services[3] and analytics.

The current treatment does not have the ambition to solve these challenges but simply to identify them. It calls for the study of techniques that are able to put the patient at the center of the design and to deal, for him or her, with the increasing complexity of health care solutions—all this for the purpose of improving the care given in terms of the quality of the care and also in terms of overall costs of care and, therefore, availability of care to the wider public. We also anticipate that the addressing of such challenges requires the design of systems that have some form of domain description: Ontologies to describe medical knowledge and also to describe the user context and daily routines. Techniques that are able to deal with open-world nondeterministic and unknown behaviors at design time will be in high demand. These are a requirement for having different systems interacting at run-time that were not specifically designed to do so: for example, insulin dispensers using the patient's mobile phone to issue reminders or alerts.

The placing of the patient at the center of the information system also requires a radical rethinking of the engineering process. The following three principles illustrate how the shift can take place:

i. *Way of thinking*: After the emphasis on e-Health, the focus is now on p-Health, providing patient-oriented health services. The collection and processing point of data is the patient, who is empowered to access and control the whole caregiving process. An approach based on the existence of remote

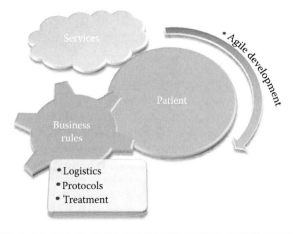

Figure 7.1   Principles for designing patient-centered information systems.

atomic services becomes then essential to be able to build patient-centric solutions.[4]

ii. *Way of governance*: Built systems need to be able to change quickly and adapt to changes in medical knowledge, available technology, and user needs. Using agile development techniques, on the one hand, and solutions that adapt at run-time appears to be essential.

iii. *Way of modeling*: Focus on "business rules" tailored to logistics, medical protocols, and treatment rather than rules made for caregiving organizations such as hospitals, nursing homes, and the like.

The cooperation of these three principles is modeled in Figure 7.1. This figure symbolizes that the patient is central; that, in order to support fast change, cloud services are required for agile development; and that agile development is managed by business rules. These business rules are based on aspects that concern the patient rather than the organization, that is, logistics, medical protocols, and treatment. In other terms, the success of e-Health systems is funded on reshuffling the importance of the stakeholders and in devising tools that will be able to cope with the increased complexity and brittleness introduced by such a shift.

7.2.1 Cases

We report next on three projects that exemplify how the challenges for e-Health systems can be addressed; in particular, we look at (1) a European project for making homes smarter, which, in turn, can provide for people to receive e-Health support directly in their own homes; (2) p-Health in Eastern Africa; and (3) a web application to provide support for people with a specific mental disease, that is, schizophrenia.

7.2.1.1 Smart Homes for All Project

The Smart Homes for All project (SM4All) is a European Union Small and Targeted Research Project funded in the context of the seventh framework.[5] It was kicked off on the first of September of 2008 with a duration of three years, having as objectives the study and development of an innovative middleware platform for the interworking of smart embedded services in immersive and person-centric environments. The challenges that SM4All addresses are those of integrating heterogeneous devices and allowing not only their interoperation, but also for them to be coordinated in such a way that they can contribute to a common goal. For instance, the user may require support for a daily activity. When this happens he or she does not have to state what actions to perform in the house but rather express a goal. The house will then proactively identify which actions need to be taken after having sensed the environment and the user context. This is possible by using state-of-the-art artificial intelligence planning techniques coupled with a software infrastructure based on the concept of service-orientation.[6] In some sense, SM4All advocates not only the heavy introduction of automation and sensors in the home, but also the shift from a purely reactive home, one that reacts to user direct commands, to a proactive smart home that can decide what to do once it identifies the user goals.[7]

Another important innovation of SM4All concerns the user interface. On the one hand, the interface adapts itself to the user context so that if he or she is in an emergency situation, he or she can rapidly issue a request for help; on the other hand, nontraditional interfaces, such as brain-computer ones (BCI), can be used. In this manner, users with the ability to concentrate but with very limited mobility can still interact with the home.[8] Figure 7.2 shows an illustration of a user using a noninvasive BCI with the possibility of issuing 16 commands (bottom left corner) and controlling a virtual four-room apartment.

In summary, SM4All is a state-of-the-art example of a system that is adaptive and resilient. It achieves these two properties by controlling the home based on the current situation of the user and the home and not by using precompiled instructions, thus achieving truly dynamic behavior to support the user best.

7.2.1.2 p-Health in Eastern Africa

There are a growing number of developed countries, and few people in developing countries are using mobile phones for health-related services. Linking the persistent problem of chronic diseases and the newly available technology in a developing country is an innovation. Many developing countries like Tanzania are eager to explore the use of mobile phones and other ICT to promote health. However, the lack of a comprehensive model, knowledge base, and published data on the health benefits poses significant barriers. These include the lack of planning for a sound, strategic health needs assessment; lack of planning for sustainability of (proven) solutions;

Figure 7.2 BCI to control a virtual home. (Courtesy of the SM4All Project.)

lack of consideration for and mitigation of change management issues; lack of sound evaluation planning or execution; limited or no dissemination (formal or informal) of findings; and no significant or structured knowledge translation and transfer to influence decision- or policy-making around future e-health implementations.

One initiative focuses on the use of Mobile-health services to promote healthy lifestyles among the elderly with chronic diseases in rural Tanzania. Mobile health services will help the society and the elderly in particular to solve the problem of compensating for age-related mobility functions; p-Health will greatly facilitate bringing patients under a doctor's care, help find problems faster, reduce response time, and cut down on the number of trips to specialists by helping older people to keep in touch with their doctors through mobile health consultative services. It is assumed that elders will be connected using mobile services to get health advice from specialists on how to live healthily with diseases and to get information on self-managed care at home while allowing patients to stay in contact with specialists. Second, urgent lines will be set to connect them right away with the specialist for first aid.

The other initiative starts from the premise that an identification card will become compulsory in the East African community. This will provide a virtual platform for the delivery of patient-centered services within dedicated environments.

Mobile p-services are foreseen to provide financial incentives for health insurance. This requires a portal platform for local service delivery.

7.3 WEGWEIS for Schizophrenia Management

People suffering from a psychotic disability (e.g., schizophrenia) have various care needs. They make use of a combination of services, including psychiatric care, rehabilitation, social services, and housing facilities. Dutch mental health services generally provide a broad array of evidence-based interventions for this group of service users, but a central problem is that the organization of care is mainly driven by supply instead of by the needs of users. The delivery of care and the selection of specific interventions are insufficiently based on individual demands and needs. In general, service users with psychotic vulnerabilities are not seen as partners in medical decision-making. The WEGWEIS project aims to investigate whether people with a psychotic disability can be supported in the self-management of health and illness by means of e-Health technology.[9] Empowerment, experiential expertise, and shared decision-making are central concepts in this project. Furthermore, the project puts emphasis on individual risk assessment and personal recovery goals, dynamically linked with knowledge systems of illness, treatment, and local availability of interventions.[10]

One of the key functionalities of WEGWEIS is a web application for personalized advice. After logging in, the patient can view personalized advice and suggestions that were compiled based on information in their electronic medical records. The information in the advice texts can include images, embedded video, or links to other websites. The presentation of personalized information is essentially a three-step process in which the system first determines the set of applicable advice, which is then prioritized or filtered by a ranking based on popularity and specificity, and finally the advice contents are customized per-user by allowing dynamic variable evaluation at run-time.

WEGWEIS can adapt to changing circumstances because it allows users to add or remove information on the fly. The advice it outputs is triggered by problem concepts instead of direct medical data. This allows the medical data as well as the advice to be changed independently of each other.[11] The problem concepts are categorized in an ontology, allowing the system to recognize rare problems as instances of more generic problem concepts. The search for applicable advice can then be extended in a logical way even if there is no advice for the exact problems that a patient might have. Resilience of the system in cases in which there is little advice or medical data available is thus guaranteed.

By making reliable, important information for individual patients accessible in an understandable format, web applications such as WEGWEIS are expected to perform a key role in improving care for patients with severe mental disabilities in the nearby future.

7.3.1 Outlook

With the aging of the population in industrialized countries and with health being underdeveloped in most third-world ones, e-Health has the potential to deliver quality health care to most, if not all. Basing itself on the ease of spreading of health knowledge and recording information of individuals, for being able to transparently deliver advice anywhere on the globe, the challenges lay mostly in the construction of general solutions that must be robust and adapt to the context of the end-user. In general, the guidelines for developing agile and resilient e-Health applications can be summarized under the label p-Health: develop applications that are patient-centric, personalized, preventive, process-driven, and participatively designed.

While the last decades have seen a great increase in the quality of solutions delivered, we have noticed the following trademarks. The solutions proposed so far tend to address one very specific health need. The solutions do not adapt to the patient but tend to put the caregiver at the center of the design with the drawback of being often hardly acceptable by the patient and only moderately adaptive to his or her real needs. For example, from SM4All we learned that users appreciate the extra functionalities of a proactive home, but they do have concerns. In an experiment ran with two groups, one of the elderly and one of university students, two issues emerged. The elderly feared the loss of control over the home, and the youngsters were worried about their own privacy.[12] These problems can be overcome by involving the users in the early phases of the design of the system and by making them participate in the design of the behavior of the home. In the SM4All project, this can be done by letting the users define what they intend to happen when a goal is given to the home and having a domain expert translate the goal into the formal language used to control the home.[13] As for the youngsters, they need to be made aware of what information is available to the system and how it is stored and where so that they feel in control of such information. And from WEGWEIS, we learned that patients should be involved from the very start of the design process. They should express their requirements, test prototypes, and give input in meetings. Involving patients can lead to higher acceptance and better usability. The role of patients is often overlooked—even in e-Health projects. For example, it is not uncommon for e-Health websites targeted at patients to only be tested by health care staff.

The solutions do not interoperate well with one another; actually, most often they do not interoperate at all, thus making it impossible to compose solutions for patients based on existing subcomponents. In the development of the SM4All prototype, we had to invest a considerable amount of time to make sure that device interoperability was guaranteed. This resulted in a great effort to automate the creation of device proxies and shared ontologies for message exchanges as well as for application interoperation. Such effort can be huge but justified if large deployments of the systems will eventually take place. The lesson learned from WEGWEIS is that interoperability for e-Health applications relies heavily on the use of standards (standards for medical communication, patient summaries, and messaging). We recommend using such standards not only for storing medical data, but also for storing internal domain knowledge. Domain knowledge stored in this fashion, that is, in ontological structures that weave in with concepts from standardized medical ontologies, is easily exported and interpreted by other health care applications.

These considerations call for a new generation of e-Health solutions that must build on the existing knowledge and also must address the new challenges of dealing with complexity, being resilient to faults and unexpected situations, and being adaptive to the actual user needs, and they are being designed and tested with and for the patients. We have overviewed three state-of-the-art e-Health projects that go in this direction. Although we claim, in no way, that we are close to a global robust and adaptable e-Health solution for all. In fact, many sub-issues arise due to the proposed vision.

Will patient-centered applications impact the way clinicians operate? Will clinicians deliver health care differently by being responsible more for the configuration of the e-Health solutions and taking responsibility for the final decisions rather than having to bear the whole diagnosis-prognosis–treatment–post-treatment cycle? Will they be able to scale their professional work to deliver care to larger sets of patients?

From the opposite point of view, the patients also will face new, challenging situations. Although they are likely to have easier and faster access to care, on the other hand, they will have to accept new technological solutions. Some of these may invade their personal spaces, such as their homes and personally worn accessories. Their privacy will be irremediably lost.

There are also growing ethical concerns about who will be ultimately responsible for decisions made and for care given. Clearly, the role of the caregivers will change although technology providers will hardly be considered solely accountable for any mistake the system makes. But because they may, it is in their utmost interest to build e-Health solutions that will be resilient to failures and adapt as much as possible to the actual users' health needs.

References

1. A.D. Black et al. (Jan 2011). The impact of eHealth on the quality and safety of healthcare: A systematic overview. *PLoS Medicine*, 8(1):e1000387.
2. G. Eysenbach (2001). What is e-health? *Journal of Medical Internet Research, 3*(2):e20.
3. P.G.W. Keen and H.G. Sol (2008). *Decision enhancement service: Rehearsing the future for decisions that matter*. IOP Press, Delft.
4. J. Weber-Jahnke and J. Williams (2010). The smart Internet as a catalyst for health care reform. *Lecture Notes in Computer Science*, 6400, 27–48.
5. The Smart Homes for All project. www.sm4all-project.eu (accessed October 27, 2012).
6. E. Kaldeli, E.U. Warriach, J. Bresser, A. Lazovik and M. Aiello (2010). Interoperation, composition and simulation of services at home. *Int. Conference on Service-Oriented Computing, ICSOC*, LNCS 6470, 167–181.

7. M. Aiello, F. Aloise, R. Baldoni, F. Cincotti, G. Guger, A. Lazovik, M. Mecella, P. Pucci, J. Rinsma, G. Santucci and M. Taglieri (2011). Smart homes to improve the quality of life for all. *33rd Annual International IEEE Engineering in Medicine and Biology Conference*, IEEE, 1777–1780.

8. F. Aloise, F. Schettini, P. Arico, S. Salinari, C. Guger, J. Rinsma, M. Aiello, D. Mattia and F. Cincotti (2011). Asynchronous P300-based BCI to control a virtual environment: Initial tests on end users. *Clinical EEG and Neuroscience*, 42(4):219–224.

9. The WEGWEIS project. Available at development.wegweis.nl and www.wegweis.nl (accessed October 27, 2012).

10. L. van der Krieke, A. Emerencia and S. Sytema (2011). An online portal on outcomes for Dutch service users. *Psychiatric Services*, 62(7):803.

11. A. Emerencia, L. van der Krieke, S. Systema, N. Petkov and M. Aiello (2013). Generating personalized advice for schizophrenia patients. *Artificial Intelligence in Medicine*, 58(1):23–36.

12. E. Kaldeli, E.U. Warriach, A. Lazovik and M. Aiello (2013). Coordinating the web of services for a smart home. *ACM Transactions on the Web*, ACM, 7(2):10.

13. E. Kaldeli, A. Lazovik and M. Aiello (2011). Continual planning with sensing for web service composition. *AAAI Conference on Artificial Intelligence*, 1198–1203.

DIFFERENT SYSTEMS-LEVEL APPROACHES

Chapter 8

Agile Computing

Niranjan Suri and Andrew C. Scott

Contents

8.1 Introduction

Agile computing is an innovative metaphor for distributed computing systems and prescribes a new approach to their design and implementation. The fundamental underlying assumption is that complex distributed systems are prone to failure for a variety of reasons. Individual nodes that are part of the system may fail. Communication links between the individual nodes may be degraded beyond acceptable limits (or fail completely). Nodes as well as links may be subjected to a variety of external influences, ranging from accidental human-induced errors to kinetic and cyber attacks. Even more likely, the system may be subjected to legitimate but unexpected or unplanned demands, resulting in degraded performance or failure. The agile computing approach to addressing these challenges is to design systems to be highly adaptive—going beyond current, more limited approaches to adaptivity. The rest of this chapter explores agile computing as an enabling capability to engineer and build adaptive and resilient distributed systems.

The chapter focuses more on agile computing as a metaphor and also briefly describes some specific middleware components that have embodied the principles of agile computing. Note that although the middleware components described target tactical military environments and networks, they are applicable to a wide variety of domains, including disaster recovery and humanitarian assistance.

Agile computing may be defined as opportunistically discovering, manipulating, and exploiting available computing and communication resources in order to improve capability, performance, efficiency, fault tolerance, and survivability. The term *agile* is used to highlight the desire to both quickly react to changes in the environment as well as to take advantage of transient resources only available for short periods of time. Agile computing thrives in the presence of highly dynamic environments and resources in which nodes are constantly being added, removed, and moved, resulting in intermittent availability of resources and changes in network reachability, bandwidth, and latency.

From a high-level perspective, the goal of agile computing is to facilitate resource sharing among distributed computing systems. At this broad level of description, agile computing draws together several other areas of research, including distributed processing, peer-to-peer resource sharing, grid computing, cluster computing, and cloud computing. The focus of agile computing is on the following set of challenges:

Transient Resources: The goal of agile computing is to exploit highly transient resources. Indeed, one of the performance metrics for agile computing is defined as a function of the minimum length of time that a resource must be available in order to be utilized productively. The expectation is to support environments in which resources are available on the order of seconds or minutes as opposed to hours, days, or longer. Therefore, agile computing differs from grid computing, cluster computing, and cloud computing, which target environments in which resources are more stable.

Limited Communications: The networks used to interconnect resources in the target environment are expected to be transient and potentially wireless and ad hoc. The implication for agile computing is that the middleware must be able to support and operate in intermittent, low-bandwidth, high- and variable-latency, and unreliable networks. Again, this differentiates agile computing from grid computing, cloud computing, and cluster computing, in which the network links tend to be high-performance, reliable, and relatively stable. Cloud computing, for example, assumes that the connection between the edge node and the cloud may be limited (e.g., a cellular phone connecting to a cloud service), but the interconnection between nodes in the cloud is expected to be fast and reliable.

Opportunistic Resource Exploitation: Another goal of agile computing is to take advantage of unexpected resources that happen to be available. In particular, the goal is to take advantage of resources that were not originally intended to be tasked but happen to have spare capacity for transient periods of time. Some peer-to-peer systems provide limited forms of the same capabilities. Agile computing extends this capability to be proactive and potentially manipulate the resources in the environment to satisfy the requirements. For example, manipulation can extend to physically moving resources (case in point, robots or other autonomous vehicles) in order to provide communications or processing capabilities when required. These additional resources may be uncommitted resources assigned for use by the middleware or other resources that can be manipulated without interfering with their original task assignments. This aspect differentiates agile computing from other peer-to-peer systems as well as from grid, cluster, and cloud computing.

8.2 Related Research Areas

One approach to defining agile computing is to relate it to other research areas in computer science. Agile computing builds upon a number of capabilities, such as discovery, mobile code, resource control, and process migration, which address the technical requirements and challenges. These and other capabilities are used by agile computing to opportunistically discover and dynamically exploit transient resources. Figure 8.1 shows the relationship between agile computing and other

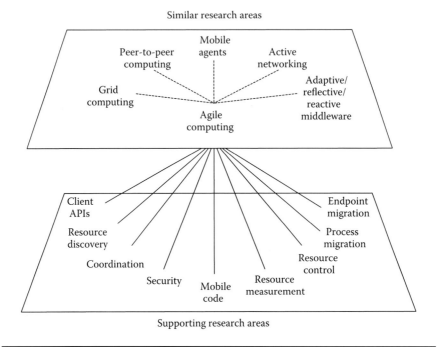

Figure 8.1 Agile computing and related research areas.

related and supporting research areas in computer science. As shown in the figure, there are a number of related research areas whose ideas can be incorporated into agile computing: grid computing, peer-to-peer computing, mobile agents, active networking, and adaptive-reflective-reactive middleware. While it differs from each of these areas, some of the appropriate concepts from each of these areas have been integrated into agile computing. It also integrates a number of other enabling capabilities, drawing from resource discovery, coordination, security, mobile code, resource measurement, resource control, process migration, and endpoint migration.

While a detailed comparison with each of these topics is beyond the scope of this single chapter, more details can be found in [1].

8.3 Lifecycle

Figure 8.2 shows the lifecycle for an agile computing system. Upon startup, the system begins by sensing the environment to find available resources. Users generate requirements that are placed on the system, which then does an allocation of the resources available to satisfy the user requirements. Any subsequent changes in either the resources available in the environment or the user requirements result in a

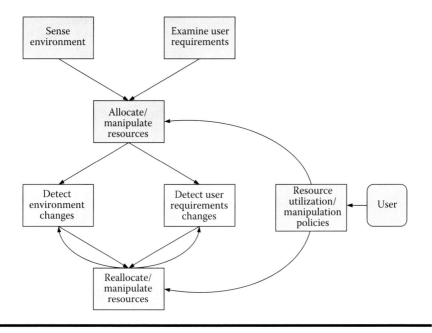

Figure 8.2   Agile computing lifecycle.

reallocation of resources to requirements. This process continues repeatedly for the duration of the system's execution. The allocation of resources is also constrained by policies that are placed on the system by the user. These policies may assign priorities to users, tasks, and resource allocations and place limits on the nature and number of resources that are allocated by the system.

Note that there are two potential approaches to the behavior of the system. One approach, illustrated in the state transition diagram in Figure 8.3, is to serialize each change that is considered by the agile computing system. A change may be to one of the nodes in the system, the connectivity between the nodes in the system, or to a client request. While the system is in the normal running state, these changes are identified by the system (or presented to the system) one by one. For each change, the system transitions from the running state into the adapting state, in which the system identifies the best possible allocation, reallocation, or manipulation of the resources available. Once this process is complete, the system returns to the normal running state.

An alternative approach is to consider a set of changes at a time. For example, a set of allocation requests could be processed together, resulting in resource assignments to all the requests in one step. Similarly, a set of changes to nodes could be processed together, resulting in a collection of resource reassignments that are executed simultaneously. Both approaches are possible, and it would be up to each implementation to select an approach.

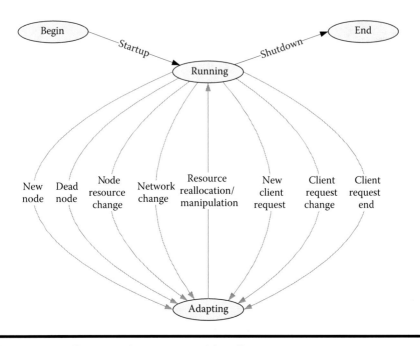

Figure 8.3 Agile computing state transition diagram.

8.4 Generic Architecture

The generic architecture described in this section serves to define a reference model for agile computing. The purpose of the generic architecture is to describe the basic and essential elements of any implementation that realizes the goals of agile computing. The generic architecture also serves as a useful mechanism to map different implementations in order to facilitate better comparisons between them.

The three fundamental technical requirements for any agile computing system are resource discovery, coordination, and APIs for clients to interact with the system. The system must support exploitation of different types of resources, including generic resources, such as CPU, memory, storage, and network bandwidth, as well as other specialized resources. Therefore, at a minimum, the system must address advertisement and discovery of resources, coordination between nodes, and tasking of nodes. The high-level architecture addresses these fundamental requirements. The architecture may be enhanced by supporting additional technical requirements, such as security, mobile code, resource measurement, resource control, strong mobility, endpoint migration, architecture independence, and environmental independence. These additional requirements are discussed later in the chapter. This generic architecture provides a starting point for any designer to begin refinement and creation of a specific instantiation. A more detailed architecture will be presented later as well.

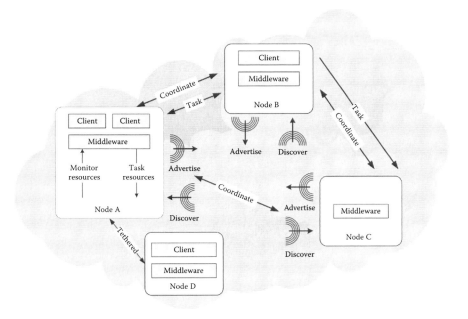

Figure 8.4   Generic architecture for agile computing.

Figure 8.4 shows the generic architecture for any agile computing system. Note that for clarity and simplification, aspects such as the operating system and other applications that would also be executing on the nodes have been excluded. As shown in the figure, nodes must carry out a minimum set of tasks that allow them to interact with each other to realize the notion of agile computing. To accomplish this, each node must execute components of the middleware that allow the node to participate in the overall system. A client is typically an application that is using one or more client APIs to interact with the middleware. Not all of the nodes need to have clients. The figure shows the activities of *Node A* in detail although *Node B* is identical to *Node A* and performs the same set of activities. The functions nominally performed by each node are listed below. Every node does not have to perform all of these functions. For example, a resource-constrained node that is only looking to exploit other resources but not share anything would not be performing any monitoring and perhaps not even advertising.

- **Monitoring** – to sense availability of local resources, typically done via monitoring components that use local operating system APIs or other libraries to determine the capabilities of the local node as well as resources available on the node (CPU, memory, storage, communications)
- **Advertising** – to make other nodes aware of the existence of the node, the capabilities of the node, and the availability of resources at the node

- **Discovery** – to look for other nodes with desired capabilities, either done proactively in order to maintain a continuous awareness of the state of the system or reactively in response to a client request
- **Coordination** – to determine the allocation of resources by nodes to satisfy the requirements of clients on other nodes
- **Tasking** – to handle the actual sharing of resources, which varies depending on the type of resource being tasked

Node C differs from *Node A* and *Node B* in the sense that it does not have any local clients but is sharing resources with other nodes. Finally, *Node D* differs from the other nodes by having a simplified version of the middleware. In this case, the middleware supports the notion of tethering, in which *Node A* acts as a proxy for *Node D*. Tethering may be used to support nodes that are resource-constrained (and hence unable to support all of the capabilities listed above), nodes that have different network technologies, or for policy reasons.

It is worth noting that there is a tradeoff between advertisement and discovery. In particular, some architectures may choose to operate passively, and there is no advertisement at all. In this case, when a node wishes to find resources, it initiates a discovery process specifying the desired resources, which will cause other nodes to respond if they can satisfy the request for resources.

These six elements identified by the abstract architecture are notional. A particular instantiation may choose to perform these tasks in a variety of ways. For example, one design might choose to only advertise the existence of the node and not the capabilities or the availability of resources, letting other nodes reactively discover them when desired.

The architecture identifies the generic operations of monitoring, advertising, discovering, coordination, and tasking of resources but does not specify the nature of the resources. Computational power, memory, storage, and communications form the basic set of resources that can be shared and exploited via an agile computing architecture. In addition, there might be customized resources, such as specialized hardware, that can also be supported. The mode of exploitation differs for each type of resource. For example, computational power is exploited by executing a thread or a process on a node that performs a computation to service another node. Remote procedure call (RPC), mobile agents, and service invocation are all mechanisms to exploit computational capabilities.

Exploitation of memory and storage involves transferring data to and holding data on a remote node. For example, a sensor with limited storage capacity that is acquiring data over an interval of time may hold or archive the data on a remote node with extra memory or storage. Note that such exploitation may be facilitated by means of an RPC or service-invocation model.

Resources are often interdependent, which implies that exploiting one type of resource may require also using another type of resource. For example, service invocation to use CPU will require network communication between the node using the service and the node hosting the service. Such interdependencies imply that a node that has idle CPU time may not be used if the network connectivity to the node does not have the necessary capacity or if the node does not have sufficient free memory.

8.5 Metrics

Given the goal of agile computing as being opportunistic in discovering and using resources, metrics that measure agility are necessary. These metrics can be used in addition to standard performance metrics, such as execution time, to characterize the performance of a particular implementation.

The degree of agility may be informally defined as the length of time for which a node needs to be available in order for its resources to be effectively exploited. The shorter the length of time, the higher the degree of agility. The degree of agility that can be realized is a direct function of the overhead involved. A node becomes available when it is reachable over a network link. There is overhead in the presence of the node being recognized; in discovering the resources available at the node; in the decision-making process to utilize the node; in setting up communication channels; and in moving code, data, and computations to the node. Before a node leaves an environment, there is potentially further overhead in moving active computations from the node onto another node.

The degree of agility may also be defined in terms of the minimum time required in order to reconfigure the system when one or more nodes are under threat. A system that has a higher degree of agility will be more survivable.

Figure 8.5 shows the stages that the system goes through when a node becomes available. As indicated in the figure, the duration of time for which the node is actually available is $t_5 - t_0$. Initially, the system goes through a discovery phase (from t_0 to t_1) when the availability of the resource becomes known. Then, the system needs to make a decision about using the node and allocating new tasks to the node or reallocating existing tasks between nodes (time t_1 to t_2). Once the allocation decisions have been completed, the new node needs to be set up for use, which

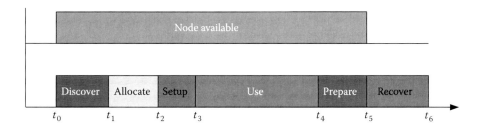

Figure 8.5 Different phases of resource exploitation.

takes time from t_2 to t_3. Setup includes such activities as pushing new code to the new node and/or migrating computations to the new node.

Once the setup process is complete, the new node is fruitfully utilized by the system. This time is represented from t_3 to t_4. This is the actual length of time during which the new resources are being productively utilized.

Before the node becomes unavailable, the system goes through a preparation phase, which typically involves moving computations out of the node about to go offline. This phase takes time from t_4 to t_5. Once the node disappears, the system goes through a recovery phase (from t_5 to t_6). This recovery phase might involve activities such as finding other systems to distribute the computations that were removed from the node that went offline.

In an ideal environment, the discovery, allocation, setup, preparation, and recovery times would be close to zero. The goal of any implementation is to try to minimize these times as much as possible. The agile computing middleware implementation developed by us has helped to better understand the current state of the art with respect to the overhead associated with each of these phases. An important analysis is to predict the impact of technology changes on each of these phases, thereby predicting the degree of agility of future systems.

Another interesting tradeoff involves preparation time versus recovery time. The expectation is that they are inversely related. That is, the greater the preparation time, the lesser the recovery time. For a system to be highly survivable, the required time to prepare should be as close to zero as possible.

Within this context, the degree of agility of an implementation may be defined using the following three metrics:

1. *Start Time*: How quickly can a new resource start to be utilized?
2. *Minimum Presence Time*: How long does a resource need to be available to offset the overhead in utilizing the resource?
3. *Release Time*: How quickly can a resource be released?

The *start time* metric measures the overhead with respect to discovery, allocation, and setup of a newly discovered resource. It is the length of time between a node becoming available and the node's resources being used. The *minimum presence time* accounts for the overhead of the adaptation on the computations and on the other nodes.

For example, consider a node N_1 that is running two computations. The arrival of a new node N_2 might result in one of the computations being moved from N_1 to N_2. The *start time* metric measures the time interval between N_2's arrival and N_2 starting to run one of the computations. The *minimum presence time* metric measures the minimum length of time for which N_2 needs to be available so that the overall execution time of the two processes is improved. Therefore, in this example, it factors in the overhead of stopping one of the computations that was running on N_1, migrating it to N_2, and restarting the computation. The more efficiently this can be done, the shorter the length of time for which node N_2 needs to be available.

The *release time* metric is also important, given that the goal of agile computing is to use resources opportunistically. In particular, one of the goals is to take advantage of resources at nodes that are currently being underutilized. However, if the resources at an opportunistic node are needed for their original purpose, then the middleware needs to release the resources so that they can be allocated back to their originally intended purpose. The *release time* metric measures the length of time it takes to release a node so that it can return to its pre-exploited state.

As an illustration of the *release time* metric, consider that the opportunistic node N_2 in the example above is a user's personal laptop. The middleware might have decided to exploit the resources on the laptop when the laptop was idle. However, if the user returns to the laptop and starts using the resources, the middleware should release the resources as quickly as possible. In the example, this would involve either killing the computation that was migrated to N_2 or, preferably, stopping it and migrating it back to N_1. The *release time* metric measures the impact on the owner of the opportunistically exploited resource.

To summarize, the agility of the system may be defined as follows:

$$Agility = (P_1, P_2, P_3) = \left(\frac{1}{Start\ Time}, \frac{1}{Minimum\ Presence\ Time}, \frac{1}{Release\ Time} \right)$$

where the ordered triple (P_1, P_2, P_3) represents the agility of the system. Note that for the purpose of comparing different implementations and systems, the times should be expressed in milliseconds. Also note that it would be possible to compute a single value using a weighted average as follows:

$$Agility = \frac{\sum_{i=1}^{3} W_i * P}{3}$$

Such a single measure would be useful only if the target environment was sufficiently well understood to determine the weights for each of the three measures.

The agility of a system may also be defined in terms of the lifecycle of the system. Figure 8.3 showed a state transition diagram that identified two primary states for an agile computing system: running and adapting. The system goes into the adapting state when a change has been identified (e.g., arrival of a new node or a change in a client request). Once the adaptation has been completed, the system transitions back into the normal running state. Therefore, the degree of agility of the system may also be defined as a function of the time to adapt given a set of changes. More specifically, the shorter the adaptation time, the better the agility of the system.

$$Agility_{Alt} = \frac{1}{\sum W_{AT} {}^* T_{AT}}$$

where AT is the adaptation type (ranging from arrival of a new node to the termination of a client request as shown in Figure 8.3), W_{AT} is the weight assigned to that type of adaptation, and T_{AT} is the time taken by the system for that type of adaptation.

Note that this discussion has assumed that changes occur one at a time (i.e., changes are presented in a serialized manner to the system). Some applications and target environments may require that the system handle a set of changes atomically. An example is a request to allocate a set of resources across multiple nodes to support a workflow. Another example is a sub-network of nodes all becoming available at the same time. These types of allocation requests have not yet been considered in our implementation of the agile computing middleware. They are currently handled one at a time. However, the agility metrics could be extended to take into account reacting to multiple simultaneous changes.

8.6 Detailed Technical Requirements and Challenges

Several technical requirements need to be satisfied in order to realize the goals of agile computing. These requirements may be divided into two categories: those that are essential and those that are beneficial. There are three key aspects that must be present in any architecture to realize the notion of agile computing:

- Client APIs
- Resource discovery
- Coordination

Furthermore, the architecture may be improved by the addition of the following eight aspects. Although strictly not necessary for an implementation, addressing the following additional requirements will improve the overall performance of the agile computing system.

- Security
- Mobile code
- Resource measurement
- Resource control
- Process migration
- Endpoint migration
- Architecture independence
- Environmental independence

Lastly, in addition to the above technical requirements, the following two miscellaneous challenges should be considered by agile computing architectures.

- Overcoming dedicated roles/ownership
- Achieving a high degree of agility

The following subsections describe each of these aspects.

8.6.1 Client APIs

Client APIs are the means by which user requirements are conveyed to an agile computing system. Resources that may be requested are processing (CPU), communications, memory, and storage as well as any specialized resources attached to nodes. Applications would typically use a set of client APIs to request resources or to perform actions that indirectly request resources. For example, using a communications library to open a connection and send and receive data from node A to node B would result in a request for communications resources between A and B.

Client requests may be more complex and involve multiple types of resources. For example, a service invocation request involves communications (between the node running the client and the node hosting the service) for sending the request and receiving the reply as well as processing and memory for the actual service execution.

Requests for resources may also be annotated with meta-information and accompanied by quality of service (QoS) requirements. Annotations may include information that facilitates allocation of resources by the middleware. For example, a request for a communications link might be accompanied by an expected duration for the connection and the amount of data that will be transmitted and received. QoS requirements indicate desired properties for the resource that is allocated. Using the same example of a communications link, the QoS requirements may indicate the minimum bandwidth and the maximum latency that would satisfy the application.

8.6.2 Resource Discovery

The fundamental goal of agile computing is to be opportunistic and take advantage of dynamically found resources. As defined earlier, resources may be CPU, memory, storage, communications, or any specialized hardware present at or attached to a node.

This goal cannot be realized without a resource discovery mechanism, which makes it possible for the middleware to recognize a new node and the resources on that node when the node becomes available. A node becomes visible when it is reachable over a network link, which happens when a node with a network interface is turned on and comes within range of another node. A node that becomes visible still may not have (or may not be willing) to share any of its resources. The resource discovery mechanism is responsible for both discovering the existence of the node as well as the resources available at the node that might be leveraged.

Approaches for resource discovery fall into two primary categories: proactive and reactive. Proactive approaches rely on nodes periodically advertising their presence and their resource availability to inform other nodes. Reactive approaches rely on nodes searching for other nodes or querying for resources at other nodes when there is a requirement. Proactive approaches use more bandwidth but potentially reduce the latency involved when a resource is needed, given that information about resources has already been advertised to other nodes. Reactive approaches only use bandwidth when a resource is required, thereby being more bandwidth-efficient but increasing latency when a resource is needed.

Hybrid approaches that combine proactive and reactive mechanisms are also possible and provide a good compromise. Hybrid approaches may advertise selected resources and search or query for other resources, depending on the frequency of use, the rate of change, and the latency that is acceptable.

8.6.3 Coordination

Coordination addresses the decision-making that is required to take advantage of resources found via the resource discovery mechanism. The coordination component takes into account the requirements being placed on the middleware by client applications and maps those to available resources. As new nodes are discovered and new resources become available, the coordination component determines how to best take advantage of the available resources and appropriately allocates resources to the client demands. When nodes no longer have resources available (or when nodes disappear), once again, the coordination component is responsible for reassigning the tasks to other existing resources.

Coordination may use one of three different models: centralized, distributed, or zone-based. In the centralized model, a single process at a single node is responsible for receiving all the requests for resources, gathering all the information about resource availability, and performing the allocation. In the distributed model, there is no single node that performs the decisions regarding resource allocation. Instead, each node operates in concert with other nodes to perform the resource allocation. Finally, the zone-based approach combines the two approaches and selects a subset of the nodes to perform the coordination.

8.6.4 Security

Agile computing facilitates sharing of resources among networked systems. Security is an important requirement that should be addressed for agile computing to be used in real-world environments. A node that contributes resources should be protected from abuse. If computations are pushed onto a node, the computations must be executed in a restricted environment to guarantee that they do not access private areas of the host node or abuse the resources available on the host. Similarly, the computations themselves need to be protected from a possibly malicious node that becomes available. Furthermore, agile computing is not immune from the traditional set of security challenges that face distributed and networked systems, including nodes and computer links being compromised.

Security challenges unique to agile computing arise from its fundamental goals: discovery and sharing of resources. One approach to addressing a part of the security problem is creating and enforcing resource-sharing groups. Policies can be used in conjunction with groups to constrain the nature and extent of resources shared. Therefore, some of the security requirements can be achieved by regulating the formation of groups. Groups may be formed through static configuration or through ad hoc discovery. For example, all of the workstations in a department may be configured to be part of a group. Groups may be restricted based on ownership, geographical location, mission, assigned task, etc. Restricting groups based on ownership or task is one effective way of addressing some of the security requirements. For example, a user may not object to his or her node's resources being used by a colleague in the same organization. As another example, a group may be formed when four laptop computers in a meeting room discover each other. In this case, resource sharing may be allowed on the basis of all the participants being part of the same meeting. Group formation needs to be controllable through policies for security reasons so that the systems that are allowed to share resources can be regulated.

The subsection on resource control further addresses the issues related to limiting the extent to which a shared resource may be used by a remote system.

8.6.5 Mobile Code

Mobile code allows new functionality to be deployed onto a node and is an important requirement to satisfy the opportunistic use of resources. The goal of agile computing is to take advantage of resources that were not originally planned to be available or to be part of the system designed or configured to solve a particular problem (or portion of a problem). Therefore, it is unreasonable to expect that these opportunistically found nodes will have the necessary code to provide the capabilities that clients are requesting. Mobile code allows the middleware to migrate new capabilities (such as services) to a newly discovered node that has resources to spare. Once the code has been deployed to the new node, it may then be activated or instantiated and used on behalf of the client application.

Dynamic and late-binding environments, such as Java, make it fairly easy to support mobile code. The virtual machine architecture also facilitates multiple CPU and OS architectures as there is no need to have multiple versions of the executable code, one for each platform. In the Java environment, source code is compiled into an intermediate representation called bytecode. At runtime, a Java virtual machine uses a mechanism called a classloader to dynamically load classes that have been compiled into bytecode. Therefore, mobile code can be easily realized using one of two approaches. The compiled code (bytecode) can simply be packaged into a JAR* file, transferred to the destination machine, and loaded into the VM using a classloader. Alternatively, Java also supports remote classloaders that can load classes directly over a network connection. One example of such a classloader is the *URLClassLoader*, which accepts a URL (uniform resource locator) to a JAR file located on a web server and loads the classes on demand. The former method increases the initial latency as all of the code has to be deployed to the node before it can be used. However, initial deployment will ensure that a network disconnection will not prevent a necessary class from loading midway through a computation.

Virtual machines or interpreters are not an absolute necessity for mobile code. Natively compiled mobile code is also possible with operating systems that support dynamic libraries (for example, DLLs, dynamically linked libraries, in Microsoft Windows systems and shared libraries in UNIX-based systems). The file containing the compiled binary code can be transferred to the destination host and then loaded by an application, thereby invoking the new code.

8.6.6 Resource Measurement

Resource measurement is the process by which the system can keep track of the utilization of resources by various components. For example, the resource measurement for a service could include the CPU time, memory, storage, and communications bandwidth utilized by the service for a given invocation by a client. The resource utilization will vary depending on the client and the nature of the operation. Resource measurement can be used to construct a resource utilization profile for services. This information, in turn, is extremely useful to the coordination component when making decisions about allocating available resources to requirements. For example, if resource measurement shows that the average invocation of a service A by any client results in the transfer of x bytes to the service and y bytes from the service, the coordination component can make an informed decision about the nodes to which service A may be deployed, depending on the bandwidth available between client nodes and server nodes.

8.6.7 Resource Control

Resource control is the notion of limiting the number of resources that any process or computation may use on a node. For example, a computation may be limited in the maximum percentage of CPU time that could be utilized (say 40%), the maximum runtime (such as 10 min), the maximum amount of memory or storage, the

---

* JAR stands for Java ARchive and is a standard way of collecting several related files, each containing bytecode for a class, together into a single file.

maximum amount of data transmitted and received, the transmit-receive data rate, and other parameters.

While not strictly necessary, resource control is an important part of agile computing. When a node is opportunistically discovered and used, the owner of that node will be more willing to share resources if the system can guarantee that the resources will not be abused. For example, the owner of a laptop might be willing to allow colleagues to take advantage of the laptop when it is idle, provided that the resources used by the dynamically deployed services can be controlled and limited to some acceptable maximum.

The resources a node is willing to contribute could be expressed using policies and may be a function of parameters such as local load, available capacity, power level, cost, and even compensation if there is any involved.

8.6.8 Process Migration

Process migration is the notion of capturing the execution state of a process that is currently executing on one node, moving the state information to a new node, and restarting the execution of the process. The migration itself is usually transparent to the process—the computation is stopped on the original node and migrated to the new node, where it resumes exactly at the next instruction. Process migration is also referred to as strong mobility in some contexts, such as mobile agents. In the case of a multi-threaded system, such as Java, individual threads may be migrated instead of the complete process. Thread migration is similar to the notion of process migration except that it may be done at the level of individual threads (or thread groups) within a process and hence supports a finer level of granularity.

While not strictly a requirement for agile computing, process and thread migration has the potential to significantly improve the performance of an implementation (as defined by the degree of agility of the implementation). Consider, for example, a long-running service A that has been deployed and activated on node N_1. If a new node N_2 (with a much smaller load than N_1) becomes available, the system will have to wait until the current invocation of service A completes before the next invocation is directed to node N_2 (presuming that the service has been deployed and activated on N_2 in the mean time). However, if process migration is available, the service can be paused midstream and migrated to node N_2, thereby improving the performance of the current invocation. Moreover, this decreases the time required from when a new resource becomes available to the time when the resource is put to use—one of the measures that contributes to the degree of agility.

Another component of the degree of agility is the time required to release an opportunistically tasked node so that it may return to its original purpose. In this example, if the resources at node N_2 are no longer free, then the system needs to stop using that node's resources. There are three options: (a) the computation (service invocation) can be terminated immediately and restarted elsewhere, (b) the computation can be allowed to finish, or (c) the computation can be migrated to another node using process migration. In the case of option (a), the work done up to the point at which the computation is terminated is wasted. In the case of option (b), the system is not being fair to node N_2 because it is allowing an opportunistic computation to disrupt the normal operation of the system (which breaks the notion of agile computing as simply exploiting the wiggle room or excess capacity of a node). Option (c), which requires process migration, provides the best alternative. By migrating the execution of the service, the system neither loses the work done up to that point nor does it maintain control over the resources at node N_2 for longer than it should.

8.6.9 Endpoint Migration

Endpoint migration allows a process that is communicating with another process to migrate the communication endpoints to another node in a manner that makes it transparent to the first node. As an example, consider a client on node N_1 invoking a service on node N_2. Endpoint migration allows the communication endpoint that is currently on node N_2 to be moved to another node N_3 without any disruption to the communication channel (and transparently to the client on node N_1). Endpoint migration goes hand in hand with process migration, in which a service can be migrated from one node to another and take any active communication endpoints with it.

It is possible to design and implement an application-level communication protocol between a client and a service so that, when the service has to migrate, it simply closes the connection and forces a reconnect between the client and the service on the new node. Endpoint migration raises the level of abstraction to the point at which clients and services do not have to worry about such additional logic. Moreover, endpoint migration makes it easier to handle connections to other preexisting components that cannot be redesigned to handle the disconnect and reconnect logic.

8.6.10 Architecture-Independence

Ideally, agile computing implies that computations must be able to take advantage of any available resources independent of the underlying hardware architecture. If a system fails unexpectedly, the infrastructure should be able to compensate by running those computations on another system, independent of architecture. Similarly, if an ad hoc group consists of computing devices of different architectures (say, ARM-based smartphones running Android and Intel-based laptops running Windows, Linux, or MacOSX), any one of the devices should still be able to take advantage of any available resources in the other devices. Limiting one type of device to exploit resources only in similar devices reduces the agility of the overall implementation. Architecture-independence improves the ability of the system to be opportunistic by allowing any resources, regardless of the type of computing device, to be exploited.

8.6.11 Environmental Independence

Architecture-independence alone is not sufficient to support migration of computations between systems. If the environment on each system is different, the computation will fail after migration due to the sudden change. Environmental factors include the operating system, data resident on a system, and configuration of the software on the system as well as any specialized hardware in the system.

8.6.12 Overcoming Dedicated Roles/Owners for Systems

One of the problems with current systems is that they are often dedicated to certain tasks or assigned to particular users. Such a priori classification of systems prevents exploitation of available resources. Agile computing relies on the notion that any available resource should be utilizable. In order to make this a reality, hardware should be generic and ubiquitous with the specialization being derived through software. If this were the case, then one system can easily be substituted for another by means of moving the software functionality as needed.

Similarly, if systems are assigned to individual owners who are protective about their systems, then resource sharing will be ineffective. One solution to this problem lies in resource accounting, which will allow the owner of a system to contribute resources but then to quantify the contribution in order to receive compensation in some manner.

8.6.13 Achieving a High Degree of Agility

The degree of agility of a system may be defined as a measure of how agile an architecture and implementation might be. A detailed measure of the degree of agility was developed in the section on Metrics. The higher the degree of agility, the better the system would be in reacting to and opportunistically taking advantage of transient resources.

Not all of the technical requirements and challenges listed above are a must for realizing the notion of agile computing. For example, if the architecture-independence challenge is not addressed, it might reduce the number of systems that can be exploited in a mixed environment consisting of different types of systems. Therefore, the degree of agility of the system will be reduced.

Many of the technical requirements and challenges outlined above fall into this category; they may not be essential but contribute to achieving a high degree of agility for the overall system. The following section, which refines the generic architecture for agile computing, elaborates on the features that would be essential to realizing the notion of agile computing versus those that are optional but would help to achieve a high degree of agility.

8.7 Refinement of the High-Level Architecture

This section extends the high-level architecture described earlier by incorporating additional components that would be necessary to address the full set of technical requirements outlined in the previous section. It should be noted that the architecture detailed in this section, while still being generic, is not a requirement for a given implementation. Rather, it is the architecture that has guided the implementation of our agile computing middleware. This refinement introduces the notion of a kernel component: the agile computing kernel, which integrates many of the capabilities necessary to support agile computing. Figure 8.6 shows an abstract view of the kernel component. The kernel should provide a platform-independent way to execute processes on heterogeneous networks. Moreover, the kernel should support discovery, local resource measurement, resource control, resource redirection, policy enforcement, and coordination.

8.7.1 Discovery and Grouping

A dynamically formed group is a fundamental structural notion in agile computing. A group is essentially defined as a set of hosts that have joined together to share resources. That could be, for instance, a set of laptops in a conference room during a meeting. Figure 8.7 shows one possible arrangement for a set of hosts. Note that groups may be disjointed or overlapping. An overlapping group is created by the

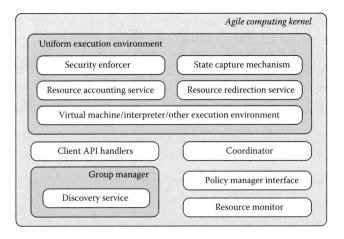

Figure 8.6 Generic architecture of a kernel for agile computing.

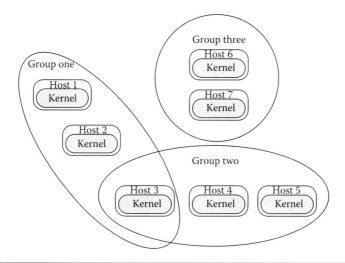

Figure 8.7   Runtime discovery and grouping of hosts.

existence of one or more shared hosts. A host may join multiple groups. Various grouping criteria are possible, but likely candidates are physical proximity, network reachability, and ownership. Sometimes, multiple groups may be created for different purposes: for example, a group based on the task or function of resources and a second group based on ownership for administrative purposes.

8.7.2 Execution Environment

The uniform execution environment should provide a common abstraction layer for code execution that hides underlying architectural differences, such as CPU type and operating system. The execution environment should also support dynamic deployment of code, dynamic migration of computations between kernels, secure execution of incoming computations, resource redirection, resource accounting, and policy enforcement.

Java-like languages provide many of the desired features for a mobile code–based framework. Moreover, the virtual machine architecture of Java provides platform-independence, another technical requirement identified earlier. Other possibilities include an approach like Microsoft's common language runtime (CLR) [2], which dynamically translates platform-independent code for execution on a specific hardware architecture or interpreted languages such as Tcl [3].

Besides the Java-compatible VM, the execution environment should also include a set of software components that supports interaction between the kernel and locally running processes. These components are (a) the security enforcer, (b) the resource accounting service, (c) the state capture mechanism, and (d) the resource redirection service.

The security enforcer should ensure that running processes have limited access to system resources to avoid denial of service (DoS) attacks. The resource usage restrictions for each process running in the VM should be obtained from the policy manager component. The restrictions may be established during process migration or even at runtime after the process execution has started. This component should also provide authentication and encryption services for secure data and state transfer.

The resource accounting service should provide a facility to track resource utilization at the process level inside the VM. This service can be used by the security enforcer and the resource monitor components to estimate overall kernel load and resource availability.

The resource redirection service should provide a means to transparently move links to local or remote resources when a process is migrated between kernels. Consider, for example, a scenario in which a computation has two socket connections open to remote hosts. Due to an imminent power failure, the computation needs to move to another intermediate host. In this case, the resource redirection service of each kernel should negotiate a redirection of the resources (the socket connections) to transparently move the computation with no apparent interruption of the links. For the computation in this example, the migration should happen seamlessly, and the socket connections with the remote hosts should be maintained during the migration.

The state capture mechanism should provide the necessary means to capture the execution state of one or multiple processes running in the execution environment. The state information can then be persisted or moved to another host on which to resume execution.

All these components should work in concert with the policy manager interface and resource monitor components that are also part of the kernel but not directly integrated with the execution environment. The policy manager interface and the resource manager are primarily concerned with higher-level interactions with the coordinator and other kernels, but they rely on the execution environment components to locally perform and enforce most of their tasks.

8.7.3 Policy Manager Interface

An implementation of an agile computing system will likely require a policy-management infrastructure, which can be used to control the behavior of the overall system. The policy infrastructure should be responsible for the specification, conflict resolution, and distribution of policies. The policy manager interface should provide a facility for other components in the kernel to query and determine policies and restrictions that apply to local and remote processes and nodes. Policies are an important requirement in terms of being able to regulate the runtime behavior of the middleware. Policies may be used to control discovery and group formation, thereby controlling the sharing of resources by the middleware. Policies may also be used to control the nature and extent of resources shared on a node—for example, limiting the amount of CPU, memory, storage, or network bandwidth that may be utilized by a remote node.

8.7.4 Resource Monitor

The resource monitor should provide an interface for the coordinator and remote kernels to query and provide information about local resource utilization. Resource availability is one of the metrics considered by the coordinator when calculating allocation of resources. The resource monitor should act as a bridge between the resource accounting service in the execution environment and the coordinator. It monitors local resource utilization in the execution environment and interacts with the coordinator, which can request migration of local computations or to notify it of local resource availability for the sharing. The resource monitor can also store historical information about node resource availability and resource requirements, which may be useful to the coordinator to make better decisions about resource allocation.

8.7.5 Client API Handlers

Clients will use one or more APIs to request services from the agile computing system. These requests need to be conveyed to the kernel by means of one or more client API handlers. For example, a client might request to use memory or storage on a node, which will require a component in the selected kernel to receive data from a client, archive the data, and later retrieve the data when requested by the client. The exact nature of the handlers will vary depending on the types of resources being shared. For example, the agile computing middleware has a number of handlers that support services because the middleware supports a dynamic service-oriented architecture (SoA).

8.8 Coordination Models

The coordinator is the logical entity that manages the overall behavior of the agile computing system. The coordinator should monitor node and resource availability as well as network connectivity and bandwidth. Client requests are received by the coordinator, which handles allocation of resources. The coordinator is also responsible for any proactive manipulation of nodes, such as moving a node to act as a relay in order to restore a lost communications link. Coordination is a continuous process as the resource allocation needs to adapt to changes in the environment.

The coordinator may be realized using a centralized approach, a zone-based approach, or a fully distributed approach. Figures 8.8, 8.9, and 8.10 show these three different approaches.

In the centralized approach, a single node behaves as the coordinator at any given point in time. This coordinator communicates with the local coordinator at each of the nodes. Requests from clients are sent to the coordinator, which, in turn, issues commands to the other nodes. If the coordinator (or the node) fails, then a new coordinator can be selected through an election process.

Like any other centralized approach, the centralized coordinator is the simplest but is not scalable and introduces a bottleneck as well as a single point of failure.

In the zone-based coordination approach, nodes are grouped into zones, each of which has the equivalent of a centralized coordinator. Zones are typically created based on network proximity (which, in wireless environments, also implies physical proximity). One of the nodes in a zone is elected to be the zone coordinator. This zone coordinator interacts with all the other local coordinators within the zone (much like the centralized approach) and also with other zone coordinators.

Two structural arrangements are possible with the zone-based approach: peer-to-peer and hierarchical. Figure 8.9 shows a fully connected peer-to-peer arrangement between the zones although it is not necessary for all of the zone coordinators to be in direct communication with each other.

The last possibility is a fully distributed coordination approach as shown in Figure 8.10. In this approach, there is no centralized or partially centralized

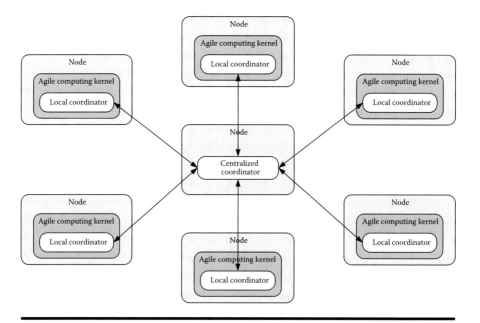

Figure 8.8　Centralized coordination approach.

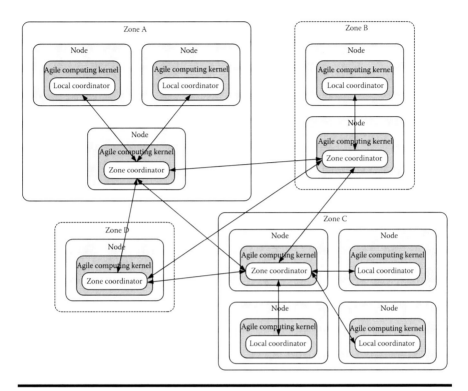

Figure 8.9   Zone-based coordination approach.

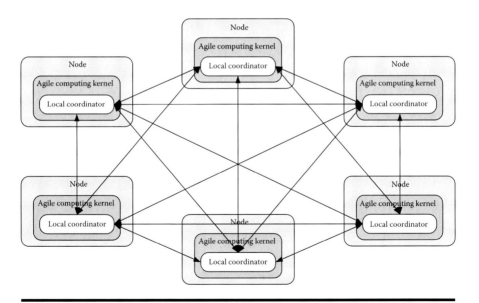

Figure 8.10   Distributed coordination approach.

coordinator at all. Each of the local coordinators directly communicates with other local coordinators as needed. Again, Figure 8.10 shows a fully connected arrangement, but that is not a requirement. The fully distributed coordination approach does not have a single point of failure, but like most distributed algorithms, it is the most complicated approach.

In realizing one of these three approaches, a number of different coordination algorithms are possible, based on the context and the problem to which the agile computing system will be applied.

Just as the kernel interfaces with a policy-management infrastructure to guide its behavior, the overall goals and desired behavior of the coordinator can also be regulated via policies. Policies may specify the coordination strategy itself or just the runtime constraints on the strategy. For example, policies can be used to specify that only 50% of a node's CPU could be used by a remote node or that a node with less than 1 h of battery life should not be exploited. The policy infrastructure should allow the administrators or maintainers of the system to dynamically change behavior at runtime.

8.9 Agile Computing Middleware

This section describes our concrete implementation of the agile computing middleware (ACM). Given the space limitations for this chapter as well as the focus on describing the general notion of agile computing, we provide only a cursory description of the middleware implementation with references to papers that contain more detailed descriptions. Figure 8.11 shows the primary components of the ACM. The following subsections provide a brief description of each of these components.

8.9.1 Mockets

Mockets provide the transport capability to applications (and components of the middleware) and are a replacement for TCP and UDP sockets. The design and capabilities

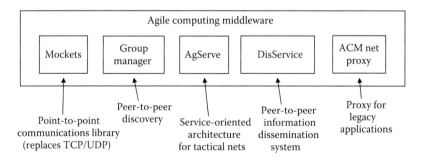

Figure 8.11   Components of the agile computing middleware.

of Mockets were motivated by observations of problems experienced in tactical network environments, which are typically wireless and ad hoc with low bandwidth, intermittent connectivity, and variable latency. The Mockets library addresses specific challenges, including the need to operate on a mobile ad hoc network (where TCP does not perform optimally), provides a mechanism to detect connection loss, allows applications to monitor network performance, provides flexible buffering, and supports policy-based control over application bandwidth utilization.

Mockets support both stream and message-based abstractions. The stream-based abstraction supports exchange of a reliable and sequenced stream of bytes, such as TCP, which simplifies adapting existing applications that use TCP to use Mockets.

The message-based abstraction provides many enhanced capabilities that are only possible when transmitting individual, self-contained messages as opposed to a continuous stream. These capabilities are

- Different classes of service: unreliable/unsequenced, reliable/unsequenced, unreliable/sequenced, and reliable/sequenced
- Tagging of messages to group messages into different categories (for example, for different types of data, such as video, voice over IP, and control)
- Prioritization of messages, either individually or by category (based on the tags assigned to messages)
- Replacement of messages in which old and outdated messages can be replaced by newer messages in order to reduce network traffic

Other key features of Mockets include the following:

- Application-level implementation of the communications library in order to provide flexibility, ease of distribution, and better integration between the application and the communications layer
- Transparent mobility of communication endpoints from one host to another in order to support migration of live processes with active network connections
- Interface to a policy-management system in order to allow dynamic, external control over communications resources used by applications
- Detailed statistics regarding the connection and data transfer, which allow the application and the middleware to observe the status of a connection and adapt as necessary to observed problems, such as connection loss, data accumulation, or significant retransmissions

The statistics gathered by Mockets are provided to the other components of the middleware so that they may adapt to changing network conditions. All the network communication between the components of the agile computing middleware is performed via Mockets. The many additional features as well as the implementation details of the Mockets library is further described in [4,5] and [6].

8.9.2 Group Manager

The group manager component supports resource and service discovery. It enables the agile and opportunistic exploitation of resources by optimizing queries to find nodes in network proximity and/or nodes that are resource rich or have excess capacity.

The group manager supports proactive advertisement, reactive search, or a combination of the two. The search is realized using a Gnutella-like [7,8] probabilistic search mechanism. In addition, the radius (in terms of network hops) of the advertisement or search can be controlled on a per request basis, providing a powerful mechanism to control how strongly or weakly a service may be advertised and how far a search request may travel. In the currently realized implementation, distance is defined as the number of hops in a MANET (Mobile Ad-hoc Network) environment but can be a more scenario-relevant parameter, such as bandwidth or latency.

Propagation of the advertisement and search messages occurs via one of three mechanisms: UDP broadcast (the simplest case), UDP multicast, or via a heuristic approach that provides bandwidth-efficient flooding. In addition, tunneling via TCP supports bridging multiple networks. In all of these four cases, each node may selectively rebroadcast an incoming message to provide control over the distribution radius.

These capabilities enable applications to make tradeoffs between discoverability, bandwidth, and latency. Proactive advertisement uses more bandwidth but reduces the latency when a client needs to find a service and vice versa. On another dimension, a service that is widely present in a network does not need to advertise strongly (or, consequently, a client looking for such a service does not need to search widely) as opposed to services that are scarce.

Groups may be used to partition network nodes into different sets, thereby restricting advertisements and queries. The group manager provides support for two different group types: peer groups and managed groups. Peer groups are completely decentralized; they do not have an owner or manager but instead maintain node membership independently from the perspective of each node. This design choice also implies that there is no attempt at maintaining a consistent group view for all nodes that are members of a peer group. In addition, no special mechanism is required in order to join a peer group. Nodes can simply query or register resources and services in the context of a specific peer group and will implicitly be treated by the group manager as being members of the same peer group.

The decentralized and dynamic nature of peer groups makes them very well suited for resource and service sharing in MANET environments. However, for additional flexibility, the group manager also supports centralized resource and service management by means of managed groups. A managed group is created by a particular node and is owned by that node. Other nodes need to explicitly join a managed group by sending a membership request to the group owner node.

Access to groups (of both peer and managed types) may optionally be restricted using a password, thereby preventing nodes that do not possess the necessary authorization credentials from joining a specific group. Password-protected groups are called private groups. Group manager is further described in [9] and [10].

8.9.3 AgServe

The AgServe component supports a SoA on top of the ACM. AgServe supports dynamic definition, instantiation, invocation, relocation, and termination of services. The underlying middleware monitors service resource utilization, invocation patterns, and network and node resource availability and uses that information to determine optimal initial placement and subsequent migration of services.

AgServe supports dynamic deployment of services by exploiting mobile code. Clients may define new services dynamically, thereby injecting new capabilities into the network. These services are packaged as self-contained archives that include all the necessary Java class files and other related JAR files. These service archives are dynamically distributed (pushed) to other nodes using the ACM kernel functions for remote service installation. Policies may be used to control and manage service implementations as required for security purposes.

AgServe also allows activating services on remote nodes. Service activation is normally triggered by application-driven service invocation requests. AgServe takes advantage of the ACM kernel functions to instantiate the specified service on the target node. Once activated, the service is identified by a unique identifier that is returned to the client stub, which is subsequently used for service invocation.

In addition, AgServe leverages the ACM kernel service container functions to provide transparent service migration. A service running inside the service container can be asynchronously stopped, its execution state captured and moved to a new service container (usually on a different node), and then restarted. This migration process is transparent to both the service itself and the client utilizing the service. AgServe takes advantage of the two Java-compatible ACM kernel service containers, respectively based on the Aroma VM [11] and the Mobile Jikes RVM [12], which support transparent service migration for services implemented in Java.

Service migration differentiates AgServe from other SoAs. Service migration allows AgServe to react to environmental changes and is crucial to realizing the goal of agility. For example, as nodes with free resources become available, service instances might be migrated from heavily loaded server nodes to other nodes, thereby improving the overall performance of the system (self-optimizing behavior). Service migration also allows the system to react to accidental events, such as a power loss or an incoming attack (survivability behavior).

AgServe uses a service manager to support dynamic service definition, registration, lookup, and invocation. The dynamic activation and migration of service instances raises the need to augment traditional service descriptions with meta-information, such as location of the service implementation code, the resource utilization profile, and the communications profile. Currently, a resource tracker inside each ACM kernel maintains a database of services and their resource utilization information. While this aspect has not yet been realized, the resource utilization profile and other meta-information could be embedded into the WSDL [13] description for the service and packaged as part of the service archive. In addition, service registration, lookup, and invocation operations have been modified in order to support dynamic changes in service locations.

AgServe is further described in [10] and [14].

8.9.4 *DisService*

DisService is a peer-to-peer information dissemination system and embodies the principles of opportunistic resource exploitation. DisService is integrated into the ACM, much like Mockets and AgServe, and provides a third entry point into the middleware for applications. DisService supports store and forward delivery of data and caches data throughout the network, thereby making it disruption-tolerant and improving availability of data. In keeping with the philosophy of agile computing, DisService opportunistically discovers and exploits excess communications and storage capacity in the network to improve the performance of information dissemination.

DisService also supports the notion of hierarchical groups to organize the information being disseminated and to be efficient about delivery of information. Subscriptions allow clients to express interest in particular groups. Information is published in the context of a group and may also be tagged to differentiate between multiple types of data (for example, blue-force tracking, sensor data, logistics, or other runtime information).

Each node in the network running DisService operates in a distributed, peer-to-peer manner while processing and communicating the published information and requested subscriptions from neighboring nodes. Information is disseminated using an efficient combination of push and pull, depending on the number of subscribers, the capacity of the network, and the stability of nodes in the network.

One of the goals of the system is to dynamically adapt to dissemination patterns that are combinations of {one | few | many} nodes to {one | few | many} nodes. Observations of information needs in tactical environments identified three primary modes of dissemination. Situation awareness (SA) data, such as blue-force tracking information, is one-to-many because it needs to be disseminated to most, if not all, nodes in the network. Directed data, on the other hand, is by nature of interest only to a small subset of nodes. Hence, it is one-to-one or one-to-few. Examples of this type of data include sensor data being transferred to a node for processing or fusion. Finally, the last mode is for on-demand data, which includes large objects, such as maps, pictures, videos, and other multimedia objects. Given the limited bandwidth of the network, this type of data should not be delivered until explicitly requested by some node in the network.

DisService is further described in [15] and [16].

8.9.5 ACM NetProxy

While the new Mockets and DisService protocols developed as part of the ACM are effective in supporting applications in dynamic environments, there are many legacy applications and existing SoA implementations that continue to use TCP and UDP protocols. Modifying all of these legacy applications would be cost- and time-prohibitive. Therefore, the solution is to create a proxy for the TCP/IP network protocols—NetProxy—to address the problems outlined above. NetProxy is part of the ACM, and its primary goal is to provide a mechanism to transparently remap TCP/IP protocols to protocols that provide better performance for tactical networks. In particular, NetProxy remaps TCP to Mockets, which is designed for mobile ad hoc and other wireless networks. NetProxy also remaps unicast UDP to Mockets and multicast UDP to DisService.

NetProxy also provides transparent compression and decompression capabilities to address the problems and inefficiencies that arise from bandwidth-intensive Web protocols. The flexible architecture of NetProxy, along with the flexibility of Mockets and DisService, enable NetProxy to be used to address many other requirements, such as remapping other protocols, bandwidth allocation, prioritization, data transcoding, and other policy-based behaviors.

Further technical details about NetProxy, including experimental results, are described in [17].

8.10 Conclusions

This research on agile computing began with a simple idea: It should be possible to extract better performance out of middleware operating in a highly dynamic environment if the middleware could react quickly and opportunistically exploit resources, given the observation that nodes often have some excess capacity—some *wiggle room*. In the process of answering this question, the notion of agile computing was defined, an abstract architecture developed, and a comprehensive middleware solution designed and implemented. The middleware includes components for communications (Mockets), resource and service discovery (Group Manager), peer-to-peer information dissemination (DisService), service-oriented architectures (AgServe), and a network proxy to support legacy applications and protocols (NetProxy). Extensive experimental analysis has shown that these components perform better than existing standards or popular solutions.

However, the novel contribution of agile computing is in combining the above components into integrated middleware and leveraging their capabilities to be opportunistic. Experimental analysis has also answered the question of extracting improved performance from the middleware by being opportunistic. While this analysis was done in a limited context, the results are promising. The degree of agility has been determined to be on the order of a few seconds for the current middleware implementation, and this can certainly be improved with further optimization.

Finally, components of the middleware have been integrated into systems being designed and built by the U.S. Army Research Laboratory, the U.S. Air Force Research Laboratory, and the U.S. Navy and Marine Corps and continue to be used in a variety of contexts.

References

1. Suri, N. 2008. Agile Computing. PhD Thesis. Lancaster University, Lancaster, UK.
2. Box, D., and Sells, C. 2002. *Essential.NET. Volume 1: The Common Language Runtime* (Microsoft.NET Development Series). Microsoft Press.
3. Ousterhout, J. K. 1994. *Tcl and the Tk Toolkit.* (Fourth ed.), Reading, MA: Addison-Wesley.
4. Suri, N., Benvegnù, E., Tortonesi, M., Stefanelli, C., Kovach, J., and Hanna J. 2009. Communications Middleware for Tactical Environments: Observations, Experiences, and Lessons Learned. *IEEE Communications Magazine*, Vol. 47, No. 10, pp. 56–63.
5. Benvegnù, E., Suri, N., Tortonesi, M., Esterrich III, T. 2010. Seamless Network Migration Using the Mockets Communications Middleware. In *Proceedings of the 2010 IEEE Military Communications Conference* (MilCom 2010), San Jose, CA, pp. 1604–1609.
6. Tortonesi, M., Stefanelli, C., Suri, N., Arguedas, M., and Breedy, M. 2006. Mockets: A Novel Message-Oriented Communication Middleware for the Wireless Internet. In *Proceedings of International Conference on Wireless Information Networks and Systems* (WINSYS 2006), Setúbal, Portugal.
7. Gnutella. 2002. Online reference. http://www.gnutella.com/.
8. The Gnutella Developer Forum. The Gnutella Protocol Specification Version 0.4. Online reference: http://rfc-gnutella.sourceforge.net/developer/stable/index.html.
9. Suri, N., Rebeschini, M., Breedy, M., Carvalho, M., and Arguedas, M. 2006. Resource and Service Discovery in Wireless Ad-Hoc Networks with Agile Computing. In *Proceedings of the 2006 IEEE Military Communications Conference* (MILCOM 2006), Washington, D.C.
10. Suri, N., Marcon, M., Quitadamo, R., Rebeschini, M., Arguedas, M., Stabellini, S., Tortonesi, M., and Stefanelli, C. 2008. An Adaptive and Efficient Peer-to-Peer Service-Oriented Architecture for MANET Environments with Agile Computing. In *Proceedings of the Second IEEE Workshop on Autonomic Computing and Network Management* (ACNM'08).
11. Suri, N., Bradshaw, J. M., Breedy, M. R., Groth, P. T., Hill, G. A., and Saavedra, R. 2001. State Capture and Resource Control for Java: The Design and Implementation of the Aroma Virtual Machine. In USENIX JVM 01 Conference Work in Progress Session. Extended version available as a Technical Report.

12. Quitadamo, R., Leonardi, L., and Cabri, G. 2006. Leveraging Strong Agent Mobility for Aglets with the Mobile Jikes RVM Framework. In Hexmoor, H., Paprzycki, M., and Suri, N. (Eds). *Scalable Computing Practice and Experience — Special Issue: Software Agent Mobility*, pp. 37–52. Online reference: http://www.scpe.org.
13. World Wide Web Consortium. Web Services Description Language (WSDL) 1.1. Online reference: http://www.w3.org/TR/wsdl.
14. Suri, N. 2009. Dynamic Service-Oriented Architectures for Tactical Edge Networks. In *Proceedings of the 4th Workshop on Emerging Web Services Technology* (Eindhoven, The Netherlands, November 9, 2009). W. Binder and E. Wilde, Eds. WEWST '09, vol. 404. ACM, New York, 3–10.
15. Suri, N., Benincasa, G., Tortonesi, M., Stefanelli, C., Kovach, J., Winkler, R., Kohler, R., Hanna, J., Pochet, L., and Watson, S. C. 2010. Peer-to-Peer Communications for Tactical Environments: Observations, Requirements, and Experiences. *IEEE Communications Magazine*, Vol. 48, No. 10, pp. 60–69.
16. Benincasa, G., Rossi, A., Suri, N., Tortonesi, M., and Stefanelli, C. 2011. An Experimental Evaluation of Peer-to-Peer Multicast Protocols. In *Proceedings of the 2011 IEEE Military Communications Conference* (MilCom 2011), Baltimore, MD, pp. 1015–1022.
17. Morelli, A., Kohler, R., Stefanelli, C., Suri, N., and Tortonesi, M. 2012. Supporting COTS Applications in Tactical Edge Networks. In *Proceedings of the 2012 IEEE Military Communications Conference* (MilCom 2012), pp. 1717–1723.

Chapter 9

A Pattern-Based Architectural Style for Self-Organizing Software Systems

Jose Luis Fernandez-Marquez,
Giovanna Di Marzo Serugendo, Paul L. Snyder,
Giuseppe Valetto, and Franco Zambonelli

Contents

In this chapter, Jose Luis Fernandez-Marquez et al. propose an approach to engineering self-organizing software systems toward self-adaptation and resilience from an architectural point of view. They argue that the adaptation of complete systems is different from the adaptation of single components within the systems and propose an architectural approach based on patterns. Besides adaptation, the authors also claim to achieve resilience by leveraging the capabilities of self-organizing systems, which are able to modify themselves in order to continue providing their functionality even in the face of unexpected situations.

9.1 Introduction

With the continuous increase in runtime scale and complexity of software systems, self-adaptivity has assumed a central role in the software engineering discourse and has been often mentioned as one of the defining challenges for the discipline [1–3].

Self-adaptivity is the ability of a software system or application to automatically modify its structure and behavior at runtime in order to ensure, maintain, or recover some functional properties even in the face of unexpected changes to operating conditions or user requirements. This kind of resilience to change is a typical property of the software architecture. Moreover, self-adaptive provisions are often in charge of nonfunctional properties and end-to-end quality factors of the software (such as availability, reliability, response time and performance, robustness, etc.) that are architectural in nature. It is therefore quite natural to approach the engineering of self-adaptive software from an architectural perspective, and a large body of research has indeed addressed this topic.

Architecture-centric approaches to self-adaptation rely upon an explicit representation of the system at runtime (its architectural configuration), obtained through monitoring and reflection; they reason about that configuration and its properties and enact adaptations deemed necessary or useful by manipulating the configuration or architectural elements (either components or connectors). This approach assumes a high degree of accuracy and completeness in the information about the runtime architectural layout and global, top-down control of the makeup of that layout and its adaptation.

Although this approach has been successfully used in many cases, there are some forces in a number of systems and domains that may test them severely. A major force of this type is uncertainty with respect to either the collection of information about the runtime system or the ability to carry out its adaptation as planned [4]. A domain in which high levels of uncertainty are intrinsic is that of systems that are strongly decentralized and have open boundaries. By "open boundaries" we mean that the number or even the type of participating elements in the system may vary at any given time, making impossible to maintain an accurate snapshot of the architectural configuration or other global knowledge; rather, every individual element may hold its own partial view of the system, together with other local information that may be useful for adaptation. By "decentralized" we mean that no one element can take far-reaching control actions that influence many or all of the other elements in the system; rather, every individual element may only take local action that affects itself and possibly a limited number of other system constituents (e.g., its neighbors).

Systems with these characteristics are becoming common as the increase in scale, complexity, and pervasiveness of information technologies accelerates, inducing real-world examples of ultra-large-scale software systems [5], such as peer-to-peer networks, ubiquitous computing, and transport control and energy distribution systems.

Our interest focuses on this kind of system because, in order to enable self-adaptivity, researchers have resorted to *self-organization principles and algorithms* [6–9], often derived from biological or otherwise natural phenomena that can be observed in complex adaptive systems (CAS) [10]. Software self-organization can then be regarded as a fundamentally different kind of self-adaptation in which control is exerted in a bottom-up, emergent fashion by a plurality of individual system elements that take actions with only local effects and local interactions.

Although several self-organization mechanisms have been identified and have proven effective in producing system-wide adaptation in a number of applications, it is commonly recognized that it is quite difficult to represent and reason about this type of adaptation. For example, it is typically extremely hard to capture the local-to-global linkage [11], that is, how local behaviors in a particular system combine up to produce the desired global effects or indicate under what set of conditions that can successfully happen. That, in turn, hinders the disciplined reuse of these mechanisms and the engineering of self-organizing software in a repeatable way with predictable outcomes. The ability to frame self-organizing software systems within an architectural setting would therefore be an important advancement as underlined also by the self-organized architecture (SOAR) workshops* recently held at conferences on self-adaptive systems (ICAC)† and software architecture (WICSA).‡

In this chapter, we present a pattern-based architectural style to address adaptivity, resilience through systematic reuse of self-organizing mechanisms. It proceeds from the observation that, in self-organizing software, a desired system-wide adaptation and resilience is often achieved through the combination of a number of self-organization mechanisms that operate at different levels of abstraction. Therefore, our vision is twofold. First, we need to organize existing, well-understood mechanisms as a set of building blocks that can be composed together. To that end, we have developed a catalog of design patterns that capture self-organization mechanisms that recur in many existing systems [12–14]. The catalog systematizes these patterns by identifying the interrelationships that exist among them, including the composition of simpler, lower-layer patterns in more abstract, higher-layer ones. Second, we need an architectural style in which those patterns can be implemented and can operate as modular primitives or operators in ways that are explicitly specified and produce predictable effects. In this chapter, we discuss, for the first time, such an architectural style, which we have devised on the basis of the insight gained by using the patterns in a piecemeal fashion. We discuss the merits and limitations of this style, show the use of different patterns in an example, and describe an implementation framework.

The rest of the chapter is organized as follows: We discuss the relevant literature in Section 9.2 and our own pattern catalog in Section 9.3. We introduce our architectural style in Section 9.4 and an example of its use in Section 9.5. We present a possible technological incarnation of the style in Section 9.6. We conclude with a discussion of the properties of this approach and outline the direction of our future work.

9.2 Related Work

The literature on architectural approaches to developing self-adaptive software is vast. A thorough overview is beyond the scope of this chapter, and several recent surveys on the state of the art of software engineering for self-adaptive systems devote substantial attention to architectural aspects, challenges, and techniques [2,3,15].

---

* http://distrinet.cs.kuleuven.be/events/soar/.
† http://nsfcac.rutgers.edu/conferences/icac/.
‡ http://www.wicsa.net/.

9.2.1 Architectural Approaches to Self-Adaptation

Here, we simply point out a few seminal papers and how they have outlined a very common model for architectural adaptation, based upon the use of an adaptation engine, that is, an architecture evolution manager that operates at run time [16]. A classic example is found in Rainbow [17], which uses a centralized adaptation engine that reasons about an up-to-date representation of the runtime target system and uses primitives that manipulate its architectural configuration based on quality considerations. Georgiadis et al. [18] propose a multiply instantiated adaptation engine together with the individual elements of the architecture. That idea goes in the direction of distributed—but not decentralized—control because, to correctly enact adaptation, the distributed engine still relies on a complete and consistent view of the target architecture across all instances. Building on that work, Kramer and Magee [19] also elaborated a layered architectural reference model for self-adaptive systems that organizes distributed adaptation in a hierarchical fashion: Elements within a higher layer work at a higher level of abstraction and drive the layer below. Although made of physically distributed and functionally distinct elements, an adaptation engine built in accord with this reference model is conceptually a single entity that still exerts centralized control on a target system.

Although this kind of approach can be quite successful, it has recognized limitations. From Cheng et al. [2]: "… the centralized control loop pattern … may suffer from scalability problems. There is a pressing need for decentralized, but still manageable, efficient, and predictable techniques for constructing self-adaptive software systems." Two years later, in 2011, de Lemos et al. [3] state: "While promising work is emerging in decentralized control of self-adaptive software there is a dearth of practical and effective techniques to build systems in this fashion," which we take as a lack of engineering and, more specifically, architectural understanding of how existing decentralized control mechanisms effectively lead to self-organization in a predictable and repeatable fashion.

Because our contribution, with this chapter, tries to address exactly that issue, we focus below on design patterns and architectures for self-organizing systems and how they have been used to attack it.

9.2.2 Design Patterns for Self-Organizing Software

In the last decade, self-organizing mechanisms have been applied in different domains, achieving results that may go beyond what is possible with more traditional self-adaptation approaches as documented, for example, in the SASO conference series.* However, researchers usually apply these mechanisms in an ad hoc manner, preventing these mechanisms from being systematically reused to solve

---

* http://www.saso-conference.org/.

recurrent problems. Thus, the issue of systematizing those self-organization techniques lends itself to a pattern-based approach.

As with other design patterns, the main promise of self-organization patterns is to support the understanding of the correct scope of a particular solution in terms of what problems it can successfully address and under which set of conditions. Abstracting the problems and the solutions is particularly important because the same self-organization technique can be, at times, used in very different application contexts, and conversely, very similar problems can often be addressed via several diverse techniques.

Among the works that attempt to define design patterns for self-organizing software, some focus on the discovery and definition of a single pattern [20,21]. Others propose a catalog of multiple patterns [6,7,9,22]. Our previous work [12], discussed in Section 9.3, is most similar to the latter. One thing that sets apart our catalog of bio-inspired patterns from previous efforts is our organization of the patterns into layers and the documented composition relationships among them. This helps differentiate the patterns and scope them correctly. As pointed out by Parunak and Brueckner [23], the definition of composition and decomposition relationships is quite important because, although it seems clear that certain self-organization mechanisms can be obtained from those finer-grained, which primitives to use and how they should be combined is often not clear and almost never explicitly codified.

9.2.3 Architectural Approaches to Self-Organization

Although self-adaptive architectures have been discussed extensively in the literature, self-organization is less frequently addressed directly, and even then, it is often as an adjunct consideration. Parunak and Brueckner [23] extensively examine a number of important issues related to the software engineering of self-organization, including architectural concerns. One of the key challenges that architectures for self-organizing software need to address is the composability of multiple self-organization mechanisms in a single system. In that direction, Cuesta and Romay [24] present a taxonomy of self-adaptive architectural elements in a distributed context and an algebraic approach to analyzing composition of adaptive elements in such a system. Specific architectural proposals are primarily in the multi-agent system (MAS) vein as in Wenkstern et al. [25], which proposes a two-level agent model with higher-level agents acting as explicit controllers, or SodekoSV [26], which pairs agents with a coordination medium hosted on an agent middleware.

In previous work, we proposed MetaSelf, a preliminary description of an architecture that combines both self-management through the enforcement of predefined system-wide policies and self-organizing through rules locally applied by entities [27]. That proposal, however, does not focus on the use of self-organizing mechanisms as primitives provided by the environment. More recently, we proposed

BIO-CORE [28], a middleware built around a data repository and chemical rules that also exploit low-level self-organizing mechanisms as basic primitives provided by the computational environment although a precursor of the architectural style proposed in this chapter, BIO-CORE, is not explicitly described as such.

9.3 Design Pattern Catalog

To facilitate the systematic engineering of self-organizing software systems, in our previous work, we constructed a catalog of bio-inspired modular and reusable design patterns, organized into layers. We report here the main characteristics of each of these patterns (i.e., the problems they solve and the solutions they provide). For a full description of each of these patterns (including class diagrams, sequence diagrams, and examples of use) the reader should refer to [12–14].

By analyzing the behavior and properties of bio-inspired mechanisms applied to self-organization of software that recur in the literature, we have isolated several mechanisms, which are basic and atomic, plus others that are composed from the basic ones. As a result, we classified our patterns into three layers. At the bottom layer are the *basic* mechanisms that can be used individually or in composition to form more complex patterns. At the middle layer, there are the mechanisms formed by *combinations* of the bottom-layer mechanisms. The top layer contains even more complex patterns that capture more complex and abstract bio-inspired self-organization mechanisms and show different ways to *exploit* the basic and composed mechanisms from the bottom and middle layers.

Figure 9.1 shows the different design patterns collected in the catalog and their relationships. The arrows indicate how the patterns are composed. A dashed arrow indicates that the use of the pattern is optional. One important observation about this classification is that basic and composed patterns capture behaviors that correspond to processes that occur within the environment in biological systems. For example, SPREADING, AGGREGATION, and EVAPORATION are processes that change

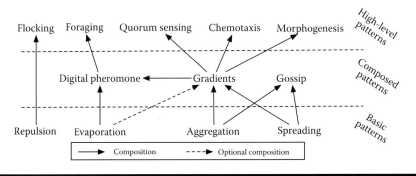

Figure 9.1 Patterns and their relationships.

the environment in which self-organizing agents are immersed. Moreover, these environmental mechanisms become essential building blocks for more complex behaviors on the part of those agents.

A good and well-known example is ant foraging. In ant colonies, the ants are able to achieve very complex self-organized behaviors; this coordination is performed via pheromones as ants create, follow, and reinforce pheromone gradients that form trails through the environment. This mechanism, to be effective, assumes an environment in which pheromones are spread (allowing ants to sense pheromone trails), aggregated (increasing pheromone concentration and reinforcing certain trails), and evaporated (allowing ants to adapt to changes as old trails that are no longer relevant are forgotten).

In line with the importance accorded by Weyns et al. to the concept of an agent's environment [29], we consider the two lower layers, that is, the basic and composed patterns, as primitives for bio-inspired self-organization algorithms embedded within the execution environment. The high-level patterns capture instead more complex processes that are the prerogative of each individual agent and can be implemented either within the agent itself or as services that agents can invoke. This insight is the basis for the software architecture proposed in the next section that integrates these design patterns in a modular fashion, hence promoting their reuse.

9.4 Architectural Style

An architectural style defines the common architectural traits of a family of software systems by describing a structural organization. Architectural styles can be used on their own or combined together to form new architectural styles.

Following Shaw and Garlan [30], we will discuss our proposed architectural style for self-organizing software systems in the following terms: structural organization for the architecture (including components, connectors and data); invariants of the components or connectors; advantages, highlighting adaptation and resilience features; disadvantages; and the underlying computational model.

9.4.1 Architectural Style for Self-Organizing Systems

As observed in [31], the main design elements of a self-organizing system are the *environment* in which the system evolves; the *agents*, autonomous, individual, active entities of the system; the *self-organizing mechanisms* governing the behavior of the agents and their interactions with the environment or other agents; and the *artifacts*, which are the passive data entities maintained by the environment and created, modified, and/or sensed by the agents (e.g., digital pheromones spread in the environment or information exchanged among agents).

The main idea behind the architectural style we propose here is the provision by the environment of low-level, atomic self-organizing mechanisms under the form of *core services* or primitives, which are leveraged by more sophisticated self-organizing mechanisms, or *services*, which, in turn, are requested on demand by *agents* acting in the system to effect their own, complex self-organizing behaviors. It is then possible to build self-organizing applications using these agents, which activate individual behaviors with the purpose of creating a convergence to the kind of emergent order that would satisfy the application requirements.

We want such an architecture to support reuse (e.g., a primitive, such as evaporation, provided by the environment can be used independently by multiple different services and activated by all agents that make use of those services) and separation of concerns (e.g., the programmer of the agent or the application concentrates on the functionality to provide and relies on the underlying self-organizing mechanisms provided by the environment).

9.4.2 Structural Description

The architectural style we propose for self-organizing systems is a combination of three other architectural styles: layered, blackboard, and implicit invocation.

Our architectural style follows overall a *layered* organization (see Figure 9.2). At the bottom layer, we find the computational environment, in which a set of *core services* is executed and provided as primitives. Those primitives correspond to basic or composed patterns in our design pattern catalog. This is a parallel with the natural environment in which many biological self-organization processes take place, in particular with respect to production and distribution of information. Therefore, the data that represent the properties of the computational environment and the information deposited in it are managed and manipulated in a shared repository. Zooming into the computational environment layer, we see that it is composed of (1) *core services*, that enact low-level self-organizing mechanisms corresponding to the basic or composed patterns of Figure 9.1 (see

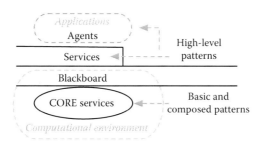

Figure 9.2   Overall layered architecture.

Section 9.3), such as SPREADING, AGGREGATION, EVAPORATION, and GRADIENT, and (2) the *blackboard component*, which hosts the shared repository of data among agents and services and is responsible to trigger the execution of core services as necessary.

In the higher layers of our architecture, we have *services* and *agents*.

Services are an abstraction that enables us to situate and enact complex self-organizing mechanisms, corresponding to the high-level patterns in our catalog, such as CHEMOTAXIS or QUORUM sensing (see Section 9.3), which agents make use of. A major difference between a service and a core service resides in the fact that a core service is of an atomic nature and triggered by the environment, and a service represents a composed self-organization mechanism that may also need some additional logic to be performed. Also, unlike core services, services are not necessarily available at all locations where a blackboard is instantiated.

Agents stand for any application-related computational component that runs some self-organization algorithm as its own individual behavior. Agents may implement some of the high-level patterns themselves, but for the sake of reuse, they may even simply delegate them to services and invoke them as needed. Both services and agents can make use of the core services provided by the computational environment by interacting with the blackboard, but they can also interact with one another.

It is through the injection of appropriate data that agents or services trigger the activation of core services by the *blackboard*. Notice that, unlike traditional blackboard architectures, the blackboard component we consider is not centralized for the whole system, but multiple blackboard components are distributed through the system. Thus, each computing node that hosts some services or agents will also host one blackboard component. Agents and services may only access the blackboard running on the same computing node, which thus restricts them to use local information only. However, core services, such as SPREADING, which are present in all blackboards, provide a way for information to travel in controlled ways and effectively connect the different blackboards (Figure 9.3).

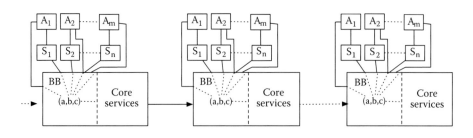

Figure 9.3 Implicit invocation of core services.

The self-organization primitive mechanisms representing core services are "embedded" within the blackboard by means of the *implicit invocation* architecture (Figure 9.3). Upon arrival of appropriate data in a blackboard repository, the corresponding core services are activated by that blackboard component.

The distributed blackboard component of our style thus allows the coordination between different software agents even though they belong to different applications or they are hosted in different nodes. Moreover, the blackboard provides an elegant way for implementing indirect communication, which is a key communication model for nature-inspired, self-organizing systems, such as those based on ant foraging, gradient propagation, or chemotaxis.

9.4.3 Components and Connectors

Following the description above, the *components* of the architectural style for self-organizing systems are the agents, the services enacting high-level self-organizing mechanisms, the blackboard (repository and computational component), and the core services enacting basic self-organizing mechanisms.

There are different *connectors* as shown in Figure 9.4:

- Between agents/services and blackboard repository: Data inserted/read/removed from the blackboard repository.
- Between agents and services: Local services are directly invoked by agents without needing to use the blackboard. That provides a higher-level programming interface that eases the programming tasks (i.e., services can be provided as libraries on which developers rely).
- Between agents/services and core services: Implicit invocation by the blackboard component upon an event generated by the arrival of an appropriate data in the blackboard repository.

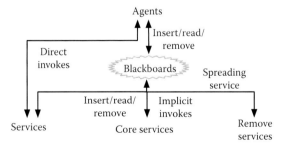

Figure 9.4 Components and connectors.

- Between agents/services and remote services: The SPREADING core service acts as connector for invoking services that are available through remote blackboards.
- Between blackboards and core services: The blackboard component activates the appropriate core service as a result of the corresponding event arising in the system (e.g., the arrival of data in the blackboard repository requesting to be spread, evaporated, etc.).

Data inserted, read, and retrieved from the blackboard repository consists of tuples of data with specific tags recognized by other agents or services or acting as events for triggering core services.

Components and connectors are summarized in Table 9.1.

Table 9.1 Primitive Vocabulary

Primitive Vocabulary	Informal Description
Components	
Agent	Autonomous software component, usually pertaining to an application; can be mobile or permanently hosted in one node
Service/remote service	Software component enacting a high-level self-organizing mechanism; invoked by agents possibly remotely
Core service	Software component enacting a basic self-organizing mechanism; invoked by agents or services
Blackboard	Components hosting a shared repository where data is injected, observed, and modified by agents, services and core services; triggers core services when necessary
Connectors	
Agents/services – blackboard repository	Data inserted/read/removed from the blackboard repository
Agents – services	Exchange of data through the blackboard repository or direct invocation of services by agents
Agents/services – core services	Generation of an event arising from the insertion of a data in the blackboard repository
Agents/services – remote services	Exchange of data through blackboard repositories, use of SPREADING core service to propagate data at remote nodes
Blackboard – core services	Invocation of core services in response to the arrival of expected data

9.4.4 Invariants

The invariants of a software architecture refer to those features that should remain constant for different implementations. The invariants of our proposed architecture are as follows:

- Nodes participating in the system have only local connections, and the blackboard component stored in each host is only directly accessible by agents and services running in the same node.
- The blackboard component should provide at least three primitives: (1) injection of data, (2) observation of data, and (3) removal of data.
- The invocation of core services is done by specific tags (e.g., keywords) present in the data injected into the blackboard repository.
- The invocation of remote services can be done by tags in the data or by using a service description. In both cases, the invocation is done by injecting some data into the blackboard repository.
- Agents and services announce themselves in the system by injecting suitable data into a blackboard. This data will contain an agent or service description or information relative to the application or service they provide.
- The order of executions of invoked services cannot be established at design time. However, core services could have priority because they will be provided together with the blackboard.
- Core services are available at all nodes through the blackboard components. In contrast, high-level services may not be available at all nodes.

9.4.5 Advantages

Implicit invocation provides strong support for reuse of low-level self-organizing mechanisms. Additional low-level self-organizing mechanisms can be introduced in the system by simply registering them for the appropriate events.

By their very nature, self-organizing mechanisms naturally provide robustness and adaptability to well-identified dynamic changes. For instance, evaporation dynamically gets rid of outdated information, or chemotaxis allows following a gradient direction despite unexpected changes of direction.

Besides the adaptability and resilience characteristics naturally provided by the self-organizing mechanisms, the proposed architectural style provides to the applications the additional following adaptation and resilience features.

9.4.5.1 Adaptation

Services can be published on the blackboard and spread or made available at remote locations using the Spreading core service. Moreover, services can also be published using gradients (i.e., a GRADIENT service), providing routes and information about distance (i.e., number of hops) to potential users (i.e., higher-level services or agents). Thus, the application can adapt dynamically, taking into account the current services available in the system.

Our approach favors scalability in two ways: (1) Applications can easily increase their functionality in a modular way, and (2) both services and core services proposed as a basis for all the applications developed following this architecture are very scalable because they are based only on local interactions and partial knowledge of the system. Their scalability has been properly demonstrated in many applications as discussed in the coordination and self-organization systems literature.

9.4.5.2 Resilience

The composition of implicit invocation and blackboard provides a way to publish services at remote nodes. Applications can thus be composed of services even though those services are not hosted in the same node. This type of composition allows the system to increase available functionality, replace a service in case of failure, or if there is a decrease in the quality of service, to provide or find a replacement and hence to increase the robustness and resilience of the system.

9.4.6 Disadvantages

The main disadvantage of this architecture is that high-level components, such as agents or services, cannot rely on any order of computation in which the core services are invoked. Agents and services cannot rely on a predefined core services flow of control.

9.4.7 Underlying Computational Model

The computational model is fully described in [28]. An abstract view of a setup for an individual computing host or node that complies with our description of the style is shown in Figure 9.5a. A concrete realization, including the interfaces and main interaction channels between the various architectural elements found in a host, is shown in Figure 9.5b. Agents are autonomous and proactive software entities running in a host; they can be programs in traditional computing environments but can also be, for example, mobile robots or smartphones with sensors and actuators for controlling their physical environment. Our architecture is distributed on a set of connected computing nodes (or hosts) that host agents, services, and the blackboard.

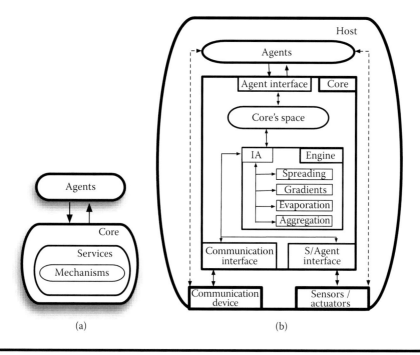

Figure 9.5 System's structure. (a) Abstract view and (b) concrete view.

9.5 Case Study

We present hereby the implementation of a self-organizing algorithm in accord with our architecture. For illustration purposes, the example is kept very simple. It consists of the decentralized computation (sum) of numbers distributed among multiple hosts. In this example, a node starts the computation and queries remote data by diffusing a gradient; nodes reached by the gradient reply by injecting their data into the blackboard. This data is then moved along the gradient toward the node that started the computation through chemotaxis. At each intermediary node, data is aggregated (sum) before being sent forward. The purpose of this example is to show how multiple self-organization mechanisms in our pattern catalog are accommodated and executed as needed within the proposed architecture.

9.5.1 Description

Let N be a set of nodes with local connections to other nodes. Each node contains a blackboard and the core services of SPREADING, AGGREGATION, GRADIENT, and EVAPORATION. As discussed in Section 9.4, these Services will be implicitly invoked based on the properties of the data injected into the blackboard.

Every node hosts an instance of a GRADIENT core service gs. Upon invocation, this service creates a data element that (a) will SPREAD from node to node, increasing its hop count by one every time it spreads to a new node; (b) will AGGREGATE the neighboring hop-count values to determine the lowest; and (c) will EVAPORATE, removing it from the system after a certain period of time. The service then injects this element to the local blackboard.

Each node also hosts a CHEMOTAXIS service cs, which monitors the local blackboard for data elements associated with a GRADIENT. If such an element is found, it determines the neighbor with the lowest hop-count value and modifies it so that the SPREADING core service will transfer it to the appropriate neighbor blackboard.

Each node $n_i \in N$ has an instance of an agent a_i. This agent stores an integer value v_i, which is a random number [0...2,000] (Figure 9.6a). The process is initiated by a single node n_{start}, and this node will contain the final sum.

1. The initiating agent a_{start} (running on node n_{start}) makes a request to the GRADIENT core service gs to establish a new gradient for this execution of the sum algorithm. To do so, agent a_{start} formats a data element representing a request that a gradient g be created and injects the request to the blackboard.
2. The GRADIENT core service running on the blackboard diffuses the gradient throughout the network.
3. Once the gradient reaches a node n_i, (Figure 9.6b), the node's agent a_i injects a data element to the blackboard containing the value of v_i with flags indicating that it is associated with the gradient g and that it should aggregate with other data elements of the same type.
4. When the CHEMOTAXIS service cs observes a data element associated with g at the node, it determines the lowest hop-count neighbor and marks the data element to be transferred there.
5. The SPREADING core service actually transfers the data element to the indicated neighbor. If multiple data elements associated with g are present, the AGGREGATION core service will combine them by summing their associated values (Figure 9.6c and d).
6. When a data element associated with gradient g arrives at n_{start}'s blackboard, the agent a_{start} removes it and adds the contained value to its local sum. The algorithm converges when all values have reached this node (Figure 9.6e).

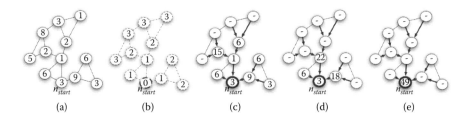

Figure 9.6 Self-organizing sum algorithm. (a) Initial values, (b) gradient, (c) progressive aggregation I, (d) progressive aggregation II, and (e) final result.

9.5.2 Adaptation and Resilience

As shown in [32], this type of progressive aggregation reduces bandwidth and memory usage, thus increasing scalability. Indeed, simulations show that progressive aggregation can be done on top of gradients, reducing the bandwidth usage and memory consumption, in spite of the existence of many different paths for reaching the source.

Simulation results show that our approach is tolerant to intermittent communication failures without a significant increment of the bandwidth usage nor memory consumption.

In mobile scenarios, we replaced the gradient with a dynamic gradient (i.e., using evaporation to regularly update the gradient). Simulations show that the system also adapts to changing node positions, the sum algorithm being able to get 100% accuracy even when the gradient is not properly updated.

Additional quantitative results showing resilience to failure in mobile scenarios are discussed in Chapter 11.

9.6 Implementation Framework

The proposed architectural style could be implemented in a number of different ways. In this section, we discuss one possible approach that has been developed as part of the SAPERE European Project.* This approach is a reification of the vision of the architectural style that is discussed in Section 9.4 and provides a framework for the style, which can thus support the development of decentralized, self-organizing applications from the pattern primitives discussed in Section 9.3.

We identify the main components and attending decisions that have been adopted within the SAPERE framework. For a more extensive discussion of this framework, readers are referred to [33,34]. In the SAPERE implementation the main concepts are:

- Live Semantic Annotation (LSA): An LSA is a tuple that represents any information about an agent or service. LSAs are injected in the LSA's space representing the (updated) state of its associated component. *An LSA represents the data described in the proposed architecture for self-organizing systems.*
- LSA's space: It is a distributed, shared space in which context and information are provided by a set of LSAs stored at given locations. *According to our proposed architecture, the LSA's space corresponds to the blackboard component and its repository.*

---

* www.sapere-project.eu.

■ Eco-laws: Eco-laws are rules acting as chemical reactions that implement core services. These rules act on the LSAs stored in each LSA's space by deleting, updating, or moving LSAs between LSA's spaces. *Eco-laws correspond to core services in our architecture.*

■ LSA bonding: An LSA bond acts as a reference to another LSA and provides fine-tuned control of what is visible or modifiable to each agent and what is not. If an LSA of a given agent includes a bond to an LSA of another agent, the former agent can inspect the state and interface of that other agent (through the bond) and act accordingly. Bonds are a specific type of eco-laws.

■ Agents: They are autonomous software that executes in hosts and are able to locally access the LSA's space available in that host.

Eco-laws reside in the LSA's space, and they are invoked following an implicit invocation pattern. An LSA can be subject to different eco-laws, depending on the specified properties or tags of the tuple. Table 9.2 shows the properties that one LSA should contain to trigger the core services.

The SAPERE infrastructure encompasses a middleware [35] running on stationary and mobile platforms, crowd-steering applications [34], situation-awareness [36], and semantic resource discovery [37].

To give the flavor of a translation into the SAPERE framework of the proposed architectural style and in line with the case study of Section 9.5, we show here the SPREADING, AGGREGATION, and GRADIENT core services as well as the CHEMOTAXIS service.

Table 9.2 LSA's Properties

Property	Description
ID	Unique identifier
SPREAD	Activate the SPREADING core service
EVAP	Activate the EVAPORATION core service. Automatically, this property sets up the relevance attribute (REL)
AGGREGATE	Activate the AGGREGATION core service
GRADIENT	Activate the GRADIENT core service. Automatically, this property also enables the SPREAD and AGGREGATE properties and sets up the distance attribute (DIST)
CHEMOTAXIS	Activate the CHEMOTAXIS service
DATA	Actual information stored in the LSA

9.6.1 Spreading Core Service

The SPREADING service retains a copy of the information received or held by an agent and sends this information to the neighbors so that it propagates over the network. The eco-law corresponding to the SPREADING service is as follows (1):

$$\langle \mathtt{ID,SPREAD,DATA} \rangle \mapsto \\ \langle \mathtt{ID,DATA} \rangle, bcast \langle \mathtt{ID,SPREAD,DATA} \rangle \tag{1}$$

When an LSA with SPREAD property is processed, the LSA is identified as a spreading LSA (i.e., the LSA is subject to the SPREADING eco-law [1]). The LSA is sent to the neighboring LSA's spaces (with a broadcast *bcast*), and a local copy of the information remains. This process is repeated by neighboring LSA's spaces, causing the LSA to be propagated over the whole system, making the information contained in the LSA available to all the agents participating in the system.

A variant of the SPREADING core service, called direct spreading, allows the specification of the node where the information should be spread. This is very useful in many cases to optimize higher-level services (e.g., optimizing bandwidth by sending information to a set of neighboring nodes instead of all of them) or simply to directly propagate information to specific nodes inside the communication range. The CHEMOTAXIS service uses direct spreading.

9.6.2 Aggregation Core Service

The AGGREGATION core service consists in locally applying a fusion operator to synthesize information. When an LSA with the AGGREGATE property is processed, the LSA is identified as an aggregation LSA (i.e., the LSA is subject to the AGGREGATION eco-law [2]). In each node, the AGGREGATION core service retrieves all the LSAs and aggregates them. These LSAs are those with the AGGREGATE property and that share the same application identifier (provided by the field $\mathtt{APP_{ID}}$).

The eco-law for the AGGREGATION service is as follows:

$$\langle \mathtt{ID_1, AGGREGATE, APP_{ID}, DATA_1} \rangle, ..., \\ \langle \mathtt{ID_n, AGGREGATE, APP_{ID}, DATA_n} \rangle \mapsto \\ \langle \mathtt{ID_{new}, AGGREGATE, APP_{ID}, Sum(DATA_1, ..., DATA_n)} \rangle \tag{2}$$

The above eco-law for the AGGREGATION service assumes the sum as a default operator (sum).

9.6.3 Gradient Core Service

The GRADIENT core service is a composition of the SPREADING and AGGREGATION core services with which the information is propagated (spread) in such a way that it provides additional information about the sender's distance and direction. Additionally, the GRADIENT core service uses the AGGREGATION core service to merge different gradients created by different agents or to merge gradients coming from the same agent but through different paths. Finally, in the dynamic version with mobile nodes, the GRADIENT core service also exploits EVAPORATION for removing outdated information.

The eco-laws for the gradient service are as follows:

$$\langle ID, GRADIENT, APP_{ID}, DATA \rangle \mapsto$$
$$\langle ID, GRADIENT, APP_{ID}, 0, DATA \rangle \tag{3}$$

$$\langle ID, GRADIENT, APP_{ID}, DIST, DATA \rangle \mapsto$$
$$\langle ID, SPREAD, GRADIENT, APP_{ID}, DIST + \Delta, DATA \rangle \tag{4}$$
$$\langle ID, AGGREGATE, GRADIENT, APP_{ID}, DIST, DATA \rangle,$$

When an LSA with GRADIENT property is processed, the LSA is identified as a gradient LSA. This automatically activates the SPREADING and AGGREGATION core services by creating two appropriate LSAs with SPREAD and AGGREGATE properties, respectively.

Moreover, when the LSA is injected in the LSA's space, an additional distance attribute called DIST, initially equal to 0, is added to the LSA. When the LSA is processed, the LSA is spread, incrementing the distance attribute by a default value (Δ) and aggregated (by default – sum).

Additionally, the EVAP property can also be set, enabling dynamic gradients. In this case, the information is subject to the EVAPORATION eco-law, allowing the gradient to be adapted automatically when network topology changes occur.

9.6.4 Chemotaxis Service

The CHEMOTAXIS service is a high-level service provided as a library. The CHEMOTAXIS service allows information to be sent to remote nodes following gradients paths. As a library, the CHEMOTAXIS service is directly invoked by agents. The CHEMOTAXIS service has been implemented exploiting the SPREADING core service–direct spreading variant. When the CHEMOTAXIS service is invoked, a gradient ID is passed as a parameter. The CHEMOTAXIS service checks that the gradient exists in the LSA's space (as an LSA) and retrieves from the LSA's space the ID of the previous node in the gradient path (uphill). The CHEMOTAXIS service then injects an appropriate LSA, specifying that it wants to use the direct spreading variant of the SPREADING core service to send the data to that previous node in the path.

9.7 Conclusions

In this chapter, we have presented an architectural approach suitable to pattern-based engineering of self-organizing software systems. Our contribution comprises a catalog of reusable design patterns capturing mechanisms for self-organizing dynamics and a composite architectural style that supports the composition and reuse of such mechanisms into full self-organized applications. This architectural style has a layered organization. A key aspect is represented by the computational environment layer, in which core services reside to enact basic self-organizing mechanisms that describe processes of information management and manipulation. In this way, the architecture clearly separates the responsibility of the computational environment from that of higher-level services and application-level agents, which make use of the processes in the computational environment level to implement more complex individual behaviors. This style can facilitate the design and implementation of many self-organizing applications or services and fosters reuse of well-understood self-organizing mechanisms that recur in the literature in a unified way.

The use of blackboards and implicit invocation within our composite style facilitates service composition and easily supports the entry and exit of services and application agents to and from the system. Overall, it enables a dynamic infrastructure composed of a potentially very large number of nodes and is applicable to application domains such as mobile ad-hoc networks, P2P, wireless sensor networks, ubiquitous computing, and ecosystem of services.

In addition to the natural robustness and adaptability provided to the applications through the self-organizing mechanisms at work in the architecture, the architecture itself favors adaptability to available services (e.g., to improve performance) and supports resilience to service failures or to degraded quality of services by providing dynamic composition with remote services.

Even though the architectural style proposed in this chapter has been implemented using a tuple based and chemical-inspired approach, it can be implemented in different ways.

Our next steps will entail further validation of this architectural style in practice. In particular, we intend to assess how well it facilitates the use and reuse of self-organizing patterns all the way up to the agent level and how effectively it eases the development of self-organizing applications comprised of many distinct agents with additional logic and individual parameterization.

Acknowledgment

G. Di Marzo Serugendo and J. L. Fernandez-Marquez are partially supported by the EU FP7 project "SAPERE—Self-Aware Pervasive Service Ecosystems" under contract No. 256873.

References

1. P. Horn, "Autonomic computing: IBM's perspective on the state of information technology," *Computing Systems*, vol. 15, no. Jan, pp. 1–40, 2001.
2. B. H. Cheng, R. Lemos, H. Giese, P. Inverardi, J. Magee, J. Andersson, B. Becker et al., "Software engineering for self-adaptive systems: A research roadmap," in *Software Engineering for Self-Adaptive Systems*, ser. Lecture Notes in Computer Science, B. Cheng, R. de Lemos, H. Giese, P. Inverardi, and J. Magee, Eds. Springer, Berlin, 2009, vol. 5525, pp. 1–26.
3. R. de Lemos et al., "Software engineering for self-adaptive systems: A second research roadmap," in *Software Engineering for Self-Adaptive Systems*, ser. Dagstuhl Seminar Proceedings, R. de Lemos, H. Giese, H. Müller, and M. Shaw, Eds., no. 10431. Dagstuhl, Germany: Schloss Dagstuhl - Leibniz-Zentrum fuer Informatik, Germany, 2011.
4. N. Esfahani, E. Kouroshfar, and S. Malek, "Taming uncertainty in self-adaptive software," in *Proceedings of the 19th ACM SIGSOFT Symposium and the 13th European Conference on Foundations of Software Engineering*, ser. ESEC/FSE '11. New York: ACM, 2011, pp. 234–244. [Online]. Available: http://doi.acm.org/10.1145/2025113.2025147.
5. L. Northrop, P. Feiler, R. Gabriel, J. Goodenough, R. Linger, T. Longstaff, R. Kazman et al., "Ultra-large-scale systems: The software challenge of the future," *Software Engineering Institute*, 2006.
6. T. De Wolf and T. Holvoet, "Design patterns for decentralised coordination in self-organising emergent systems," in *Proceedings of the 4th International Conference on Engineering Self-Organising Systems*, ser. ESOA '06. Berlin, Heidelberg: Springer-Verlag, 2007, pp. 28–49. [Online]. Available: http://portal.acm.org/citation.cfm?id=1763581.1763585.
7. O. Babaoglu, G. Canright, A. Deutsch, G. A. D. Caro, F. Ducatelle, L. M. Gambardella, N. Ganguly et al., "Design patterns from biology for distributed computing," *ACM Transactions on Autonomous and Adaptive Sys*, vol. 1, pp. 26–66, 2006.
8. M. Mamei, R. Menezes, R. Tolksdorf, and F. Zambonelli, "Case studies for self-organization in computer science," *Journal of Systems Architecture*, vol. 52, pp. 443–460, August 2006. [Online]. Available: http://portal.acm.org/citation.cfm?id=1163824.1163826.
9. L. Gardelli, M. Viroli, and A. Omicini, "Design patterns for self-organizing multi-agent systems," in *2nd International Workshop on Engineering Emergence in Decentralised Autonomic System (EEDAS) 2007*, T. D. Wolf, F. Saffre, and R. Anthony, Eds. ICAC 2007, Jacksonville: CMS Press, University of Greenwich, London, June 2007, pp. 62–71.
10. F. Heylighen, "The science of self-organization and adaptivity," *The Encyclopedia of Life Support Systems*, vol. 5, no. 3, pp. 253–280, 2001.
11. T. De Wolf and T. Holvoet, "Towards a methodology for engineering self-organising emergent systems," 2005. [Online]. Available: http://www.cs.kuleuven.be/tomdw/publications/pdfs/2005soas.pdf.

12. J. L. Fernandez-Marquez, G. Di Marzo Serugendo, S. Montagna, M. Viroli, and J. L. Arcos, "Description and composition of bio-inspired design patterns: A complete overview," *Natural Computing*, pp. 1–25, 2012.
13. J. L. Fernandez-Marquez, J. L. Arcos, G. Di Marzo Serugendo, M. Viroli, and M. Sara, "Description and composition of bio-inspired design patterns: The gradient case," in *Workshop on Bio-Inspired and Self-* Algorithms for Distributed Systems (BADS '2011)*. ACM, 2011, pp. 25–32.
14. J. L. Fernandez-Marquez, J. L. Arcos, G. Di Marzo Serugendo, and M. Casadei, "Description and composition of bio-inspired design patterns: The gossip case," in *Proc. of the Int. Conf. on Engineering of Autonomic and Autonomous Systems (EASE '2011)*. IEEE Computer Society, 2011, pp. 87–96.
15. M. Salehie and L. Tahvildari, "Self-adaptive software: Landscape and research challenges," *ACM Trans. Auton. Adapt. Syst.*, vol. 4, no. 2, pp. 14:1–14:42, May 2009.
16. E. M. Dashofy, A. van der Hoek, and R. N. Taylor, "Towards architecture-based self-healing systems," in *Proceedings of the First Workshop on Self-Healing Systems*, ser. WOSS '02. New York: ACM, 2002, pp. 21–26. [Online]. Available: http://doi.acm.org/10.1145/582128.582133.
17. D. Garlan, S. Cheng, A. Huang, B. Schmerl, and P. Steenkiste, "Rainbow: Architecture-based self-adaptation with reusable infrastructure," *Computer*, vol. 37, no. 10, pp. 46–54, 2004.
18. I. Georgiadis, J. Magee, and J. Kramer, "Self-organising software architectures for distributed systems," in *Proc. of 1st Wkshp. on Self-Healing Systems*. ACM, 2002, pp. 33–38.
19. J. Kramer and J. Magee, "Self-managed systems: An architectural challenge," in *Future of Software Engineering, 2007. FOSE '07*. IEEE, 2007, pp. 259–268.
20. H. Kasinger, B. Bauer, and J. Denzinger, "Design pattern for self-organizing emergent systems based on digital infochemicals," in *Proc. of the Int. Conf. on Engineering of Autonomic and Autonomous Systems (EASE '2009)*. IEEE Computer Society, 2009, pp. 45–55.
21. H. Parunak, S. Brueckner, D. Weyns, T. Holvoet, and P. Valckenaers, "E pluribus unum: Polyagent and delegate mas architectures," in *Proc. of 8th Intl. Workshop on Multi-Agent-Based Simulation (MABS07)*. Springer, 2007, pp. 36–51.
22. J. Sudeikat and W. Renz, "Engineering environment-mediated multi-agent systems," D. Weyns, S. A. Brueckner, and Y. Demazeau, Eds. Berlin, Heidelberg: Springer-Verlag, 2008, ch. Toward Systemic MAS Development: Enforcing Decentralized Self-organization by Composition and Refinement of Archetype Dynamics, pp. 39–57. [Online]. Available: http://dx.doi.org/10.1007/978-3-540-85029-8_4.
23. H. Parunak and S. Brueckner, "Software engineering for self-organizing systems," in *Proc. of the Twelfth International Workshop on Agent-Oriented Software Engineering (AOSE 2011)*, 2011.
24. C. Cuesta and P. Romay, "Elements of self-adaptive architectures*," in *Proceedings of the Workshop on Self-Organizing Architecture (SOAR 2009)*. Springer, Heidelberg, 2009.
25. R. Wenkstern, T. Steel, and G. Leask, "A self-organizing architecture for traffic management," *Self-Organizing Architectures*, 2009.
26. J. Sudeikat, L. Braubach, A. Pokahr, W. Renz, and W. Lamersdorf, "Systematically engineering self-organizing systems: The Sodekovs approach," *Electronic Communications of the EASST*, vol. 17, no. 0, 2009.

27. G. Di Marzo Serugendo, J. Fitzgerald, and A. Romanovsky, "Metaself: An architecture and development method for dependable self-* systems," in *The 25th Symposium on Applied Computing (SAC 2010)*. Sion, Switzerland: ACM, 2010, pp. 457–461.

28. J. L. Fernandez-Marquez, G. Di Marzo Serugendo, and S. Montagna, "Bio-core: Bio-inspired self-organising mechanisms core," in *Bio-Inspired Models of Networks, Information, and Computing Systems*, ser. Lecture Notes of the Institute for Computer Sciences, Social Informatics and Telecommunications Engineering, E. Hart, J. Timmis, P. Mitchell, T. Nakamo, and F. Dabiri, Eds. Springer Berlin Heidelberg, 2012, vol. 103, pp. 59–72. [Online]. Available: http://dx.doi.org/10.1007/978-3-642-32711-7_5.

29. D. Weyns, A. Omicini, and J. Odell, "Environment as a first class abstraction in multiagent systems," *Autonomous Agents and Multi-agent Systems*, vol. 14, no. 1, pp. 5–30, Feb 2007.

30. M. Shaw and D. Garlan, *Software Architecture: Perspectives on an Emerging Discipline*. Prentice-Hall, 1996.

31. G. Di Marzo Serugendo, "Robustness and dependability of self-organising systems: A safety engineering perspective," in *Int. Symp. on Stabilization, Safety, and Security of Distributed Systems(SSS)*, ser. LNCS, vol. 5873. Lyon, France: Springer, Berlin Heidelberg, 2009, pp. 254–268.

32. J. L. Fernandez-Marquez, A.-E. Tchao, G. Di Marzo Serugendo, G. Stevenson, J. Ye, and S. Dobson, "Analysis of new gradient based aggregation algorithms for data-propagation in distributed networks," in *First International Workshop on Adaptive Service Ecosystems: Nature and Socially Inspired Solutions (ASENSIS) at Sixth IEEE International Conference on Self-Adaptive and Self-Organizing Systems (SASO12)*. IEEE Computer Society, 2012.

33. F. Zambonelli et al., "Self-aware pervasive service ecosystems," *Procedia Computer Science*, vol. 7, pp. 197–199, 2011.

34. M. Viroli, D. Pianini, S. Montagna, and G. Stevenson, "Pervasive ecosystems: A coordination model based on semantic chemistry," in *27th ACM Symp. on Applied Computing (SAC 2012)*, S. Ossowski, P. Lecca, C.-C. Hung, and J. Hong, Eds. Riva del Garda, TN, Italy: ACM, March 26–30, 2012.

35. F. Zambonelli, G. Castelli, M. Mamei, and A. Rosi, "Integrating pervasive middleware with social networks in SAPERE," in *Mobile and Wireless Networking (iCOST), 2011 International Conference on Selected Topics in*, Oct. 2011, pp. 145–150.

36. G. Stevenson, J. L. Fernandez-Marquez, S. Montagna, A. Rosi, J. Ye, A.-E. Tchao, S. Dobson et al., "Towards situated awareness in urban networks: A bio-inspired approach," in *First International Workshop on Adaptive Service Ecosystems: Nature and Socially Inspired Solutions (ASENSIS) at Sixth IEEE International Conference on Self-Adaptive and Self-Organizing Systems (SASO12)*. IEEE Computer Society, 2012.

37. G. Stevenson, J. Ye, S. Dobson, M. Viroli, and S. Montagna, "Self-organising semantic resource discovery for pervasive systems," in *First International Workshop on Adaptive Service Ecosystems: Nature and Socially Inspired Solutions (ASENSIS) at Sixth IEEE International Conference on Self-Adaptive and Self-Organizing Systems (SASO12)*. IEEE Computer Society, 2012.

Chapter 10

Adaptation and Resilience of Self-Organizing Electronic Institutions

David Sanderson, Dídac Busquets, and Jeremy Pitt

Contents

Modern computing systems are composed of different, possibly autonomous, components that form a collective, which must provide its functionalities in the face of unexpected and unpredictable perturbations in the environment or in the collective itself. In this chapter, David Sanderson, Dídac Busquets, and Jeremy Pitt show how such resilience can be achieved by the use of institutions to record the value of conventionally agreed-upon variables, together with an institutionalized consensus formation algorithm to reach and maintain agreement on the value. To illustrate the approach, they propose an example related to platooning in intelligent vehicular networks.

10.1 Introduction

There are many situations in which a set of autonomous agents opportunistically form a collective and have to either sense and agree on the value of some environmental variable or negotiate and agree on the value of some conventional variable (e.g., sensor networks, mobile ad hoc networks). Subsequently, they have to maintain that value in the face of unexpected and unpredictable perturbations in either the environment or the collective itself.

A prime example of this is platooning in vehicular networks [1], in which a number of cars travel in closely spaced single file, reducing road-space occupancy and fuel consumption. In the absence of each vehicle ceding its autonomy, this requires a set of "intelligent vehicles" to form a cluster, agree on a number of parameters (speed, spacing, lead vehicle, duration in lead, etc.), and maintain or adapt those values in the light of unexpected events, such as fragmentation or aggregation of the cluster, role failure, environmental factors such as congestion, and so on.

Autonomous entities negotiating an agreement about conventional facts without a central controller is a feature of institutionalized activity in human society. Therefore, we propose to use the idea of self-organizing electronic institutions [2], in which a number of agents use a dynamic norm-governed system specification to self-organize their activities to achieve some individual or collective purpose. We then use a norm-governed specification of an institutionalized consensus formation algorithm to reach and maintain agreement on cluster values and so provide resilience to dynamic, unexpected, and unpredictable events.

However, these negotiated agreements within any one vehicular cluster do not occur in isolation. For example, in intelligent transportation systems (ITSs), platooning or, rather, the opportunistic formation of temporary structures to achieve some transient collective goals is a manifestation of intelligence at a meso-level. There is also intelligence to be considered at the micro-level, residing in the highly instrumented and interconnected individual vehicle, which might want to join or form a cluster to satisfy its own individual goals (e.g., journey time), and intelligence to be considered at the macro-level, residing in an equally highly instrumented infrastructure, which may try to encourage cluster formation to satisfy macro-level goals (e.g., emission minimization). Meso-level intelligence then needs to be situated in a micro-meso-macro approach. Such a micro-meso-macro approach can also be found in other domains, such as meso-economics [3], self-governing institutions [4], or planned decentralization [5].

Accordingly, this chapter is structured as follows. In Section 10.2 we give an example of a system whose entities need to agree on the value of some variables, namely the case of vehicle platooning, and identify the need for resilience. Section 10.3 defines an institutionalized consensus-formation algorithm, IPCon, as a convention-based variation of the well-known Paxos algorithm for fault-tolerance in distributed databases [6]. The formal notion of a self-organizing electronic institution is introduced in Section 10.4, and Section 10.5 gives an executable specification of IPcon for use in such institutions. Finally this work is contextualized in Section 10.6 through a micro-meso-macro structure, in particular linking to the work of Elinor Ostrom on self-governing institutions for enduring and resilient collectives [4].

10.2 Example: Platooning in Intelligent Vehicular Networks

Note that although we use the motivating example of vehicular networks throughout this chapter, there are problems that are common to any electronic institution, self-organized system, or multi-agent system requiring groups to mutually agree on values.

One example of such networks is the platooning of vehicular ad hoc networks in ITSs, in which vehicles are organized into a platoon, or "road train," that travels along the roadway in an agreed-upon order at a constant agreed-upon speed, maintaining a constant agreed-upon spacing. Advantages include synchronized braking to reduce congestion waves, reduced spacing between vehicles, and improved fuel economy due to reduced air resistance owing to a slipstreaming effect, and semi-unattended driving allowing drivers to do other things.

In order to maintain a platoon, the vehicles must agree on a number of values; examples range from the physical (speed, spacing, ordering, etc.) [1] to matters such as security [7] and road safety [8]. These agreements must be resilient and not be compromised by changes in the platoon or external environment. This, then, is a problem of maintaining consistency in a distributed system.

Consistency in distributed systems requires that values are agreed upon by a set of nodes in the system and that these values are not changed after they have been agreed upon. In addition to the problems of maintaining consistency in static ("closed") distributed systems, our dynamic open system results in the problem of clusters of nodes aggregating and fragmenting. We use electronic institutions [9] for consensus on conventionally agreed-upon institutional facts to be preserved in cooperative open systems. We address three challenges in addition to the problems of maintaining consistency in static distributed systems; they are given below and indicated diagrammatically in Figure 10.1 when taken in the context of vehicular networks:

- Fragmentation/aggregation—Dynamic open systems of the sort we are interested in will entail cluster aggregation and fragmentation as agents join and leave the clusters and clusters merge and split over time. In any mobile ad hoc network, nodes in the network may physically move out of range of some networks and into range of other networks. Even in non-mobile networks, there may be reasons for the membership of the network to change.
- Revision of conventionally agreed-upon values (parameter change)—The value that is chosen for an institutional fact may not always be the best value, so we must deal with its mutability in a consistent manner. In vehicular networks, for example, the vehicle platoon may agree on a speed at which to travel and then come to an area of road that is more congested; this would then require them to alter the previously agreed-upon speed to something lower.
- Role failure—We must ensure robustness against role failure as the loss of a role-fulfilling agent may impede the proper functioning of the institution. In a voting system in which only one member has the authority to declare that a motion has passed, no motions can be passed if that member fails; having a single point of failure is undesirable in many different types of systems.

A source for a potential solution to the issues of resilience and fault tolerance in open systems is in the domain of distributed systems. For example, a well-known algorithm for designing fault-tolerant distributed systems under certain conditions is Leslie Lamport's Paxos.

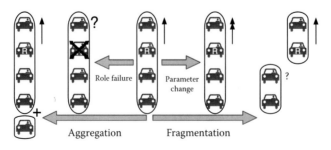

Figure 10.1   Problems to address.

10.3 Algorithm for Fault Tolerance in Conventional Systems

In this section, we first describe Paxos, a well-known algorithm from distributed database research, as a mechanism for designing fault-tolerant distributed systems. We then give details of IPCon, an Institutional Paxos Consensus algorithm for open, dynamic, institutional systems. In the next section we will give a specification of it in the event calculus.

10.3.1 Paxos

Paxos is an algorithm for implementing a fault-tolerant distributed system using the state machine approach [6] as shown in Figure 10.2. Fault tolerance is achieved by executing an infinite sequence of separate independent *instances* of the Paxos algorithm (in which each instance is a run of the algorithm to decide a command to be sent to the distributed state machine), which contain numbered *ballots*. Each ballot decides on the value of an instance; it is orchestrated by a *leader*, voted on by *acceptors*, and *learners* act as redundant storage. Values are proposed by *proposers*. Each leader may choose ballot numbers from an unbounded but individual set of natural numbers; it is unbounded to allow "infinite" ballots to occur and individual so that it is impossible for another leader to begin a ballot that has already taken place. The Paxos consensus algorithm operates on the undecided commands before they are sent to the "state machine" in order to ensure that all the distributed nodes in the fault-tolerant system agree on the value of those commands. Once they have been decided, they are unalterable, and each node can execute the same commands in the same order, thus ensuring that each distributed node has the same state.

Different versions of Paxos exist to deal with different sets of constraints and to address different types of fault; here we explain the most basic form, "classic Paxos" [6,10]. The algorithm assumes that agents operate asynchronously at arbitrary speed, may fail by stopping, and may restart, but they may not lie or impersonate other agents. Furthermore, messages can take arbitrarily long to be delivered, can be duplicated, and can be lost, but they cannot be corrupted—this is termed "non-Byzantine communication" by Lamport.

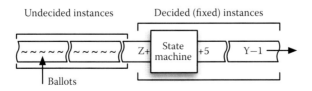

Figure 10.2   Paxos state machine approach.

As a consensus algorithm, Paxos has three safety requirements:

S1 Only a value that has been proposed may be chosen.
S2 Only a single value is chosen.
S3 A process never learns that a value has been chosen unless it actually has been.

To maintain safety, all versions of Paxos rely on the fact that all possible *quorums* have at least one member in common [6, Section 2]. This results in the assurance that f failures can be tolerated in a system of $2f + 1$ processes. This defines the term *quorum* in the context of all Paxos derivatives.

Paxos maintains the following "normative" properties:

P1 An acceptor can vote for value v in ballot b only if v is *safe at b*.
P2 The acceptors may not vote for different values in one ballot.
P3 A value v is *safe at* ballot number b if no value other than v has been or ever can be chosen in any ballot numbered less than b.

Once a quorum of acceptors has voted for a value v in a ballot b, the value is said to have been *chosen* and should be communicated to any additional non-acceptor *learners* if this is relevant. Once a value has been chosen, then, in any further proposals in that instance, the leader will be restricted to the chosen value by P3 and the restriction that all quorums must have at least one acceptor in common. This means that further proposals in that instance will retrieve the chosen value rather than choose one.

Although Paxos provides fault-tolerant consensus in a static distributed system, current versions are limited in their application in open systems insofar as they presume a static set of nodes in which the set of decision-makers (that is, the cluster) does not change, aims for consensus on a single value for each instance that will not and cannot change, and presumes temporary failures with the possibility of restart where the cluster size remains constant. Our solution is to design a new algorithm to overcome these limitations by explicitly representing different dynamic clusters of agents, allowing conventionally agreed-upon values to be changed in a coherent manner without causing confusion between different nodes and providing resilience against permanent role failures and the departure of an agent, which is equivalent to a permanent failure. This new algorithm is IPCon, an algorithm for institutionalized Paxos Consensus.

10.3.2 IPCon

Paxos operates by deciding commands that are to be sent to a state machine; the static set of nodes means that all of the decision-makers are also maintaining a copy of the state machine, so the state of the machine itself (the values in the database) doesn't need to be expressed explicitly by the Paxos algorithm. This is because all

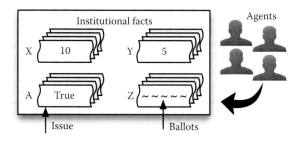

Figure 10.3   IPCon institutional approach.

the nodes are guaranteed to start with the same state and have the same changes applied to them.

Our requirement for dynamic clusters means that we work on two levels of abstraction at the same time: the changes to send to the database and the database itself. For this, we require a new algorithm for IPCon. We therefore explicitly represent the values and operate on them, not on the commands to be sent. We begin by mapping an *instance* of the Paxos algorithm to an *issue* in the IPCon algorithm to allow multiple separate instances of Paxos to be related and reasoned about together in one run of IPCon.

Changing a value in the database is no longer simply a case of issuing a new command to change it; it now requires modifying an agreed-upon value, so we wrap it in a new concept of a *revision* of an issue. This adds to our mapping between Paxos and IPCon by relating a Paxos instance to a revision of an issue in IPCon. This will require a new method for revising a previously agreed-upon value.

While Paxos is designed to deal with temporary node failure, the dynamic membership of clusters and the explicit representation of institutional power will require new methods for dealing with cluster aggregation and fragmentation and permanent role (node) failure.

In contrast to the presentation of Paxos in Figure 10.2, we provide that of IPCon in Figure 10.3. Rather than instances of Paxos operating on commands that are yet to be sent to the "machine," the IPCon issues are now institutional facts inside the "machine" that the agents operate on. Despite this change, the independence of Paxos instances from each other means that we maintain the properties that are internal to each instance of Paxos for each issue in IPCon.

10.3.3 Definitions

Cluster: Agents are organized into clusters that represent self-organizing electronic institutions. In our work on ITSs, these clusters are groups of vehicles traveling along the road together for some period of time.

Quorum: Any set of agents in the cluster is considered a quorum if it has at least one member in common with every other possible quorum. In a cluster of N agents, any set of agents that has more than $\dfrac{N}{2}$ members is therefore a quorum.

Issue: Clusters have a number of institutional facts to agree on; in our vehicular networks, examples could be the agreed-upon speed of the group, the space to be left between vehicles, the order in which the front vehicle in a road-train should rotate, and so on. These issues are specific to each cluster.

Value: Each issue has a value associated with it at any given time that is agreed upon by the members of the cluster. These are the values that we use consensus to agree on. In our example, for the issue of speed, the value could be 70 mph.

Ballot: In Paxos and IPCon, the values for each issue are decided using a sequence of numbered ballots that are orchestrated by a leader. In each ballot, the leader submits a value for the issue, and the acceptors may either vote for it or abstain. The process by which a ballot proceeds is given in Section 10.3.5.

Roles: Leader, acceptor, learner, and proposer are the roles in Paxos, and they are also required in our institutions; each role has different associated powers, permissions, and obligations. We have introduced powers to open the system and constrained the actions of the agents with permissions and obligations.

We introduce the following concepts not present in any existing Paxos variant:

Revision: The value that is chosen for an issue may not always be the most suitable value. For example, over time, the ideal speed for a group of vehicles may change due to any number of reasons. As Paxos is explicitly designed to prevent agreed-upon values changing, we introduce the concept of a revision to allow values to change safely. Each revision is unaffected by the previous revisions and represents a "clean slate." We provide a new action to create a new revision of an issue. See Section 10.3.5 for more detail.

Self-organization: The cluster requires a leader to coordinate the process of the algorithm; this central role should be appointed from within the system by the autonomous components themselves using self-defined rules. The leader needs to manage the cluster by granting or removing roles from agents as well as adding or removing them from the cluster. We provide actions allowing any agent to arrogate or resign the role of leader. See Section 10.3.5 for more details.

10.3.4 Properties and Assumptions

IPCon maintains the assumptions of asynchronous communication, possibility of failure, and non-Byzantine communication. Indeed, as our motivating example is that of ITSs, it does not make sense to consider the possibility of deceit in systems that are clearly safety-critical. It is another problem to consider a Byzantine IPCon algorithm.

Unlike Lamport, we give a precise specification of the requirements for liveness and progress; the dynamic nature of our application means that clusters may change frequently, but we aim for a system that will eventually result in a chosen value being learned. A non-faulty quorum is required for liveness as the algorithm cannot change state without a quorum of agents. Progress is guaranteed by ensuring that, eventually, a single leader exists as the only one trying to issue proposals, and it can communicate successfully with a quorum of (non-faulty) acceptors [10, Section 2.4]. The liveness and progress requirements are given as L1 and L2. We assume that if a non-faulty leader exists, it can communicate successfully with the non-faulty quorum of acceptors by virtue of them all being non-faulty.

L1 At least one non-faulty quorum of acceptors must exist.
L2 Eventually, (only) one non-faulty leader must exist.

Lamport describes the design of the Paxos consensus algorithm as following "almost unavoidably from the [Safety] properties we want it to satisfy" [10, Section 1]. We design IPCon such that it maintains institutional "normative" versions of the properties maintained by classic Paxos. The clear difference is that we explicitly include multiple issues and clusters and allow values to be *revised*. The properties are derived from Paxos by mapping "ballot number b" to "ballot number b on revision r of issue i in cluster c" and terming it "b'". P1 and P2 are now normative rules but remain the same while P3 is broken into three in translation to normative terminology. P3a, 3b, and 3c are a definition of a *safe* value expressed through normative states.

P3a If no empowered acceptor in the quorum voted in a ballot numbered less than b', all values are *safe at* b'.
P3b If an empowered acceptor in the quorum has voted, let c be the highest-numbered ballot less than b' that was voted in. The value voted for in ballot c is *safe at* b'.
P3c A value v is *safe at* b' if no empowered acceptor in the quorum has voted for any value other than v in a ballot less than b in revision r of issue i in cluster c. Likewise, all values are *safe at* b' if no empowered acceptor has voted for any value in any revision equal to r or greater.

We use a conceptual construct of the "highest numbered ballot" (**hnb**) to explain which values are *safe* for given (usually quorum-sized) sets of agents. The **hnb** for any set of agents Q is the ballot and value that was voted for in the highest-numbered ballot in which an agent in Q has voted. The set of **hnb** for a set C of overlapping quorums (i.e., a cluster) is simply the set containing the **hnb** of the quorums $Q_{1...n}$ in C. It is important to note that as both the ballot number and the value that is voted for determine which values are safe, the **hnb** encapsulates a $\langle ballot, value \rangle$ pair.

10.3.5 IPCon Algorithm

In this section, we will describe the functioning of IPCon by detailing the parts of it algorithmically. Lamport observes in [11, Section 8] that voting is a refinement of consensus. We take advantage of this to modify classic Paxos to create a single collective choice algorithm, IPCon.

> **Main message flow:** This gives the standard flow of the protocol that decides on a value for an issue as adapted from classic Paxos [10, Section 2].
>
> **0A** The proposer sends a `request0a` message to the cluster leader requesting to know the value of an issue if it has one or to propose a value if it does not.
>
> **1A** The leader sends a `prepare1a` message to all empowered acceptors with a ballot number b' it has chosen.
>
> **1B** On receipt of a `prepare1a` message from the leader, an empowered acceptor responds with a `respond1b` message containing the number of the **hnb** that they voted in and the value they voted for. If they have not yet voted, then they indicate this by replying with a special `noVote` value. This message also represents a promise from the acceptor to not participate in any ballot numbered lower than b'. If the leader receives `respond1b` messages stating that some acceptors have voted in higher-numbered ballots than b', this indicates that it should retry with a higher ballot number.
>
> **2A** Once it has received `respond1b` messages from a quorum of empowered acceptors, the leader chooses a value v that has previously been proposed and is *safe at b'* (see Section 10.3.4) and sends a `submit2a` message to the empowered acceptors asking them to vote for this submitted value.
>
> **2B** On receipt of a `submit2a` message from the b' leader, an empowered acceptor either votes for the value v in ballot b' by sending a `vote2b` message or abstains by sending no message. An acceptor cannot vote in a ballot if it has already voted in a higher-numbered ballot on the same issue. A value is chosen once a quorum of agents have voted for it.

By sending a `request0a` message to the leader, a *proposer* initiates the standard message flow for the algorithm. This "request" may either be a request to discover the currently chosen value of an issue or to suggest a value for one that has no value. If a value has already been chosen, it will be discovered by the leader before Phase 2A. Phase 2 can be executed in either case; if a value has been chosen, it is impossible for a correctly functioning leader to send a `submit2a` message with a different value. Phase 2 then redistributes the chosen value to new cluster members and allows the consensus to be strengthened. The single-issue, single-cluster interaction between leader and acceptors can be seen in Figure 10.4. Each phase corresponds to the creation of a new institutional fact representing the algorithm's state as shown on the right of the diagram.

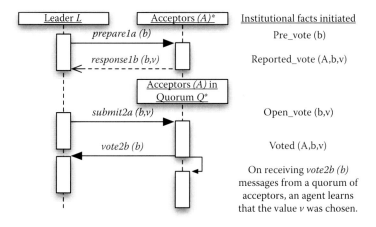

Figure 10.4 IPCon messageflow.

Clustering: Cluster fragmentation and cluster aggregation are related problems; in both cases, the set of acceptors will change, and this will cause a change in the set of **hnb** for the cluster; we serialize changes for simplicity. We provide functions for agents to leave the cluster and for the leader to grant and remove agents' roles.

Revision: By ballot ordering, the promise in 1B of the message flow and the restriction on voting in 2B, lower-numbered ballots cannot progress once a higher-numbered ballot has been interacted with by a quorum-sized group of acceptors. Our method of tagging ballot numbers with revision numbers extends the ballot numbering system so that "older" revisions of an issue are explicitly out of date. We can therefore say that "revising an issue" by creating a new revision safely "un-chooses" a value by resetting the **hnb** set so that all values are *safe*. Revision can then be seen as a reset switch that does not affect safety except in the way that we want it to: namely, by allowing us to change our minds on the chosen value.

Leadership: The details of electing a process to the role of leader are left as an implementation detail by Lamport [10, Section 2.4] under the assumption that a single agent will eventually have the role of leader. Our method has no impact on the values that are chosen as no values are referenced by the commands, so it has no effect on safety. We provide commands to allow any agent to claim the leadership of the cluster and for any agent with the role of leader to resign. We presume some mechanism or agreement whereby the agents will organize a solution resulting in only one leader at any given time. As explained by Lamport in [10], "dueling leaders" can only impede progress, and we allow agents to leave a cluster; this means that if such a situation occurs, agents may leave the cluster and reform without them in order to continue progression. Once a "new" cluster is formed by fragmentation, one of the agents should arrogate the leadership so that new values can be chosen.

10.4 Self-Organizing Electronic Institutions

A self-organizing electronic institution is essentially a collection of agents plus a specification of a dynamic norm-governed system [12]. Formally, it can be denoted by $\mathcal{IC}_t$, which is a multi-agent system at time t defined by

$$\mathcal{IC}_t = \langle \mathcal{A}, \mathcal{C}, R, \mathcal{L} \rangle_t$$

where (omitting the subscript t if clear from context):

- $\mathcal{A}$ is the set of all agents.
- $\mathcal{C}$ is the set of action situations.
- R is a binary *nesting* relation on C.
- $\mathcal{L}$ is a dynamic norm-governed system specification.

In the framework of [12], a number of degrees of freedom (DoF) are identified, so $\mathcal{L}$ defines a specification space, in which each specification instance is defined by a different set of values assigned to the DoF. For example, one DoF is the method by which the resources are allocated (e.g., at random, by ration, or using legitimate claims), and another might be which role-assignment protocol to use.

An *action situation* is any situation in which institutionalized power [9] is exercised. Institutionalized power enables and empowers an agent appointed to a role in an action situation to bring about a fact of conventional (or institutional) significance by performing a designated action in that specific context. Note that because action situations can be nested within each other, we might have a logical hierarchy of clusters, corresponding to a single physical platoon of vehicles. Therefore, some DoF may be changed by decisions made within the action situation, and some may be changed by decisions made in the action situation within which it is nested as determined by the nesting relation R.

Each action situation $C_{i,t} \in \mathcal{C}_t$ is defined by

$$C_{i,t} = \langle \mathcal{M}, l, \epsilon \rangle_t$$

where (again omitting the subscripts as clear from context):

- $\mathcal{M}$ is the set of members, such that $\mathcal{M} \subseteq \mathcal{A}$.
- l is a specification instance of $\mathcal{L}$.
- ϵ is the cluster's local environment, a pair $\langle Bf, If \rangle$.

Regarding the environment ϵ, Bf represents the set of "brute" facts whose values are determined by the *physical* state, and in a vehicular cluster, this includes speed, spacing, lead vehicle, duration as lead vehicle, etc. *If* represents the set of "institutional" facts, whose values are determined by the *conventional* state, that is, are asserted by the exercise of institutionalized power, for example, who occupies which role and so on.

In previous work, we have used the event calculus (EC) [9] as the specification language for $\mathcal{L}$. We continue to use that language in this chapter, and in the next section, we give an EC specification of IPCon.

10.5 Event Calculus Specification of IPCon

The EC is a formal language from artificial intelligence that is intended to reason about actions and changes in non-monotonic systems [9]. An EC specification of a system consists of a domain-independent part that includes axioms for determining what fluents *hold at* a specific time and an application-specific part that specifies axioms for determining what holds at the initial time, what fluents are initiated by actions at any time, and the constraints on what holds or does not hold at any given time. The latter two can be expressed axiomatically as follows:

$$\text{Action initiates Fluent} = \text{Value at Time} \leftarrow$$

$$\text{Condition}_1 \wedge \ldots \wedge \text{Condition}_m$$

$$\text{Fluent} = \text{Value } \texttt{holdsAt } \text{Time} \leftarrow$$

$$\text{Condition}_1 \wedge \ldots \wedge \text{Condition}_m$$

What follows is a formal axiomatization of the property and message specification of the IPCon protocol. The specification gives the relevant fluents, actions, powers, roles, and permissions and covers the state of the algorithm.

10.5.1 Fluents and Actions

Table 10.1 gives the fluents and actions used in IPCon. All fluents in IPCon are either true or false; actions generally correspond to agents sending a message. We use the following initialisms: Agent (A), Ballot (B or N), Cluster (C), Issue (I), Revision (R), Value (V), and `Vote = (A,N,V)`.

A typical run of the protocol begins with a "proposer" P sending a message of the form `request0a(P,V,I,C)` to the leader L of cluster C regarding the value V of issue I; this sets the fluent `proposed(V,I,C)` to *true*. L then chooses a ballot number B and, as can be seen in Figure 10.4, sends a `prepare1a(L,B,I,C)` message to the acceptors and learners of I in C, initiating the fluent `pre_vote(B,I,C)`.

An acceptor A who receives this message will reply with a message of the form `response1b(A,(A,N,V'),B,I,C)`, indicating that it has previously voted for

Table 10.1 Fluents and Actions in IPCon

Fluents	Actions
proposed(V,I,C)	request0a(A,V,I,C)
pre_vote(B,I,C)	prepare1a(A,B,I,C)
reportedVote(A,N,V,B,I,C)	respond1b(A_1,Vote,B,I,C)
open_vote(B,V,I,C)	submit2a(A,B,V,I,C)
voted(A,B,V,I,C)	vote2b(A,B,I,C)
sync(A,V,R,I,C)	syncReq(A_1,A_2,V,R,I,C)
role_of(A,Role,R,I,C)	syncAck(A,V,R,I,C)
possibleAddRevision(R,I,C)	arrogateLeadership(A,C)
possibleRemRevision(R,I,C)	resignLeadership(A,C)
quorum_size(R,I,C)	addRole(A_1,A_2,Role,R,I,C)
pow(A,Act)	remRole(A_1,A_2,Role,R,I,C)
per(A,Act)	leaveCluster(A,C)
obl(A,Act)	Revise(L,I,C)

value V′ in ballot N on I in C, which will set a reportedVote(A,N,V′,B,I,C) fluent to true to indicate that the cluster knows the agent voted previously. This is the Phase 1B action as specified in Figure 10.5. Specifically, an agent has the power if it has the role (indicated by the fluent role_of detailed later) and if a pre_vote exists, and it has permission if Ag voted for V in B (indicated by voted(Ag,B,V,I,C)), or did not vote yet in that revision (indicated in the second of the two **per** rules).

At this point, reportedVote fluents tell the leader the highest ballot number that has previously been voted on by the agents in the cluster. If it has received a quorum of reported votes with the same highest ballot number in the same revision, then the value of I has already been decided. The proposer should either find this out from the broadcasted messages, or the leader should inform them of the result. Otherwise, the leader proceeds to Phase 2 in order to start a vote.

From the reportedVote fluents, the leader finds the highest-numbered previous ballot on I. If their initial B was not high enough for safety to be ensured, they should increase it and try again before continuing with a submit2a(L,B,V,I,C) message. Once a high enough number has been chosen for B, the leader must

```
response1b (Ag₁, (Ag₂, N, V), B, I, C) initiates reportedVote (A,N,V,B,I,C)  at T ←
   (N = (R_N, B'_N)) ^ (B=(R, B') ) ^
   pow (Ag₁, response1b (Ag₁, (Ag₂, N, V), B, I, C) ) = true holdsAt T.
```

```
pow (Ag₁, response1b (Ag₁, (Ag₂, N, V), B, I, C) ) = true holdsAt T←
   ( B = (Rev, B') ) ^ pre_vote (Rev, B', C) = true holdsAt T ^
   { role_of (Ag₁, acceptor, Rev, I, C) ^ (Ag₁ = Ag₂) } = true holdsAt T.
```

```
per (Ag₁, response1b (Ag₁, (Ag₂, N, V), B, I, C) ) = true holdsAt T←
   pow (Ag₁, response1b (Ag₁, (Ag₂, N, V), B, I, C) ) = true holdsAt T ^
   voted (Ag₂, N, V, C) = true holdsAt T.
```

```
per (Ag₁, response1b (Ag₁, (Ag₂, N, V), B, I, C) ) = true holdsAt ¬T
   pow (Ag₁, response1b (Ag₁, (Ag₂, N, V), B, I, C) ) = true holdsAt T ^
   (N = (Rev, 0) ) ^ (V = noVote) ^
   (B = (Rev, B') ) ^ not (voted (Ag₂, (Rev, Bal), V', I, C) = true holdsAt T) ^
   not (V' = noVote) ^ ( Bal > B' ).
```

```
obl (Ag₁, response1b (Ag₁, (Ag₂, N, V), B,I,C) ) = true holdsAt T ←
   per (Ag₁, response1b (Ag₁, (Ag₂, N, V), B, I, C) ) = true holdsAt T ^
(Ag₁ = Ag₂) ^ voted (Ag₂, N, V, I, C) = true holdsAt T ^
   not (reportedVote (Ag₂, N, V, B, I, C) = true holdsAt T).
```

Figure 10.5   Example of specifying the Phase 1B action in the EC.

calculate what values are *safe at B* as defined in Section 10.3.4. Note that the leader can choose any quorum-sized group of acceptors in order to calculate the *safe* value; if this were not the case, then only the first value that was voted for would ever be *safe*. An acceptor casts a vote with a vote2b(A,B,I,C) message.

An example of voting progress and safety can be seen in Table 10.2. The **hnb** denotes the *highest numbered ballot* that was voted for in a set of agents. At the beginning of the round, no votes have been cast; this is indicated by "−" and "ø". Ballot 1 has x as the submitted value and is voted for by Ag_1. The **hnb** for all sets that include Ag_1 are now set to (1, x) to indicate this vote. The voting continues with Ag_2 and Ag_3 voting for y in ballot 2 and Ag_4 joining Ag_1 to vote for x in ballot 3. After ballot 3, observe that the leader could choose to either submit x or y for voting because at least one quorum-sized set of agents has it as its **hnb** (1,4,5 and 2,3,5 respectively). In ballot 4, Ag_5 votes for y, and this sets the **hnb** of all possible quorum-sized sets of agents to y. From now on, the only value that will ever be found to be safe and therefore eligible for submission is y. It is possible for an agent to vote regardless of its previous votes; for example, in ballot 5, Ag_1 decides to also vote for y, thus strengthening the consensus on the value. There may come a time when the value that was chosen is not suitable, and the acceptors will not vote for it (ballot n); this indicates that the leader may wish to revise the issue.

Table 10.2 Example Showing Voting Progress and Safety in IPCon (Quorum Size of 3)

Ballot Number	Value	Agent					*hnb* after Ballot for Agents			
		Ag_1	Ag_2	Ag_3	Ag_4	Ag_5	1–5	2–5	1,4,5	2,3,5
0	–	–	–	–	–	–	∅	∅	∅	∅
1	x	x	–	–	–	–	(1, x)	∅	(1, x)	∅
2	y	–	y	y	–	–	(2, y)	(2, y)	(1, x)	(2, y)
3	x	x	–	–	x	–	(3, x)	(3, x)	(3, x)	(2, y)
4	y	–	y	y	–	y	(4, y)	(4, y)	(4, y)	(4, y)
At this point, only y can ever be found to be safe.										
Before this, the leader could choose the quorum to allow x or y.										
5	y	y	y	y	–	y	(5, y)	(5, y)	(5, y)	(5, y)
Note that Ag_1 has now voted y; this strengthens the consensus.										
⋮	⋮	⋮	⋮	⋮	⋮	⋮	⋮	⋮	⋮	⋮
n	y	–	y	–	–	–	(n, y)	(n, y)	(n, y)	(n, y)
The reduced number of votes for y may indicate a new revision is needed.										

10.5.2 Roles and Cluster Membership

There are four possible roles for agents in IPCon: *proposer, learner, acceptor,* and *leader.* An agent has a role if a fluent of the form role_of(Ag,Role,Rev,I,C) is true. This would give agent Ag the role Role in revision Rev of issue I of cluster C.

Dynamic clusters may cause roles to be duplicated or absent. For the situation in which a cluster has no leader (for example, after the leader has left or after cluster fragmentation) or multiple leaders (after cluster aggregation), we provide the actions arrogateLeadership and resignLeadership. An agent may arrogate leadership for itself at any time; this is to allow "bootstrapping" a cluster out of nothing and to allow an agent to replace a leader that it believes may have lost connectivity or left the cluster unexpectedly. Apart from requiring it to already be a leader, an agent may likewise resign leadership at any time. We assume that, in the case of multiple agents claiming leadership in a cluster, there is some method of voting to decide who will remain leader. It is enough for our purposes that we allow a method for leadership to be gained and lost. See [10] for why the "dueling leaders" problem cannot violate safety.

An empowered leader *may* wish to add a role to any agent it wishes, provided that the agent in question does not already have that role. Similarly, the leader can remove a role from any agent it wishes, provided that the agent has the role. This allows the leader to populate a cluster with agents using the `addRole` command or to remove an agent from a cluster using `remRole`.

We model temporary failures of agents by having them not send messages. An agent can intentionally leave a cluster at any time using the command `leave-Cluster`. We leave the issue of detecting when an agent has permanently failed or left the cluster without warning as an open question.

The addition or removal of an acceptor can endanger the safety of the consensus due to the members of the cluster changing. The simplest solution to this is to make a new revision of the issue whenever the quorum size changes, but consider, for example, when an acceptor leaves and is replaced by a different acceptor; the quorum size is unchanged, but the set of **hnb** *has* changed, thus changing the set of values that are *safe* and possibly even "un-choosing" a value. We therefore take an alternative approach; for additions of acceptors, we provide the `sync(A,V,R,I,C)` fluent, which can be initiated by a leader when an agent becomes an acceptor by sending a `syncReq(L,A,V,R,I,C)` message. This message indicates to new acceptor `A` that value `V` has been chosen in `R` of `I` in `C`. The acceptor can then reply with a `syncAck(A,Reply,R,I,C)` message, and `Reply` is either `V` or any other value (typically "`no`") to indicate its agreement or disagreement, respectively. If the acceptor agrees, then the reply *initiates* both a `voted` and `reportedVote` fluent, and no further action is required to maintain the consensus. If the acceptor does not agree, then the leader may start a new revision for that issue.

As with addition, the loss of an acceptor requires our attention when a value has previously been chosen. As the acceptor is leaving, however, it is simpler to deal with as no interaction is required; we must simply check to see if the agent that is leaving had voted for the chosen value. If no value had been chosen, or if the agent had not voted for it, then no action is required at all.

In both the cases of an acceptor joining and not accepting the previously chosen value and the loss of an acceptor that had previously voted for the chosen value, there are situations in which the leader *must* start a new revision for the issue in order to preserve safety. As previously mentioned, the sets of **hnb** are the indicators of this need. We monitor the votes *for* the currently chosen value and the votes *not for* the currently chosen value. When these two groups are the same size, there is the risk that the addition or loss of an acceptor could compromise safety, so we initiate the fluents `possibleAddRevision` and `possibleRemRevision`, respectively, to indicate these situations. Specifically, `possibleAddRevision` is initiated when an agent is `synching` and will create a new quorum-sized group of agents that do not have the chosen value as their **hnb**. Likewise, `possibleRem-Revision` is initiated when the number of agents that voted for the currently chosen value is equal to the number of agents that did not vote for it as this creates a new quorum-sized group by changing the quorum size.

The `possibleAddRevision` and `possibleRemRevision` fluents are checked when an agent completes its `sync` process and when an agent that had voted for the chosen value leaves. If the relevant one is true, then an obligation for the leader to revise the issue is created. By sending a `revise` message, the leader can initiate a new revision. When a new revision is begun, all open `sync` fluents are terminated, and all active ballots are closed.

10.5.3 Powers, Permissions, and Obligations

We maintain the standard distinction in legal, organizational, and social theory between physical capability, institutionalized power, and permission [9]. We also distinguish between institutionalized power being used to change an *institutional fact* by an empowered agent performing a designated act in a specific context (in our case, a speech act in the context of the IPCon protocol) and physical capability being used to change a *brute fact* (a fact that is true by physical properties rather than by conventional agreement).

Accordingly, the EC axioms in the context of IPCon are used to determine which (institutionalized) powers, permissions, and obligations hold as follows:

Institutionalized Power is the power to perform an action that has an effect on an institutional fact. In our case, this usually means that an agent is empowered to perform an action when it occupies the relevant role and when the algorithm is in the correct state.

Permission determines the valid exercise of institutional power in accordance with the rules and expectations of the institution. In our case, power is always necessary for permission but not always sufficient.

Obligations constrain the exercise of power to ensure the "normative" properties hold. In our case, these take two forms. The first is the obligation to perform an action or send a message with the correct values, thereby ensuring that the consensus algorithm continues to function as expected; this roughly corresponds to "do not perform maliciously faulty actions." The other form is the obligation to perform an action when not doing so may cause the safety requirements of the consensus algorithm to be violated.

Figure 10.5 demonstrates the relationship between **pow**er, **per**mission, and **obl**igation for the action `response1b`. An agent has the power to send a response about itself if it is an acceptor and if it is replying to a `pre_vote`. An agent has the permission to send a response if it voted, in which case it says what it voted for, or if it didn't vote, in which case it says that it hasn't. Axioms for the other messages are similarly defined.

10.6 Micro-Meso-Macro Approach

In the previous sections, we have shown how a set of agents can form a cluster to agree on the value of some conventional facts through the use of institutions and a consensus-formation algorithm. However, such clusters are not created in isolation; they are often part of a system with other agents as well as other clusters. Usually each individual agent has its own goals, which drive its behavior. At the other extreme, the system as a whole may have macro-goals, which may be conflicting among them as well as with the goals of individual agents. Moreover, clusters of agents may have to interoperate among them, and there may be also cases in which some agents belong to more than one cluster at the same time (e.g., one for negotiating the cruise speed in a platoon with neighboring vehicles and another one for deciding what route to take, which may involve a different set of vehicles, depending on its destination); this may result in the agents receiving conflicting values from each cluster.

In order to deal with the interactions between the individual, cluster, and macro goals and inspired by evolutionary economics [3], we have taken the approach of using a three-layered framework. This results in a *micro-meso-macro* approach, in which each layer is concerned with the problem at a different level of detail. We believe that the combination of the three levels is needed to provide adaptation and resilience in highly dynamic complex systems as is the case for the transportation domain. In this section, we first describe the micro-meso-macro structure and then we give an example of how the micro and macro levels could be realized in the domain of ITSs.

10.6.1 Overview of the Three-Layered Structure

In systems with complex interactions among their components, the usual micro-macro approach does not suffice to account for the behavior of the system. This is the case, for instance, of evolutionary economics [3], complex systems [13], or organic computing [14], in which the need for a level that acts as a bridge between the micro and macro levels has been identified. While the micro level deals with the behavior of individual agents and the macro level deals with the behavior of the system as a whole, the meso level accounts for system dynamics and is responsible for changes in agents' behaviors and the relationships among them. This meso level is usually realized as a temporary structure (e.g., a cluster) composed of some of the components of the whole system having a close interaction among them. The behavior of the cluster affects the behavior of the individual components (either members or not of the cluster itself) as well as the behavior or goals at the macro level. On the other hand, the macro level could influence the formation of, as well as the decisions taken in, clusters in order to achieve some of its system-wide goals.

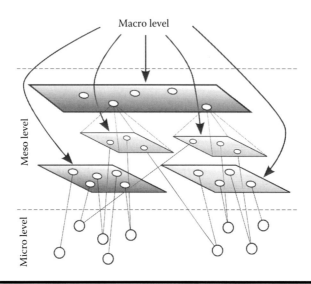

Figure 10.6 Micro-meso-macro architecture.

Although the approach is presented as having three levels, there could be multiple nested meso levels, as shown in Figure 10.6, depending on the level of detail at which the system is inspected (e.g., in the domain of transportation, this could be at the level of borough, city, county, state…). Moreover, as already mentioned, an individual could belong to more than one meso level system at the same time.

10.6.2 Example of Micro-Meso-Macro Structure: Intelligent Transportation Systems

Transportation systems are composed of a collection of individual agents (i.e., vehicles) with their own goals (e.g., minimize journey times, reach destination without having an accident, etc.) interacting in an infrastructure that has some system-wide, possibly conflicting, goals (e.g., minimize congestion and pollution, maximize network throughput, etc.).

The use of the micro-meso-macro approach in the domain of an ITS aims to provide intelligence at its different levels, tackling the particular problems of each of them. The end goal of our approach is that the combination of the intelligent behavior in each of the levels leads to a more sustainable and efficient use of vehicles and the road infrastructure.

At the *micro level*, we deal with individual behavior, providing intelligence to the vehicle. At this lower level, we design an agent architecture to make use of all the sensing equipment and on-board technology in order to put together a decision-making system concerned with the actual execution of driving. This may include safety issues (e.g., emergency breaking or steering, collision warning) as well as fulfilling drivers' and passengers' goals and preferences (e.g., driving speed, safety distance).

The *meso level* is concerned with collective decision-making within groups or clusters of vehicles. These clusters are created according to some relationship among the vehicles; for instance, we could have clusters at every junction, at which the vehicles should reach agreements on who passes first or, as shown in the previous sections, in platoons of vehicles, among which they should agree on the travel speed, among others. Moreover, due to the high dynamics of road traffic, vehicles are constantly moving and, therefore, joining and leaving different clusters, which causes them to be temporary.

Finally, the *macro level* deals with infrastructure or system-wide goals, such as reducing congestion and pollution, making an efficient use of the road network, or managing the interaction of the network and other infrastructures (e.g., energy systems, public transportation, emergency services), among others. The decision-making at this level is based on all the information gathered by sensors distributed in the infrastructure as well as by information received from (and sent to) the vehicles. This allows the macro level to have a big picture of the state of the infrastructure and thus act accordingly.

10.6.2.1 Micro Level: Affective Anticipatory Architecture

We have developed an intelligent anticipatory vehicular architecture [15] to serve as the micro level of the system. The architecture allows the agents to choose from a number of actions to fulfill their goals by modeling other agents, anticipating future states and actions, and performing cognitive evaluations of their quality. We combine a theory of intent prediction based on aspects of the HAMMER (hierarchical attentive multiple models for execution and recognition) architecture [16] with an affective evaluation of predictions [17]. This results in a method for proactive intent prediction based on affective appraisal of possible future states by using HAMMER as the basis of our predictor and a cognitive functioning of expectations as the affective dimension of the evaluation of states.

The HAMMER architecture is a framework for online action selection and intention recognition developed for the robotics and multi-agent system domains that use the *generative, simulationist* approach to understanding actions by mentally rehearsing them from a motor-based perspective [16]. It uses groups of inverse-forward model pairs in parallel, each with an additional "prediction verification" block and then feedback/competition to select the correct action as can be seen at the heart of Figure 10.7. We use HAMMER as both a "controller" to plan and control the vehicular agent and also as a "predictor" to make predictions about the actions of the neighboring vehicles.

The cognitive anatomy of expectations and their function in purposive action in [17] presents an analytical decomposition of the structure of expectations as well as

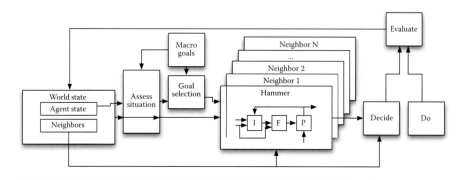

Figure 10.7 Agent architecture diagram.

their purpose, meaning, and side-effects arising from their fulfillment or invalidation. We use this to inspire the affective evaluation of predicted states. In the work, an "expectation" consists of two parts: a *belief* with strength and a *goal* with value. These are the subjective certainty and importance of the expectation, respectively. We base our method of affective evaluation on the concept of an expectation having a value to the agent and a certainty weight. As an agent moves through the world, it will have to make choices about its actions and be able to rank its goals. Any given goal of the agent will have an expectation associated with its fulfillment; these can be paired with the action(s) that can lead to its achievement and the predicted world states that will result from these actions. These predicted states will themselves have an expectation associated with them, and so they can also be ranked. Once the agent has determined the most favorable or least unfavorable possible state, it can execute the action that will lead to it.

The agent architecture is shown in Figure 10.7. The agent assesses the current state of the world, taking into account the degree to which its various goals are being fulfilled. Each possible goal is then taken into consideration, and a set of instantaneous goals are selected that would potentially lead to the main goals being fulfilled. In parallel to this, the HAMMER-style predictor blocks are used to generate possible future states based on the actions of the nearby agents; in this case, the agent in question places itself "in the shoes of" the neighbor to be simulated in order to get the best outcome for that neighbor. This set of predicted future states are then used as inputs to a HAMMER-style planning phase using the previously generated possible actions for the agent. This generates a number of predicted possible future world states, depending on which action the agent performs; the agent can then choose the preferred one.

We have experimentally demonstrated [15] that there is a social utility in exhibiting affective, anticipatory reasoning and that integrating an affective evaluation of predictions into an intelligent agent architecture provides significant global advantages, resulting in improved and stabilized system throughput, reduced congestion, and increased global agent happiness.

In our multidimensional approach to intelligent transportation, this architecture provides the micro level of the system. In general, a driver wishes to get from his or her start point A to destination B in a given time t, with *zero* accidents, and obeying a set of additional rules R that may be personal (such as fuel efficiency or a certain route) or external (such as laws or relevant social norms). These external rules may originate from the macro level of the system in the case of road laws or from the meso level on a potentially more temporary basis as in the case of agreed-upon cluster speed or spacing. This formulation can replace the set of simple instantaneous goals used so far.

Any group, or cluster, of these agents at the meso level would have to organize a consensus among themselves on a set of social normative rules to govern their behavior. To this end, we have designed the IPCon algorithm to manage consensus formation as described in Section 10.3.

10.6.2.2 Macro Level: Infrastructure for Sustainability

The macro level, having a big picture of the infrastructure's state, deals with system-wide goals. The main concern at this level is to achieve a more efficient and sustainable use of the infrastructure. This includes reducing congestion, greenhouse emissions, or noise pollution, among others. It may also manage the interaction between the road infrastructure and other systems, such as public transportation (for instance, giving priority to mass-transportation vehicles), emergency services (providing fast routes to reach accidents), or energy systems (managing the power demand of electrical vehicles).

The decision-making at this macro level may influence the behavior of the agents at the meso and micro levels. For instance, depending on the traffic conditions, the macro level may suggest or incentivize the formation of platoons to achieve a smoother stream of vehicles in certain areas of the network. Such incentives may include the generation, when possible, of green waves for the platoons. Similarly, the macro level may impose some temporary restrictions on the maximum driving speed so as to reduce pollution. For the platooning clusters, this would imply a restriction on the values the agents may agree on or the need for a revision on the already agreed-upon value. This level may also inform about alternative routes to reach the different destination points of the drivers, thus distributing the vehicles among the network and using its capacity more efficiently.

Given the high dynamics and unpredictability of transportation systems, the macro level should be constantly monitoring the state of the system in order to quickly react to any problem that may arise. It should also be able to forecast potential problems, anticipating them and taking corrective actions before the actual problem occurs. Such proactive behavior would really improve the efficiency of the system.

10.7 Summary and Conclusions

This chapter has shown how resilience can be achieved through the use of institutions to record the value of conventionally agreed-upon variables and has presented an institutionalized consensus formation algorithm to reach and maintain agreement on that value.

Resilience is the sustained operation of a group of agents in light of environmental perturbations. In our example of vehicular clusters, such as platoons, examples of these perturbations are numerous. We focus on the fragmentation and aggregation of clusters caused by the dynamic network of vehicular agents, the revision of conventionally agreed-upon values caused either by external (unsuitability of the chosen value) or internal (lack of continued consensus on the chosen value) factors, and role failure caused by the departure of role-fulfilling vehicular agents from the cluster.

Providing resilience against these perturbations requires that the agents share a method of agreeing on values. One solution to the issues of resilience and fault tolerance in distributed systems is the concept of consensus formation from distributed databases. However, dynamic open systems face additional problems when compared to static closed systems. We therefore use electronic institutions to apply our solution to open systems with roles and institutional facts. We have adapted the Paxos algorithm to an open, dynamic, institutional setting by developing IPCon.

IPCon has been presented as an adaptation of the Paxos algorithm for open, dynamic, institutional systems, and we have given a specification of it in the EC. This specification has demonstrated the methods by which we maintain the properties of classic Paxos and also the additions we have made to it.

We note that the IPCon algorithm reinforces Principles 1–3 of Ostrom's design principles for enduring institutions [4]:

1. **Clearly defined boundaries of the institution**—IPCon is built on a set of rules that clearly define institutional powers, permissions, and obligations for agents, depending on their role in the cluster. This ensures that the members of the cluster and the right of those members in the cluster are always clearly defined.
2. **Institutional rules that are adaptable to local conditions**—The cluster is able to revise conventionally agreed-upon values in response to changes in the cluster or external environment.
3. **Collective-choice arrangements**—The agents themselves are the ones that decide the organizational structure of the cluster and the institutional facts.

The consensus algorithm and the self-organizing electronic institution have been contextualized in a micro-meso-macro structure, fulfilling the duties of the meso level. The inclusion of this meso level, in combination with the micro level, dealing with the behavior of individual agents, and the macro level, in charge of system-wide goals, should enhance the resilience and adaptability of dynamic systems.

As work on self-adaptive and self-organizing (SASO) systems operating on multiple levels becomes more common, we may find that further work is required to predict and understand the interplay between SASO methods. Although two different methods may operate in isolation as expected, when they are combined or have to work side by side, they may not operate as well or even at all. The interactions between non-adaptive methods is fairly well studied, but the same cannot be said for interactions between SASO systems. Avoiding such collisions requires thorough study and understanding of the design of SASO systems from an emphatically *systems* perspective.

References

1. The SARTRE Project, http://www.sartre-project.eu/.
2. Pitt, J., Schaumeier, J., Artikis, A.: Axiomatization of socio-economic principles for self-organizing institutions: Concepts, experiments and challenges. *Transactions on Autonomous and Adaptive Systems* 7(4), 1–39 (2012).
3. Dopfer, K., Foster, J., Potts, J. D.: Micro-meso-macro. *Journal of Evolutionary Economics* 14(3), 263–279 (2004).
4. Ostrom, E.: *Governing the Commons*. CUP, Cambridge (1990).
5. Binmore, K.G.: *Natural Justice*. Oxford University Press, Oxford (2005).
6. Lamport, L.: The part-time parliament. *Transactions on Computer Systems (TOCS)* 16(2), 133–169 (1998).
7. Raya, M., Hubaux, J.: Securing vehicular ad hoc networks. *Journal of Computer Security* 15(1), 39–68 (2007).
8. Bonnefoi, F., Bellotti, F., Schendzielorz, T., Visintainer, F.: SAFESPOT applications for infrastructure-based co-operative road-safety. In: *14th World Congress and Exhibition on Intelligent Transport Systems and Services*, pp. 1–8. No. 7 (2007).
9. Jones, A., Sergot, M.: A formal characterisation of institutionalised power. *Journal of the IGPL* 4(3), 427–443 (1996).
10. Lamport, L.: Paxos made simple. *ACM SIGACT News* 32(4), 18–25 (2001).
11. Lamport, L.: Byzantizing Paxos by refinement. *Distributed Computing* pp. 211–224 (2011).
12. Artikis, A.: Dynamic specification of open agent systems. *Journal of Logic and Computation*, doi: 10.1093/logcom/exr018 (2011).
13. Johnson, J.: The future of the social sciences and humanities in the science of complex systems. *Innovation: The European Journal of Social Science Research* 23(2), 115–134 (June 2010).
14. Müller-Schloer, C., Schmeck, H., Ungerer, T. (eds.): *Organic Computing — A Paradigm Shift for Complex Systems*. Springer (2011).
15. Sanderson, D., Pitt, J.: An affective anticipatory agent architecture. *2011 IEEE/WIC/ACM International Conferences on Web Intelligence and Intelligent Agent Technology*, pp. 93–96 (2011).

16. Demiris, Y.: Prediction of intent in robotics and multi-agent systems. *Cognitive Processing* 8(3), 151–158 (2007).
17. Castelfranchi, C., Lorini, E.: Cognitive anatomy and functions of expectations. *Proceedings of IJCAI03 Workshop on Cognitive Modeling of Agents and Multi-Agent Interactions*, pp. 9–11 (2003).

Chapter 11

Assessing the Resilience of Self-Organizing Systems: A Quantitative Approach

Matteo Risoldi, Jose Luis Fernandez-Marquez, and Giovanna Di Marzo Serugendo

Contents

As described in the previous chapter, self-organizing systems exhibit adaptation and resilience features, but the assessment and measurement of these features is not trivial, even if it would be very useful in order to quantify the adaptation and the resilience of different approaches and to compare systems. In this chapter, Matteo Risoldi, Jose Luis Fernandez-Marquez, and Giovanna Di Marzo Serugendo propose a framework, called DREF, which aims to support the assessment and the measurement of these features in a quantitative manner. They use a case study to show the applicability of their proposal.

11.1 Introduction

Resilience, intended as the persistence of dependability when facing changes [1], is one of the central features of self-organizing (SO) systems due to their ability to adapt their behavior following changes and faults. The assessment of resilience is generally achieved with experiments and simulations. Robustness and adaptation to some changes is obtained through specific SO mechanisms, which have their limits and do not help overcoming any possible type of faults or change [2]. For instance, digital pheromone in ant-based systems helps overcome the appearance of obstacles in the environment or the disappearance of food but is of limited help in case of faults (malicious or not) in the agent behavior (e.g., not properly following the pheromone). Therefore, in the process of development of a SO system, a developer will often want to achieve better resilience by adding, removing, or modifying the system's behavior, then assessing the system to see whether and how it has improved. Due to the complex behaviors of SO systems, however, it is not easy to quantify how a new version of a SO system compares to the one it replaces. Informal methods of comparison are generally effective only for relatively simple and small-scale systems. As SO systems are often used to model large, complex behaviors, a structured, systematic, and repeatable way to compare the resilient properties of different versions of a system is necessary.

In this chapter, we show how the evolution process taking place during the development of a SO system can benefit from a quantification of the satisfaction of resilience-related properties by different versions of the system. To this end, we will assist the classical "trial and error" development process (i.e., varying parameters and performing simulations) with a formal framework for the quantification of resilience called DREF (*dependability and resilience engineering framework*) [3]. DREF provides a generic framework to organize resilience measures in a coherent, formally defined way that is flexible and customizable. It supports a systematic practice for assessing resilience and is applicable to any system of interest.

The main goal of this chapter is to show how a precise, quantitative definition of resilience measures helps the developer in the choice of a particular version of a system. We used DREF on a dynamic gradient case study with 20 nodes. To this end, we selected specific measurable requirements: (1) probability of message delivery,

(2) accuracy of the gradient, and (3) number of messages created. We then measured quantitatively those requirements in a mobile scenario and assessed how modifying key parameters in subsequent versions of the system impact (positively or negatively) the resilience to this type of perturbation. Resilience is directly linked to the measures above assessed across the different executions. For instance, the higher the accuracy of the gradient, the higher the resilience of that version of the system to the mobility of nodes.

The chapter is organized as follows. Section 11.2 discusses how in SO systems literature there is a growing interest in formal methods. Section 11.3 describes the SAPERE framework used to create the case study described in Section 11.4. The latter section also discusses the simulation environment used to assess identified properties. Section 11.5 briefly introduces the notions of the DREF framework essential to this chapter and how they apply to the case study. In Section 11.6, we describe the resilience assessment results. Finally, Section 11.7 draws conclusions and outlines the perspectives of this work.

11.2 Formal Methods in SO Systems

The development of SO systems, and multi-agent systems in general, is a software-engineering activity like any other. The different paradigm made it so that, during a certain time, development practices were still not as mature as in other software-engineering fields. In 2000, Jennings argued [4] that one of the pitfalls of agent-oriented software engineering (AOSE) is that "You forget you are developing software: [...] the development of any agent system [is] a process of experimentation. [S]oftware engineering good practice [is] ignored."

Since then, there has been progress on this subject. In 2005, Bernon et al. [5] reviewed several AOSE methodologies and proposed a unified meta-model in an effort to standardize the conceptual framework behind multi-agent systems (MAS). More advanced techniques, such as model-driven engineering, and formal methods slowly but surely made their way into AOSE. A 2009 survey [6], catalogued about 50 works from the previous 10 years that are concerned with the application of formal methods to AOSE. A large number among them are concerned with validation and verification of MAS. Several of them use a model-driven engineering approach. In addition to novel approaches to simulation-based verification [7], there have been more rigorous approaches to verify MAS using model checking with an accent on modeling [8], model transformation [9], and requirements enrichment with business constraints [10].

Current methodologies specifically meant for developing SO systems [11] follow the typical phases of software-engineering methodologies: requirements, analysis, design, implementation, verification, and test. Those methodologies all focus on the design phase and, in some cases, on a series of design phases [12], in which the self-organizing behavior is progressively modeled and, in some methodologies, also simulated. Some of them consider analysis of faults [13], but none of them specifically addresses assessment or quantification of resilience.

The cited works are a strong contribution to bringing AOSE and SO development methods on par with other software-engineering domains but are mostly oriented to the requirements and design stages of software development. In the domain of SO systems, however, validation and verification methods should deal with the dynamic changes in modes and contexts and address runtime assurance as stated in [14]. The authors of [14] also agree that requirements for self-adapting systems should be expressed in a way that accounts for the uncertainty characterizing this kind of systems. They argue that the very same nature of self-adapting systems calls for models and development practices that blur the line between design-time and runtime with models being "built and maintained at runtime" with evolutions of the system being reflected at the model level. The DREF framework was built, in part, with the goal of integrating requirements, assurance, and system evolutions in the model itself. DREF is thus a suitable candidate to consider for assessing resilience of SO systems as it considers both runtime aspects and design evolutions.

11.3 SAPERE Framework

In this section, we present the SAPERE framework, used for implementing the case study presented in this chapter. The SAPERE framework has been developed under the SAPERE European Project.* SAPERE provides a theoretical and practical framework for decentralized deployment and execution of self-aware and adaptive services for future and emerging pervasive network scenarios.

The SAPERE framework combines two different technologies: (1) tuple space systems, providing a shared space in which agent and services are advertised, and (2) chemical-inspired systems, providing chemical rules (called *eco-laws*) that govern the behavior of the SAPERE system [15,16].

An innovative architecture is built on top of these two technologies. The main motivation is to allow applications to rely on core services (e.g., spreading, gradient, or evaporation) provided by the environment or by specific libraries, building spatial structures across distributed (possibly mobile) nodes [17]. This architecture is organized into different layers; it provides abstractions that ease the design and implementation of new decentralized applications by relying and reusing functionalities provided by those core services. Chapter 9 discusses the corresponding architectural style [18].

In the SAPERE implementation the main concepts are the following:

- "Live Semantic Annotation" (LSA): A tuple that represents any information about an agent or service. LSAs are injected in the LSA's space representing the (updated) state of their associated component.
- LSA's space: A distributed, shared space in which context and information are provided by a set of LSAs stored at given locations.

---

* http://www.sapere-project.eu/.

- Eco-laws: Rules acting as chemical reactions that implement core services. These rules act on the LSAs stored in each LSA's space by deleting, updating, or moving LSAs between LSA's spaces. In the same way that biological systems' processes obey nature's laws, a SAPERE application is subject to eco-laws that implement its active environment.
- LSA bonding: An LSA bond acts as a reference to another LSA and provides fine-tuned control of what is visible or modifiable to each agent and what is not. If an LSA of a given agent includes a bond to an LSA of another agent, the former agent can inspect the state and interface of that other agent (through the bond) and act accordingly. Bonds are a specific type of eco-laws.
- Agents: Execute in hosts and are able to locally access the LSA's space available in that host.

Eco-laws reside in the LSA's space and are implicitly and seamlessly triggered by tags or properties appearing in the LSAs. An LSA can be subject to different eco-laws depending on the specified properties or tags of the tuple. Core services provided by the eco-laws include spread, decay, aggregate, and bond. Higher-level services, built on top of those core services, are provided either by libraries or specialized agents. They include services for building or exploiting spatial structures, such as Chemotaxis or Dynamic Gradient, and application-level services, such as crowd steering.

The SAPERE infrastructure encompasses a middleware [19] running on stationary and mobile platforms, situation-awareness [20], semantic resource discovery [21], and crowd-steering applications [16].

11.4 Case Study: Dynamic Gradient

This section describes the Dynamic Gradient service, designed and implemented using the SAPERE framework.

Gradients are spatial structures spread across nodes that provide information about distance from and direction to the gradient source. Thus, when a piece of information is propagated among nodes under the form of a gradient, additional information about the sender distance and direction is progressively computed at each node.

Figure 11.1 shows an example of gradient spread over a network of 12 nodes. The nodes represent devices with computational and connectivity capabilities (e.g., laptops, smartphones, or tablets). The edges represent the connections between nodes; two nodes are connected if they are in communication range. In this example, the node with value 0 (source of the gradient) creates a gradient that spreads across the network. The value inside each node increases at each hop. When two or more gradient paths reach a node, the lowest gradient value is kept. The number of hops provides information about the distance. The node within communication range with the lowest hop count provides the direction to the gradient source.

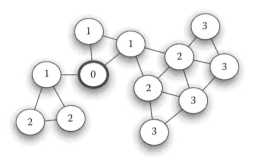

Figure 11.1   Gradient example.

This section focuses on *dynamic gradients*. A dynamic gradient provides information about the sender's (i.e., gradient source) distance and direction that is *periodically updated*. A dynamic gradient allows the maintenance (i.e., the update) of the gradient structure even when nodes are moving.

Dynamic gradients may be used in a wide range of applications, such as routing information in mobile ad hoc networks, crowd steering, or coordination in multi-agent systems.

11.4.1 Dynamic Gradient Implementation

The Dynamic Gradient service is provided by the SAPERE framework as an external library and is available in each node. The Dynamic Gradient service makes use of the spread, aggregate, and decay eco-laws.

Spread allows the information to be spread across the network, increasing hop counts at each node. Aggregate is in charge of selecting the value with the minimum number of hops when information is coming through different gradient paths. The combination of spread and aggregate eco-laws allow the creation of a gradient across the network, but the gradient is not updated; thus, the information about the sender's distance and direction becomes outdated when the nodes are moving. To avoid this problem, the dynamic gradient relies on the decay eco-law to periodically decrease the lifetime of the gradient LSAs. When the lifetime value reaches 0, the SAPERE middleware removes the gradient information LSA from the node. If the gradient information LSA is removed from the node that created the dynamic gradient (i.e., the gradient source), the gradient LSA is injected again in the system by the Dynamic Gradient service (library agent). This actually restarts a new gradient propagation thus providing updated information about the sender's distance and direction.

Eco-laws fire periodically, for example, sending information from one node to another (spreading) and aggregating them. The *frequency of eco-laws firing* is a key parameter that has an effect on the performance of the system. The decay eco-law is also involved in the dynamic gradient and fires at the same frequency as the other eco-laws.

A second key parameter in the case of the Dynamic Gradient service is the *decay value* (i.e., the lifetime) of the gradient information. Intuitively a short lifetime favors a fast adaptation but involves huge resource consumption (i.e., mainly bandwidth), and a long lifetime saves resources but does not cope with topological network changes. This parameter is exploited in the decay eco-law.

In this case study, we analyze the evolution process of setting these two parameters involved in the Dynamic Gradient service. The experimental values in the evolution process are assessed using The ONE simulator [22] introduced in the next section.

Figure 11.2a shows the map of Helsinki we have used for the case study with 20 SAPERE nodes. Figure 11.2b represents the real geographical map used in the simulations. We used a part of Helsinki city center (600 × 500 meters). We set the number of moving nodes to 20. Nodes simulate people (and their devices) walking on the streets. We used the mobility pattern called "shortest path based on map." Each node chooses a random destination, and it reaches the desired destination by following the shortest path on the map (following roads). The speed of the nodes is randomly chosen for each trajectory between 1–3 m/s. Each node has a wi-fi device with 200 m of communications range, and the transmission speed is 250 KBps. The simulation runs for 900 simulation steps, and each simulation step corresponds to 1 *s* of simulated time.

Figure 11.3a shows the simulation scenario before the gradient creation; black lines represent connections among nodes. Figure 11.3b depicts the gradient (black lines) corresponding to the initial configuration. The black circle node is the gradient source (i.e., the node that initially creates the gradient). We observe two nodes on the right that are disconnected. Figures 11.3c and 11.3d show subsequent updates of the gradient once the nodes start moving. Notice that for clarity reasons, Figures 11.3b, 11.3c, and 11.3d do not show the communication links representing the physical connections between the nodes and do not show the circles representing the communication range. They only show the gradient links.

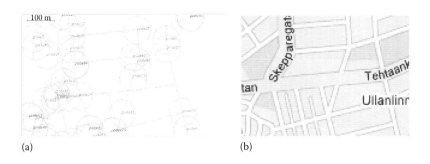

(a) (b)

Figure 11.2 Scenario map: Helsinki area with SAPERE nodes (a), real geographical map (b).

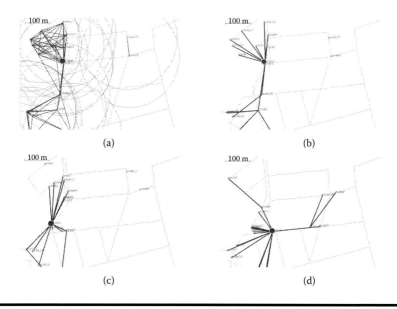

Figure 11.3 Simulation scenario before gradient creation (a), after gradient creation (b), once nodes have moved (c) and (d).

11.4.2 *The ONE Simulator*

The case study described above has been implemented using The ONE simulator [22], an open source network simulator providing a large range of realistic scenarios. Among others, it allows the investigation of simulations composed of many different entities, such as pedestrians, cars, buses, and trams, to import real mobility traces, to visualize the movement of nodes on Google maps, and to provide realistic network-related simulations of bandwidth consumptions, time to send information among nodes, or packet collisions.

We extended The ONE with the actual SAPERE middleware [19]. Consequently, the Dynamic Gradient is implemented using the SAPERE middleware (libraries, eco-laws, LSAs) running in the nodes, and the movement of nodes, visualization, performance metrics, and connectivity between the nodes are done by The ONE simulator.

Mainly, The ONE allows us to do the following:

■ Simulate network connections, providing metrics about the network behavior (e.g., message drops, messages properly received or bandwidth consumption).
■ Provide different network interfaces to the entities participating in the system, such as wi-fi or Bluetooth. This feature allows the creation of realistic heterogeneous networks.

■ Use a library with many different mobility patterns.
■ Execute the simulation with visualization or in batch mode, making it possible to use servers for very large-scale simulations.

11.5 DREF Framework

DREF [3] is a framework that formalizes the fundamental concepts used to define dependability and resilience of ICT systems. It allows quantifying how the level of satisfiability of system properties varies over an evolution axis when facing changes. DREF formalizes a rather large set of core concepts related to the assessment of resilience. While the complete definitions can be found in [3], we will summarize here the ones that are useful in the context of this chapter:

Entities and properties: An *entity* is anything of interest that is considered. An entity could be, for example, a program, a database, a person, a hardware device, or a development process. In this chapter, we consider one entity *Dyn*, a dynamic gradient system:

$$Ent = \{Dyn\}. \tag{11.1}$$

A *property* is a basic concept used to characterize an entity. It can be, for example, an informal requirement or a mathematical property. In this chapter, we consider three relevant properties of the dynamic gradient system, namely, the probability of message delivery *dp*, the average accuracy of the gradient *ac* (i.e., the distance of the actual gradient from the ideal gradient that would exist if nodes were not moving), and the number of messages created *nm*:

$$Prop = \{dp, ac, nm\}. \tag{11.2}$$

Evolution axis: An evolution axis is a set of values that is used to index a set of entities or a set of properties. In the context of this chapter, the development of *Dyn* went through different versions, which are indexed as Dyn_i^d, where i is the eco-law firing interval, and d is the decay value (i.e., lifetime of gradient information)—the two key parameters discussed in Section 11.4.1. We variated i between 1 and 3, and d between 2 and 12 in two-step intervals. Thus, there are two evolution axes indexed as

$$ev_i = \{1, 2, 3\}; \tag{11.3}$$

$$ev_d = \{2, 4, 6, 8, 10\}. \tag{11.4}$$

Satisfiability: An entity will generally have to satisfy some property or requirement. This fact can be expressed with a *satisfiability function*, defined as follows. Let *Ent* be a set of entities and *Prop* a set of properties. The satisfiability of properties by entities is a function *sat* such that

$$sat : Prop \times Ent \rightarrow \mathbb{R} \cup \{\bot\}. \tag{11.5}$$

Semantically, *sat* quantifies how much an entity satisfies (or not) a property and is defined depending on the application.

The satisfiability functions for the properties in our case study are defined as follows. The *dp* property is measured in terms of the ratio between the messages that are delivered and created:

$$sat(dp, e) = \frac{delivered messages}{created messages}. \tag{11.6}$$

The *ac* property is measured as the average accuracy (i.e., percentage of nodes with a correct value) of the gradient in the simulation:

$$sat(ac, e) = \frac{\sum_{i=1}^{n} ac_i}{n} \tag{11.7}$$

where $ac_i = \dfrac{k_i}{20}$ is the accuracy of the gradient at step i of the simulation, k_i is the number of nodes for which the gradient value is correct, 20 is the number of nodes, and n is the total number of simulation steps.

The *nm* property is measured as

$$sat(nm, e) = \frac{20,000}{created messages} \tag{11.8}$$

where 20,000 is an empirically established ideal value for the number of created messages.

Nominal satisfiability: An additional relevant concept defined by DREF is that of *nominal satisfiability*, denoted as *nsat*, such that

$$nsat : Prop \times Ent \rightarrow \mathbb{R}. \tag{11.9}$$

This is a satisfiability function used to represent the *expected satisfiability*—what minimum satisfiability value is considered to be acceptable.

The nominal satisfactions for the properties in our case study are defined as follows:

$$nsat(dp,e) = 0.99 \tag{11.10}$$

$$nsat(ac,e) = 0.8 \tag{11.11}$$

$$nsat(nm,e) = 1 \tag{11.12}$$

(i.e., the *nm* property is exactly satisfied if a system creates 20,000 messages, over-satisfied if it creates less, and under-satisfied if it creates more).

Failure: Given *sat*, a satisfiability function, and *nsat*, a nominal satisfiability function, we say there is a failure for a tuple $(p,e) \in dom(sat)$ if $sat(p,e) < nsat(p,e)$. In other words, an entity fails a property if its satisfiability is lower than the nominal satisfiability.

Let *e* be an entity, *p* a property; a failure of property *p* by entity *e* is denoted *fail*(*p*,*e*). It is defined as follows:

$$fail : Prop \times Ent \rightarrow \mathbb{R}_0^+ \tag{11.13}$$

$$fail\,(p,e) = nsat(p,e) - Min\,(nsat\,(p,e), sat\,(p,e)) \tag{11.14}$$

and it quantifies how far the actual satisfiability is from satisfying the property.

Cumulative failure level: The cumulative failure level for an entity is the sum of failures for *all* of its properties. The cumulative failure level for an entity *e* is denoted *sfail*$_e$ and is defined as

$$sfail_e = \sum_{p \in Prop} fail(p,e) \ (0 \ if \ Prop = \varnothing). \tag{11.15}$$

11.6 Experimental Results

We took the case study system through two rounds of evolution, which will explore some possible variations of the two parameters of the system. The parameters we chose are the eco-law firing interval and the decay value (i.e., the lifetime of the information in the nodes). The eco-law firing interval parameter is of major importance as it can cause network overload if its value is too high, and conversely, it can prevent the gradient from updating properly if the value is too low. The decay value is similarly important as a too-low value would prevent the information from surviving long enough to form a gradient, and a too-high value would hamper gradient updates.

As explained in the previous sections, we modeled the case study using the SAPERE framework, implemented it in The ONE simulator, and used DREF to process the results and extract useful quantitative measures about property satisfaction.

11.6.1 First Evolution: Fixing an Eco-Law Firing Interval

In the first evolution in the development of our dynamic gradient system, we want to search for an appropriate eco-law firing interval. What we are looking for is an interval that is sufficient to let the gradient establish itself while not overloading the network. In this iteration, we will stop at the first value that achieves good accuracy, leaving further optimization of the number of messages for a further iteration.

We fix the decay value at 10, which, given the size of the network, is an educated guess at a value high enough to let the information propagate through the network. We then start by considering an eco-law firing interval of 1 s, increasing it one further second for each new version (and simulation). Table 11.1 shows the satisfaction values for the first three versions of Dyn that we consider: Dyn_1^{10}, Dyn_2^{10}, and Dyn_3^{10}. We see that for all three properties there is a constant improvement (i.e., a reduction of the failure level $fail(p,e)$) for each individual property. As a consequence, the cumulative failure level $sfail_e$ decreases with each new version, indicating that it is an improvement over the previous one. What really matters here is the $fail(p,e)$ for properties dp and ac. As soon as we find a failure level of 0 for both (i.e, in version Dyn_3^{10}), we can stop this first evolution.

Table 11.1 First Evolution: Varying the Eco-Law Firing Interval

e	p	sat(p,e)	nsat(p,e)	fail(p,e)	sfail_e
Dyn_1^{10}	dp	0.8380	0.99	0.152	1.1201
	ac	0.5104	0.80	0.2896	
	nm	0.3215	1	0.6785	
Dyn_2^{10}	dp	0.9649	0.99	0.0251	0.61
	ac	0.7241	0.80	0.0759	
	nm	0.4910	1	0.5090	
Dyn_3^{10}	dp	0.9981	0.99	0	0.1462
	ac	0.8144	0.80	0	
	nm	0.8538	1	0.1462	

11.6.2 Second Evolution: Fixing a Decay Value

In this second evolution, we want to search for an appropriate decay value. Fixing the eco-law firing interval at 3 s, based on the results of the first iteration, we will simulate the system for decay values ranging from 2 to 12, in two-unit steps.

Table 11.2 shows the satisfaction values for the obtained versions of Dyn: Dyn_3^2, Dyn_3^4, Dyn_3^6, Dyn_3^8, Dyn_3^{10}, and Dyn_3^{12}. The measures here follow a slightly more interesting dynamic. The dp property is constantly satisfied by all versions, albeit with some oscillations in values. The ac property is deeply under-satisfied by Dyn_3^2, then improves with versions until reaching *nsat* with Dyn_3^6 (even staying above

Table 11.2   Second Evolution: Varying the Decay Value

e	p	sat(p,e)	nsat(p,e)	fail(p,e)	sfail$_e$
Dyn_3^2	dp	1	0.99	0	0.7473
	ac	0.0527	0.80	0.7473	
	nm	32.4	1	0	
Dyn_3^4	dp	0.9984	0.99	0	0.1287
	ac	0.6713	0.80	0.1287	
	nm	2.2629	1	0	
Dyn_3^6	dp	0.9985	0.99	0	0
	ac	0.8338	0.80	0	
	nm	1.2138	1	0	
Dyn_3^8	dp	0.9934	0.99	0	0.0575
	ac	0.8085	0.80	0	
	nm	0.9425	1	0.0575	
Dyn_3^{10}	dp	0.9981	0.99	0	0.1462
	ac	0.8144	0.80	0	
	nm	0.8538	1	0.1462	
Dyn_3^{12}	dp	0.9965	0.99	0	0.2093
	ac	0.8052v	0.80	0	
	nm	0.7907	1	0.2093	

nsat). The *nm* property instead follows a degradation pattern with the increase of decay values, going from satisfaction in the first three versions, to a gradually deeper under-satisfaction in the following ones. There is one version, Dyn_3^6, that satisfies all three properties. We can thus conclude that, in this iteration, we established an appropriate decay value to be 6.

11.6.3 Cumulative Failure Level as a Useful Overall Assessment of Versions

It is rather evident that just by using the satisfiability *sat(p,e)*, shown in Figure 11.4,* it is not immediate to the eye of the developer whether a version is better than another at satisfying properties. Even by looking at the individual failure levels *fail(p,e)* for each property, shown in Figure 11.5, the assessment may not be simple as the failure values series may vary independently from one version to another.[†] This is the main reason for the existence of the cumulative failure level indicator *sfail$_e$*: It gives a concise quantification of how much a version is failing. The hypothetical developer behind our example would easily be able to plot the values for all the created versions and quickly spot the lower points that identify the best versions (shown in Figure 11.6).

Figure 11.7 shows how the gradient continuously recovers from the mobility of nodes for the case of Dyn_3^6. The continuous line shows the evolution of ac_i. The dotted line shows the average of ac_i from time 1 to time i (note that at time 900, it is equal to $sat(ac, Dyn_3^6)$).

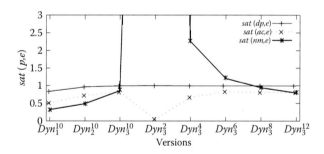

Figure 11.4 *sat(p,e)* levels for each property versus evolution for versioned entity *Dyn* (n.b.: $sat\left(nm, Dyn_3^2\right) = 32.4$ is off scale for readability).

---

* In this figure, as well as in Figures 11.5 and 11.6, the entity Dyn_3^{10} only appears once, although it is part of both evolutions.
[†] Admittedly, in this case study, there is a version that stands out for having all three failure levels at 0, that is, Dyn_3^6; however, this is far from being the general case.

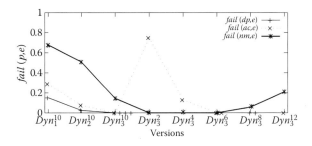

Figure 11.5 *fail(p,e)* **levels for each property versus evolution for versioned entity** *Dyn.*

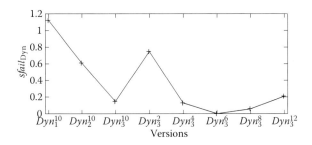

Figure 11.6 **Cumulative failure level versus evolution for versioned entity** *Dyn.*

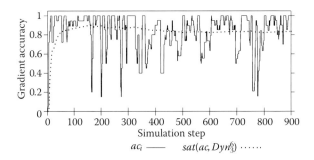

Figure 11.7 **Gradient accuracy,** Dyn_3^6.

11.6.4 *Possible Further Evolutions*

We stopped this example with the second evolution for the sake of brevity. However, it is easy to imagine that there could be a third evolution to answer the question of whether a further exploration of larger eco-law firing intervals could further

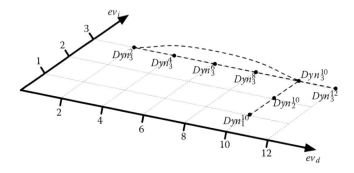

Figure 11.8   Evolution of our example over two coordinated axes.

improve the performance of the system. In this way, we could picture that the evolution of the system in our example, which is occurring over two coordinated axes (shown in Figure 11.8), could continue by adding further versions indexed on the axis (possibly adding more indexes to the axes or even additional axes). Also, instead of a manually defined evolution path like the one we followed, one could perform a complete exploration of all combinations of parameter values encompassed by given bounds if the available resources allowed for it.

11.6.5 Tolerance

In this example, we considered that a property is either satisfied (i.e., $sat(p,e) \geq nsat(p,e)$) or not. However, in DREF, it is possible to define a tolerance threshold for a property, that is, a level smaller than $nsat(p,e)$ that indicates a satisfaction value that does not quite satisfy a property but is still within acceptable limits. This is a particularly relevant concept for resilient systems as it can be used to describe tolerated failures (whether foreseen or out of the envelope). In our example, a tolerance threshold of 0.85 could be introduced for, for example, property *nm*, which would result in versions Dyn_3^8 and Dyn_3^{10} to be also considered satisfactory.

11.7 Conclusion and Perspectives

This chapter showed how the DREF framework can be used to quantitatively assess the resilience of SO systems during their development process. Through a case study, based on the SAPERE framework, we showed how the indicators defined in DREF can offer a useful synthesis of simulation results in order to compare and choose among different versions of a system.

DREF also defines other indicators, not discussed here but found in [3], that bring other useful information to the table. This chapter only focused on failures, allowing one to choose the version that "fails less." However, in other cases, it might be interesting to choose the version that "satisfies more" among several versions that are all correct. In this respect, DREF defines, for example, the concept of *global satisfiability*, which is a unique indicator measuring the cumulative level of satisfaction of *all* properties of an entity, possibly taking into account different property weights when applicable.

One of the main advantages of DREF is that we can use the cumulative failure level in order to easily compare different versions. Using the cumulative failure level, one could implement an evolutionary algorithm, such as a genetic algorithm, in order to automate the creation and selection of different versions. Thus, the system would start with random versions that would evolve through the execution of the evolutionary algorithm until the cumulative failure level of one of the versions is below a desired value (i.e., the cumulative failure level would be used as the evolutionary fitness function in case of using a genetic algorithm).

Our future perspectives for this study include the integration of DREF in The ONE simulator in order to systematically apply quantitative assessments of resilience in simulated systems. In a first instance, this will assist us in systematically investigating resilience of classified self-organizing mechanisms [23] against a range of faults and perturbations. In the long term, in conjunction with identified metrics, this will constitute a definite step toward the establishment of benchmarks and assessments for self-* systems.

Acknowledgments

M. Risoldi is supported by the National Research Fund, Luxembourg, project MOVERE C09/IS/02. G. Di Marzo Serugendo and J. L. Fernandez-Marquez are partially supported by the EU FP7 project "SAPERE—Self-Aware Pervasive Service Ecosystems" under contract No. 256873.

References

1. J.-C. Laprie, "From dependability to resilience," in *Proceedings of the 38th Annual IEEE/ IFIP International Conference on Dependable Systems and Networks (DSN)*, Anchorage (USA), 2008.
2. G. Di Marzo Serugendo, "Robustness and dependability of self-organising systems — A safety engineering perspective," in *Int. Symp. on Stabilization, Safety, and Security of Distributed Systems(SSS)*, ser. LNCS, vol. 5873. Lyon, France: Springer, Berlin Heidelberg, 2009, pp. 254–268.
3. N. Guelfi, "A formal framework for dependability and resilience from a software engineering perspective," *Central European Journal of Computer Science*, vol. 1, pp. 294–328, 2011, http://dx.doi.org/10.2478/s13537-011-0025-x.

4. N. R. Jennings and M. Wooldridge, "Agent-oriented software engineering," *Artificial Intelligence*, vol. 117, pp. 277–296, 2000.

5. C. Bernon, M. Cossentino, and J. Pavón, "Agent-oriented software engineering," *Knowl. Eng. Rev.*, vol. 20, no. 2, pp. 99–116, June 2005.

6. A. E. Fallah-Seghrouchni, J. J. Gomez-Sanz, and M. P. Singh, "Formal methods in agent-oriented software engineering," in *Proceedings of the 10th International Conference on Agent-Oriented Software Engineering*, ser. AOSE '10. Berlin, Heidelberg: Springer-Verlag, 2011, pp. 213–228.

7. M. Niazi, A. Hussain, and M. Kolberg, "Verification & validation of agent based simulations using the VOMAS (virtual overlay multi-agent system) approach," in *MALLOW*, ser. CEUR Workshop Proceedings, M. Baldoni et al., Eds., vol. 494. CEUR-WS.org, 2009.

8. J. Barjis, I. Rychkova, and L. Yilmaz, "Modeling and simulation driven software development," in *Proceedings of the 2011 Emerging M&S Applications in Industry and Academia Symposium*, ser. EAIA '11. San Diego, CA, USA: Society for Computer Simulation International, 2011, pp. 4–10.

9. R. H. Bordini, M. Fisher, W. Visser, and M. Wooldridge, "Verifying multi-agent programs by model checking," *Autonomous Agents and Multi-Agent Systems*, vol. 12, no. 2, pp. 239–256, Mar. 2006.

10. M. Montali, P. Torroni, N. Zannone, P. Mello, and V. Bryl, "Engineering and verifying agent-oriented requirements augmented by business constraints with b-tropos," *Autonomous Agents and Multi-Agent Systems*, vol. 23, pp. 193–223, 2011, http://dx.doi.org/10.1007/s10458-010-9135-4.

11. M. Puviani, G. Di Marzo Serugendo, R. Frei, and G. Cabri, "A method fragments approach to methodologies for engineering self-organizing systems," *ACM Trans. Autonomous and Adaptive System*, vol. 7, no. 3, pp. 33:1–33:25, Oct. 2012.

12. M. Schut, "On model design for simulation of collective intelligence," *Information Sciences*, vol. 180, pp. 132–155, 2010.

13. G. Di Marzo Serugendo, J. Fitzgerald, and A. Romanovsky, "MetaSelf — An architecture and development method for dependable self-* systems," in *Symp. on Applied Computing (SAC)*, Sion, Switzerland, 2010, pp. 457–461.

14. B. Cheng et al., "Software engineering for self-adaptive systems: A research roadmap," in *Software Engineering for Self-Adaptive Systems*, ser. Lecture Notes in Computer Science, B. Cheng, R. de Lemos, H. Giese, P. Inverardi, and J. Magee, Eds. Springer Berlin/Heidelberg, 2009, vol. 5525, pp. 1–26, http://dx.doi.org/10.1007/978-3-642-02161-9-1.

15. F. Zambonelli et al., "Self-aware pervasive service ecosystems," *Procedia Computer Science*, vol. 7, pp. 197–199, 2011.

16. M. Viroli, D. Pianini, S. Montagna, and G. Stevenson, "Pervasive ecosystems: A coordination model based on semantic chemistry," in *27th ACM Symp. on Applied Computing (SAC 2012)*, S. Ossowski, P. Lecca, C.-C. Hung, and J. Hong, Eds. Riva del Garda, TN, Italy: ACM, 26–30 March 2012.

17. J. L. Fernandez-Marquez, G. Di Marzo Serugendo, and S. Montagna, "Bio-core: Bio-inspired self-organising mechanisms core," in *Bio-Inspired Models of Networks, Information, and Computing Systems*, ser. Lecture Notes of the Institute for Computer Sciences, Social Informatics and Telecommunications Engineering, E. Hart, J. Timmis, P. Mitchell, T. Nakamo, and F. Dabiri, Eds. Springer Berlin Heidelberg, 2012, vol. 103, pp. 59–72, http://dx.doi.org/10.1007/978-3-642-32711-7_5.

18. J. L. Fernandez-Marquez, G. Di Marzo Serugendo, P. Snyder, and G. Valetto, "Pattern-based architectural style for self-organizing software systems," in *Adaptive, Dynamic, and Resilient Systems*, G. Cabri and N. Suri, Eds. Taylor & Francis, 2013.

19. F. Zambonelli, G. Castelli, M. Mamei, and A. Rosi, "Integrating pervasive middleware with social networks in SAPERE," in *Mobile and Wireless Networking (iCOST), 2011 International Conference on Selected Topics in*, Oct. 2011, pp. 145–150.

20. G. Stevenson, J. L. Fernandez-Marquez, S. Montagna, A. Rosi, J. Ye, A.-E. Tchao, S. Dobson, G. Di Marzo Serugendo, and M. Viroli, "Towards situated awareness in urban networks: A bio-inspired approach," in *First International Workshop on Adaptive Service Ecosystems: Nature and Socially Inspired Solutions (ASENSIS) at Sixth IEEE International Conference on Self-Adaptive and Self-Organizing Systems (SASO12)*. IEEE Computer Society, 2012.

21. J. Stevenson, G. Ye, S. Dobson, M. Viroli, and S. Montagna, "Self-organising semantic resource discovery for pervasive systems," in *First International Workshop on Adaptive Service Ecosystems: Nature and Socially Inspired Solutions (ASENSIS) at Sixth IEEE International Conference on Self-Adaptive and Self-Organizing Systems (SASO12)*. IEEE Computer Society, 2012.

22. A. Keränen, J. Ott, and T. Kärkkäinen, "The ONE Simulator for DTN Protocol Evaluation," in *SIMUTools '09: Proceedings of the 2nd International Conference on Simulation Tools and Techniques*. New York: ICST, 2009.

23. J. L. Fernandez-Marquez, G. Di Marzo Serugendo, S. Montagna, M. Viroli, and J. L. Arcos, "Description and composition of bio-inspired design patterns: A complete overview," *Natural Computing*, pp. 1–25, 2012.

Chapter 12

Leveraging ICT to Enable e-Maintenance for Automated Machines

Roberto Lazzarini, Cesare Stefanelli, and Mauro Tortonesi

Contents

Distributed systems are not limited to the Internet and data centers. Indeed, in the increasingly connected future, many commercial and consumer devices will be part of distributed systems. Traditionally, automated monitoring, notification, and adaptation have not been extended to these systems if they were not part of critical infrastructure. This is likely to change as we rapidly approach the Internet of Things. In this chapter, Roberto Lazzarini, Cesare Stefanelli, and Mauro Tortonesi describe their experiences with leveraging information and communication technologies to enable automated maintenance support for industrial machines.

12.1 Introduction

E-maintenance is a recently emerged discipline that relies on the integration of information and communication technologies (ICT) into the maintenance processes in order to enable the remote management of industrial equipment and to support proactive decision-making and efficient maintenance operation planning (Muller, Crespo-Màrquez, and Iung, 2008). In this context, e-maintenance represents a major evolution of traditional maintenance practices that expresses the synthesis of two major trends in today's society, that is, the growing importance of maintenance as a key technology and the rapid development of ICT. E-maintenance integrates and synchronizes the various maintenance and reliability applications to gather and deliver asset information where it is needed, thus improving adaptation and resiliency (Campos, 2009).

So far, e-maintenance solutions have been developed mostly for large and expensive machinery, for instance, in the energy industry and in the heavy one because of the high costs of remote monitoring technology (Kunze, 2003; Garcia, Sanz-Bobi, and del Pico, 2006). However, the cost for the implementation of e-maintenance functions is significantly decreasing thanks to recent ICT developments, such as powerful components off the shelf (COTS) hardware, the emergence of web technologies as the de facto standard for interoperable distributed applications, the evolution of data-intensive real-time processing techniques that enable the highlighting of nontrivial (cor)relations among data, and the widespread availability of (relatively) high-bandwidth Internet-based communications.

These driving factors are now fostering the evolution of e-maintenance platforms from *large plants* to *large-scale* machine installations. This enables the application of e-maintenance to *household and similar appliances*, that is, low-cost automated machines, such as those designed to be operated by untrained personnel in homes, shops, warehouses, light industries, or farms. Usually, these machines incorporate motors, heating elements, CPUs, or all of the above. Because there is no onsite technical support available, remote monitoring and diagnosis are crucial to organize and schedule timely assistance operations in order to improve the after-sales assistance of these machines.

State-of-the-art large-scale e-maintenance platforms can provide an array of remote assistance services for the management of automated machines, such as *automated monitoring, self-diagnostics and prognostics*, and *remote management.*

The evolution of large scale e-maintenance platforms is also likely to have a major impact on the design of next-generation automated machines, leading the to realization of adaptive machine-installed control software (and firmware) *that enables a dynamic change in the production workflow* in response to component wearing, to changes in environmental conditions, or to the need to change the result of the manufacturing processes (for instance, a food processing machine can produce either ice cream or chocolate by simply changing its internal thermal cycle). In addition, the maintenance data collected by large-scale e-maintenance platforms will surely become an invaluable asset for the enterprise, thus leading to the development of efficient management strategies and new business models.

The considerations presented in this chapter emerge from the authors' experience in building an innovative e-maintenance platform, called Teorema, for the management of the ice cream–making machines produced by Carpigiani (one of the leading industries in the market). Teorema is currently enabling the remote assistance of more than 4500 machines distributed all over the world, thus allowing for significant savings in maintenance operations, and represents a disruptive innovation in after-sales services within the ice cream machine market.

12.2 From Maintenance to e-Maintenance

Maintenance is the field of management that consists of skills, techniques, methods, and theories that aim at developing technical and organizational solutions to ensure the correct performance of the required functions in industrial equipment. Maintenance is an essential process for the preservation of large assets, such as factories, power plants, vehicles, and buildings, as well as of smaller items, such as vending machines, appliances, and commercial equipment in working conditions—also in terms of energy saving, environmental impact, and safety.

In recent years, engineers and researchers have developed different organizational approaches to improve the quality of maintenance, such as total productive maintenance (TPM, Nakajima, 1988), reliability-centered maintenance (RCM, Moubray, 1997), and condition-based maintenance (CBM, Crespo-Màrquez, 2007). These methods have been implemented in industry and process plants with excellent results.

TPM consists of a range of methods developed in Japan in recent times, which aim at maximizing plant and equipment efficiency. TPM is closely related to total quality management and can be seen as an application of a continuous improvement plan-do-check-act cycle, otherwise known as a Deming cycle. TPM is based on an organizational model that involves machine operators in the maintenance process. More specifically, TPM requires operators to take over some of the day-to-day maintenance tasks, for example, cleaning, lubricating, tightening, and reporting of changes observed in the operation of the equipment.

RCM follows a different approach—less pervasive than that of TPM. The goal of RCM is to identify the minimum levels of maintenance required to keep machinery in safe operating conditions as well as to implement the proper processes and policies that enable the enforcement of these maintenance standards. As a result, RCM can be defined as a technology-centric concept, in contrast to the TPM that instead represents a human-centric technique. In fact, RCM is a highly structured method for the planning of maintenance, originating from a critical analysis of maintenance requirements.

CBM marks the key conceptual shift from *preventive maintenance* strategies, such as TPM and RCM, to *predictive maintenance* ones. In fact, by leveraging condition-monitoring technologies that measure physical (temperature, pressure, viscosity, vibrations, etc.) and functional working parameters in order to determine the current status of a piece of equipment, CBM enables the planning of maintenance interventions when deterioration in equipment condition occurs. As a result, CBM allows the reduction of the number of maintenance interventions and the consumption of spare parts with respect to preventive maintenance strategies, resulting in significant savings.

However, rapid changes and developments in ICT have recently brought disruptive innovations in industrial process management and maintenance, opening the way to the implementation of *proactive maintenance* (Voisin et al., 2010). Proactive maintenance differs from predictive maintenance as it aims to implement more sophisticated monitoring techniques that go beyond the detection of equipment (or component) degradation or machine wear after they have taken place. In fact, proactive maintenance focuses on the early detection and the correction of problems that might lead to equipment malfunctions, thus enabling easier prevention of failures.

The introduction of ICT in maintenance has given rise to e-maintenance, a recently emerged discipline that aims to enable the remote management of industrial equipment and the planning of efficient maintenance operation (Muller, Crespo-Màrquez, and Iung, 2008), which represents the key component for the realization of proactive maintenance in the digital factory (Chryssolouris et al., 2009). E-maintenance platforms augment the set of maintenance tools available to technical support personnel to enable the restructuring of such maintenance processes as monitoring, diagnosis, and prognosis in order to fully reap the benefits provided by proactive maintenance (Campos, 2009; Lee et al., 2006; Levrat, Iung, and Crespo-Màrquez, 2008).

12.3 e-Maintenance: From Large Plants to Large Scale

So far, most of the research work in e-maintenance has focused on large plant machinery, in particular in the oil and gas industry (Wan Mahmood et al., 2011), in the energy industry (Kunze, 2003), and in the heavy one (Garcia, Sanz-Bobi, and del Pico, 2006). The size and cost of those machines justify and encourage the development of sophisticated e-maintenance solutions to improve efficiency. To implement 24/7 remote monitoring, e-maintenance solutions for large and critical machinery usually adopt expensive and often proprietary supervisory control and data acquisition (SCADA) technologies that leverage relatively high-bandwidth ad hoc communication stacks, for example, private radio over UHF or unlicensed bands. SCADA-based e-maintenance solutions usually leverage dedicated hardware and complex software and communication technologies to realize the remote monitoring functions and are difficult to deploy on a large scale as they require a significant effort to enable the interoperability between equipment from different vendors (Brunello, 2003) and to realize a proper communications infrastructure (Yang, Barria, and Green, 2011). The cost of SCADA-based e-maintenance solutions make them unsuited for adoption in the case of low-cost, mass-produced equipment, such as household and similar appliances.

Modern household and similar appliances are often complex and high-duty systems, which perform critical operations, for example, in food-processing applications, and have a long (often more than a decade) lifecycle. As a result, notwithstanding the quality of their components and of their manufacturing processes, during their lifetimes, these machines typically require numerous maintenance interventions. These machines can be deployed in extremely heterogeneous environments, such as restaurants, shops, offices, and homes, where they operate unattended with no onsite technical support personnel available. Household and similar appliances are typically managed with a reactive maintenance strategy, based on onsite maintenance interventions scheduled according to a combination of "run to failure" and periodic maintenance policies. Because the technical personnel working in the field have little or no knowledge about the problem before performing an onsite machine diagnostics procedure, they sometimes might not carry the spare part needed for fixing the specific component that has failed. Thus, technicians have to delay the problem resolution to a second onsite intervention, leading to high machine downtime and revenue losses or to substituting larger parts of the machine, leading to an excess consumption of spare parts. As a result, reactive maintenance strategies for household and similar appliances usually have very high costs.

These considerations suggest that e-maintenance platforms could bring significant benefits for the management of large-scale installations of household and similar appliances, enabling considerably more effective and more efficient maintenance processes based on remote operations and proactive interventions, especially with regards to the efficient scheduling of human resources that have to perform both remote and onsite maintenance interventions.

12.4 ICT Building Blocks for e-Maintenance

Several ICT trends are fostering the adoption of e-maintenance for large-scale installations of automated machines, enabling the implementation of sophisticated machine-installed monitoring functions, which were once only possible for large and costly equipment (Lazzarini et al., 2013; Jantunen et al., 2011). COTS hardware is becoming more and more powerful and affordable, thanks to the huge demand and the fierce competition in the mobile and embedded devices markets. Web-based software technology is allowing the realization of sophisticated and highly interoperable communication infrastructures for distributed applications, leveraging open source software that has low barriers to entry and no license costs. Advances in distributed data analysis techniques enable the real-time analysis of large amounts of data, thus allowing the implementation of sophisticated anomaly detection techniques to proactively identify problems within machines. Finally, the formidable improvements in Internet technology now provide several solutions, for example, Internet and mobile connectivity, which can significantly facilitate the remote access to appliances, with low costs, especially with regard to data traffic.

12.4.1 Evolution of COTS Hardware

Low-cost and low-power hardware platforms have seen a significant development recently. Following a large demand for affordable CPUs in the smartphone and tablet and in the home automation markets, relatively high-performance COTS microprocessors are now available at prices that are becoming affordable for large-scale adoption.

This increases the computational capabilities available for machine-installed (self-)monitoring equipment while keeping the hardware costs reasonably low in order not to excessively increase appliance production costs.

As an example of the kind of processing power-to-cost ratio that is currently available on the market, we mention the Raspberry Pi initiative (http://www.raspberrypi.org), which sells ARM-based credit card-sized computers running a full-fledged version of the Linux operating system for $35 with the purpose of supporting and fostering software-development education. The well-known Arduino initiative, that develops and sells a family of Open Source micro-controllers for electronic prototyping with a starting price of about $20, has also recently released significant developments, such as the powerful Pentium-based Intel Galileo board and the next-generation Arduino TRE board, which integrates an ARM-based processor running Linux on the micro-controller board.

Let us notice that the availability of cheap ARM-based solutions represents a significant advance for embedded hardware applications. In fact, these enable the running of full-fledged operating systems, such as Linux with the Xenomai (http://www.xenomai.org) extension, that are also capable of delivering (soft) real-time guarantees, thereby significantly facilitating the operation of porting existing software, for example, anomaly-detection algorithms. At the same time, this provides developers with tools and languages that provide high-level programming abstractions that are typical of traditional (desktop and web) applications and enables COTS software component reuse. This considerably increases productivity and, at the same time, lowers the barriers to entry in the embedded applications market.

12.4.2 Web-Based Software Technologies

The development of web-based software technologies is also a huge driving factor behind the realization of large-scale e-maintenance platforms. In fact, the interoperability standards developed so far in the industrial computer applications market, such as the OSA-CBM project, (open systems architecture for condition-based maintenance) (Lebold and Thurston, 2001), have not yet delivered protocols and service interfaces that are well suited for the integration of systems on a large scale.

These application scenarios could benefit from the experience in the realization of truly open and easily integrated systems developed in the Internet and web applications market. Web technologies are the de facto standard in the realization of interoperable distributed applications. Web services are the basis for service-oriented architectures that enable easily dealing with heterogeneity in the hardware and software platforms as well as the realization of complex services through the composition of simpler ones (Zhu, 2003).

In addition, web services are evolving beyond SOAP-based solutions to encompass ReST-based solutions that are easier to integrate and significantly more scalable and available on low-power devices. Solutions for the development of distributed systems originally available only for PC-class machines are now available for embedded devices (Belqasmi, Glitho, and Fu, 2011).

The modern web-based software-development technologies for distributed and embedded applications represent an ideal basis for the realization of complex management functions, such as diagnostics and prognostics.

12.4.3 Data-Intensive Real-Time Processing Techniques

The recent evolution of data-intensive real-time processing techniques enables the processing of large quantities of data and to highlight nontrivial (cor)relations among them. This is of critical importance for large-scale e-maintenance platforms, which have to process concurrently the maintenance data received from a high number of automated machines for diagnostics and prognostics purposes.

In these kinds of applications, the large scale of data to process forces the development of methods that realize smart and efficient data storage and data processing. In addition, there is the need to adopt scalable data-analysis solutions that can perform diagnostic and prognostic procedures for each automated machine, comparing the maintenance data with those coming from machines operating in similar conditions (lifecycle, installation conditions, etc.).

To this end, e-maintenance platforms can leverage modern data-fusion techniques developed in sensor network applications (Mitchell, 2012) and can take advantage of modern data-mining techniques, which enable the detection of nontrivial associations in large data sets (Reshef et al., 2011) and the performance of queries on data collected from multiple sources of both a structured as well as an unstructured type (Freitas et al., 2012). Finally, for the scalable execution of computationally intensive data-processing tasks, e-maintenance platforms can exploit the elastic and parallel computations provided by cloud-based systems (Sakr et al., 2011).

Many of the above-mentioned techniques and tools were developed in the context of the "big data" application field (Alexander, Hoisie, and Szalay, 2011), which has recently attracted a lot of interest from both the industry and the academy. The term "big data" encompasses all those issues that arise from the analysis of extremely large quantities of data, typically found in many modern applications, such as in the business intelligence and in the scientific computing fields (Bryant, 2011; Ahrens et al., 2011).

12.4.4 Ubiquitous and Large-Bandwidth Internet-Based Connectivity

The widespread diffusion of Internet connectivity significantly facilitates the deployment of large-scale e-maintenance platforms, which need to provide remote access to a high number of machines installed in a wide array of locations.

To this end, it is often possible to leverage a preexisting Internet connectivity infrastructure at the customer's premises. This enables the easy provision of remote access to appliances, usually with high-bandwidth connectivity and at negligible costs. However, some locations might not have wired access to the Internet or might not allow the exploitation of it. For instance, in some situations, it may be advisable not to convey maintenance data over the LAN installed at the customer's premises for security reasons. In those cases, mobile connectivity can be exploited instead.

Let us notice that, in the context of mobile communications, e-maintenance solutions can take advantage of commercial offers originally designed for the machine-to-machine (M2M) market (Katusic et al., 2012).

12.5 e-Maintenance Platforms for Large Scale

The realization of e-maintenance platforms for this class of machines presents some peculiar challenges because they have to deal with large-scale installations, heterogeneous machines, and network connectivity and discontinuous transmission of maintenance data.

More specifically, e-maintenance platforms in charge of controlling automated machines on a large scale need to provide a basic set of functions to support the main operations required. First of all, these machines usually operate unattended with no onsite technical support personnel available. It is therefore of utmost importance to minimize failures by adopting proactive maintenance practices. In fact, the ability to detect upcoming failures early is essential in order to reduce operation downtimes, maintenance costs, and safety hazards of appliances and therefore to preserve revenues.

To achieve this goal, a large-scale e-maintenance platform typically implements four main functions: *automated monitoring*, *self-diagnostics*, *prognostics*, and *remote management* (Muller, Crespo-Màrquez, and Iung, 2008). In addition, many of these platforms are capable of integrating with the enterprise information systems to generate value-added services, for instance, by providing the machines' production data to their customers (Lazzarini et al., 2013). Finally, large-scale e-maintenance platforms are typically designed to address the security concerns raised by the remote management of machines.

12.5.1 Automated Monitoring

The automated monitoring function is very important for any large-scale e-maintenance platform. In fact, the availability of highly informative maintenance data is essential in order to accurately capture the current state of the machines, to enable the scheduling of field service interventions only when really needed, and to realize periodical and/or on-demand (when applicable or appropriate) reporting of the operational state (including incipient and conditional failure notifications).

The proactive management of household and similar appliances demands the monitoring, as closely and as continuously as possible, of a potentially large number of variables. In fact, the monitoring should not only consider the variables that enable the evaluation of the degree of equipment degradation, but also those that might enable the detection of the presence of hidden problems before significant degradation takes place.

The automated monitoring function is in charge of collecting maintenance data from the machines, of predisposing them for further processing through the application of aggregation and manipulation procedures, and of enabling their analysis by both automated diagnostic and prognostic components as well as technical support personnel through a dedicated management interface.

More specifically, an e-maintenance platform typically implements an interface that enables the synoptic visualization of the machine state (programming table, firmware version, analysis of alarms, diagnostics of failures, events, machine access log), adopting the proper graphical and time-varying representation for critical metrics (pressure values, temperature values, etc.) and enabling the monitoring of values in real time for a live monitoring session.

In addition, e-maintenance platforms usually provide advanced reporting mechanisms that automate and facilitate the machine-monitoring processes, for example, sending to each technical support employee a daily report about the state of the machines under his responsibility, etc.

12.5.2 Self-Diagnostics and Prognostics

The self-diagnostic and prognostic functions of large-scale e-maintenance platforms realize both the automated fault diagnosis (usually implemented directly on the machine side) as well as the comparative analysis of data collected from all the machines in order to detect deviations, anomalous operations, or conditions favorable to the fault.

The early detection of anomalies and the forecast of machinery health based on the collected maintenance data are essential to enable proactive maintenance as they represent the best way to reduce both machine downtime and maintenance costs. As a result, the implementation of diagnostic and prognostic functions through online or just-in-time data analysis techniques that enable the real-time processing of the collected maintenance data is highly desirable.

Sophisticated prognostic techniques, together with ongoing advances in data-fusion techniques, and increasingly comprehensive databases of condition data hold promise for improved design of new machines and adaptive strategies of maintenance management in real-world scenarios.

It is also vital to perform timely notifications in case of unexpected behaviors detected in one ore more machines. To this end, an e-maintenance platform typically exploits a messaging infrastructure to notify the technician who is in charge of its maintenance in a timely manner.

12.5.3 Remote Management

Large-scale e-maintenance platforms also provide a remote management function, which is essential in order to implement a proactive maintenance strategy. In fact, this function allows the establishment of on-demand live monitoring and diagnostic sessions to help technicians quickly identify the cause of deviations and abnormal operations on the machines under their responsibility and in deciding how to proceed for their resolution. In addition, these platforms typically provide a remote reconfiguration function to enable the automation of firmware and configuration upgrades for the machines.

Generally, the remote management functions provided by large-scale e-maintenance platforms are specifically designed to allow nonintrusive measurements and analyze machine operation and production data in real time from different locations, thus supporting the repair companies and the staff personnel working on the field.

The remote management function enables not only the minimization of the time between the notification of failure and the beginning of the resulting service intervention, but also to plan and optimize the work time of the highly skilled specialist technicians, a critically important asset. Traditional maintenance strategies, exclusively based on onsite field service, would instead force expert technicians to spend most of their time traveling to and from the machine site locations within large geographic areas (Holmberg et al., 2010).

12.5.4 Security

Finally, let us notice that the remote management of machines raises significant security concerns. In fact, maintenance data is an important industrial secret that has become the objective of a large (and growing) number of malicious attacks (Cai, Wang, and Yu, 2008). As a result, there is the need to protect confidentiality of data transfers between appliances and external monitoring stations. It is also necessary to prevent malicious access to the machines from the communication link to avoid the unauthorized changing of critical configuration parameters of the manufacturing process and, consequently, malfunctions (and potentially damage) of machines.

These considerations suggest the opportunity to adopt a *secure-by-design* software-engineering practice when realizing e-maintenance platforms. Designing the system to be secure from the ground up is essential to hindering malicious attacks and to minimizing the impact of security vulnerabilities.

**THE TEOREMA CASE STUDY: e-MAINTENANCE
FOR ICE CREAM–MAKING MACHINES**

The considerations presented in this chapter emerge from the authors' experience in building an innovative e-maintenance platform, called Teorema, for the management of the ice cream machines produced by Carpigiani (one of the leading industries in the market). Carpigiani's ice cream–making machines are usually installed in many locations within large geographic areas; thus onsite maintenance interventions can be very expensive in terms of both machine downtimes (and consequently loss of revenues) and significant travel costs for the technical support personnel. As a result, the after-sales maintenance process is particularly expensive in the ice cream–making machines market, and the deployment of advanced and innovative e-maintenance platforms could bring a significant reduction in costs.

Project Teorema is already in production with 4550 remotely controlled machines installed in more than 20 countries around the world. Overall, the adoption of Teorema led to very significant savings. In the last months, around 20% of the assistance interventions for Teorema-controlled machines have been performed remotely. This has brought a reduction in maintenance costs of almost 25%, and we estimate that, in the long run, reductions could stabilize around 40%. Finally, from the Carpigiani's customers' perspective, the prompt improvement in maintenance intervention enabled by Teorema translated into much smaller machine downtimes and, therefore, significantly decreased loss of revenues due to faults.

Finally, Teorema provides Carpigiani customers with a reporting service for the real-time access to the production data of their machines, for example, total ice cream dispensed, daily ice cream dispensed by flavor, etc., which represent a disruptive innovation in after-sales services within the ice cream–machines market.

Teorema adopts a centralized client/server web architecture, depicted in Figure 12.1, based on a central monitoring and control station and on machine-installed remote monitoring kits that operate as clients. The central monitoring and control station, called the Teorema server, represents the core of the e-maintenance platform as it collects, stores, analyzes, and provides reports on maintenance data. The remote-monitoring kits transfer maintenance data to the Teorema server and enable the remote control of Carpigiani's ice cream machines. The kits communicate with the Teorema server through the Internet, leveraging one of the supported connectivity media to access the network.

Carpigiani technical support personnel and customers can access Teorema information and services anywhere, anytime, and from a wide range of devices with different screen sizes, CPU capabilities, and input interfaces, for example, laptops, netbooks, PDAs, and smartphones. In fact, Teorema provides a web-based, multi-modal interface that tailors data visualization according to the characteristics of the user terminal device.

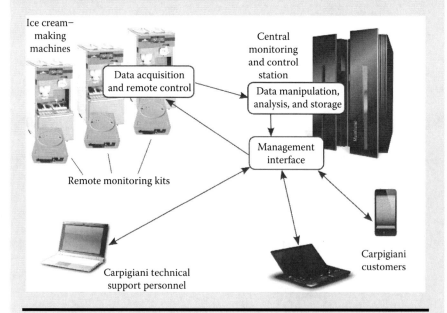

Figure 12.1   Main components of the Teorema platform and their interactions.

TEOREMA OPERATIONS

Teorema provides several services, which include automatic notification of malfunctions; synoptic visualization of maintenance data; automated reporting for monitoring and diagnostic purposes; prognostic functions to identify machines and components that are likely to fail; and remote assistance interventions, such as interactive diagnostic analysis, reconfigurations, and firmware updates.

The functions provided by Teorema as well as the technologies adopted for their implementation are proving robust and effective. Teorema-controlled machines have already proven the importance of keeping the machines under real-time control, promptly identifying faults, and in some cases, even identifying worn-down components and machines that are likely to break in the near future.

Remote Management

The remote assistance functions provided by Teorema have also allowed the performance of several remote maintenance operations. For instance, recently a UK-based customer requested the customization of about 2000 already installed Carpigiani ice cream machines. Using Teorema, the Carpigiani technical personnel managed to accomplish this task by simply upgrading the machine firmware online. The Teorema remote reconfiguration functions have also enabled the adoption of new, significantly facilitated machine installation and configuration procedures. In fact, machines now do not need to be preconfigured in the factory or onsite but are simply installed and configured online. In addition, in the last few months, we have performed hundreds of remote reconfigurations of Teorema-controlled machines to tune their working parameters, for example, scheduling of the pasteurization process, according to their specific deployment location.

Diagnostics

The Teorema diagnostic functions enable Carpigiani technicians to perform many assistance operations remotely. More specifically, Teorema allows technical support personnel to trigger real-time on-demand monitoring sessions on a specific ice cream–making machine, thus enabling them to retrieve the desired maintenance information and perform remote assistance interventions via a handy and ubiquitously accessible Web interface. In case an onsite intervention is necessary, technicians can leverage on Teorema-provided information to identify and carry with them the required spare parts, thus reducing the need for further onsite operations. In addition, real-time access to diagnostic information allows users to make sure that ice cream machines do not experience costly downtimes due to improper or untimely manage-ment by operating personnel, for example, unjustified delays in the loading of ingredients or in the cleaning of the machine. For instance, in a recent maintenance intervention, a Carpigiani technician received an alarm from a Teorema-enabled machine and managed to trace the source of a problem to an obstruction in the ventilation grid—a misuse that could have led to serious damage to the machine. The technician immediately contacted the customer, who confirmed that there was a folded cardboard box leaning on the back of the machine and removed it, thereby fixing the problem.

Prognostics

The Teorema prognostic functions have also proven capable of detecting anomalous situations and of identifying which components are likely to break in underperforming machines, especially those involved in the cooling process. For instance, in one machine, Teorema detected the upcoming fail-ure of the beater blades component by identifying a negative trend in the ratio between the number of served ice cream cones and the number of working hours for the refrigeration compressors.

Teorema prognostic information has enabled Carpigiani to adopt a pro-active maintenance strategy, which allows the optimal scheduling of onsite assistance interventions according to the machines' current operating condi-tions with significant reductions of machine downtimes. The remote detec-tion of the type of assistance intervention required for failed components allows the optimization of (in terms of specialized personnel and spare parts) onsite assistance interventions, which now are better planned and scheduled. In addition, Teorema has enabled a more effective use of work time from expert technicians, a critically scarce resource, who now can be much more productive as they do not have to spend most of their time traveling to gain onsite access to the ice cream machines. The ubiquitous web-based access to all maintenance data also facilitates the cooperation of technical support per-sonnel operating in the field with the experts of the Carpigiani management division working at the Carpigiani HQ.

12.6 Other Remote Control Systems

For their peculiar characteristics, large-scale e-maintenance platforms for auto-mated appliances present significant similarities to smart utility networks (Sum et al., 2011). Smart utility networks are telemetry systems for the efficient manage-ment of energy and other utilities. While these systems share similar remote moni-toring concerns with e-maintenance platforms, research on smart utility networks focuses on the realization of wireless mesh networks enabling access to electronic devices. Instead, e-maintenance platforms leverage a preexisting Internet connectivity infrastructure to provide access to remotely monitored machines and realize much more sophisticated services, such as the visualization of synoptical data and anomaly or fault detection based on complex event analysis.

Some aspects of the Internet-of-Things scenario, such as crowd-sensing (Ganti, Fan, and Hui, 2011), are also similar to the monitoring function of e-maintenance platforms for household and similar appliances. However, the Internet of Things focuses on the design and development of interoperable architectures that enable the collection of data from remote and extremely heterogeneous devices, usually through the realization of semantic service infrastructures (Pfisterer et al., 2011) or the deployment of M2M communication solutions on top of existing service platforms (Foschini et al., 2011). Instead, operating with machines of a single vendor, e-maintenance platforms typically have less strict device heterogeneity issues to address, but they have to deal with the coordination of maintenance data transmission, enabling users to choose which specific data to obtain and to receive it in real time. In addition, e-maintenance platforms have to provide advanced distributed management functions, such as remote device reconfiguration and automated diagnostics and prognostics.

M2M communications represent a similar application to e-maintenance, of which, to some extent, they could be considered a precursor. In fact, M2M aims at realizing distributed monitoring of devices, mostly leveraging on narrow-bandwidth GSM-based connectivity (Wu et al., 2011). The business opportunities in the M2M market that promised to enable the monitoring of electronic devices on a global scale leveraging GSM-based text-messaging solutions led many mobile operators to offer (relatively) cheap specialized data-only mobile contracts. Large-scale e-maintenance platforms can benefit from these offers to realize support for the remote monitoring and assistance of automated machines.

12.7 Evolution of Automated Machines

The considerations presented in previous sections enable the envisioning of a future scenario in which automated machines are connected through a monitoring network and a significant portion of maintenance interventions is performed remotely. Automated machines would then become an adaptive and resilient system of interconnected, self-monitoring, and self-managing nodes.

12.7.1 Next-Generation Automated Machine Design

The application of e-maintenance platforms for the large-scale remote monitoring and control of automated machines has a disruptive potential that could have a considerable impact on the way in which these machines are designed and maintained.

In fact, the design of automated machines could change significantly as a result of the integration of the (relatively) advanced ICT support required to realize e-maintenance functions. It is possible to envision that in next-generation automated machines the software/firmware will not only be capable of controlling the transitions between the steps of the (preconfigured) production process workflow implemented by the machine, as in most of the machines currently produced, but that it will enable a much finer-grained control over the entire production processes. Next-generation machines will expose a much larger portion of their internal state to the control software/firmware and enable the software/firmware not only to change from one stage to the next of a preconfigured manufacturing workflow, but *to change the workflow itself.*

More specifically, the software components running in automated machines would have access to a much larger number of observable variables (or sensors) as well as to a larger number of controllable variables (or actuators). This would make it technically possible to switch to a different manufacturing process workflow, enabling the implementation of customer-specific processes (for instance, the customization of machines to implement specific thermal cycles required by signature recipes is very important in the ice cream–making machines market). Machines would therefore change from being *mildly controllable* to being *highly controllable.*

From the maintenance point of view, the fault detection and the fault recovery components might cooperate to enable the machines to autonomously switch to a less efficient but more robust one that would work with slightly deteriorated components in case of failures.

This would also represent a significant step toward *mass product customization* in the household and similar appliances market. The adoption of e-maintenance solutions would foster the mass customization process, which is already reshaping some markets, such as the automotive one (Shamsuzzoha and Helo, 2009), in the household and similar appliances market. In fact, e-maintenance would enable manufacturers to produce identical machines for all the customers from the hardware point of view and to reconfigure their software and firmware on demand to implement the specific operations and the manufacturing processes required by each specific customer even at the single-machine level.

Another interesting evolution in automated machine design might be the possible *adoption of overengineering and/or overprovisioning practices.* In fact, as e-maintenance enables significantly cheaper after-sales assistance compared with traditional maintenance practices, for next-generation automated machines overengineering design approaches, such as the inclusion of redundant components and the automated fallback to secondary components when primary ones fail, might be more economically convenient than a traditional design that would lead to forced costly downtime in case of failures.

Finally, let us notice that the software/firmware is becoming, more and more, the most valuable part of the machine. This is likely to raise even more the problem of preserving intellectual propriety in the near future.

12.7.2 Evolution of Maintenance Data-Processing Techniques

We can also envision an evolution of the maintenance data-analysis practices. More specifically, monitoring tools for next-generation automated machines will be able to analyze and process a larger array of data streams, of significantly different type and nature, through sophisticated anomaly-detection techniques. This will enable the extrapolation of nontrivial features and ultimately build a deeper knowledge of the machine internal state, which could help in developing sophisticated diagnostic and prognostic tools.

In fact, with the increase in computational and bandwidth capabilities and the availability of more and more sophisticated sensors, e-maintenance platforms will be able to collect and analyze a significantly larger amount of maintenance data, which could be processed locally or at a central monitoring station according to the specific requirements of the monitoring process and of the data characteristics. For instance, accelerometer sensors, which generate a large amount of data that is computationally very expensive to analyze, as a result, are currently mounted only on large plant machines, and they could be installed even on low-end automated machines to monitor internal processes. At the same time, it is possible to hypothesize that the evolution of sensors could enable significantly more sophisticated process-monitoring solutions, for example, based on biological sensor data acquisitions.

In addition to a larger amount of data, next-generation monitoring solutions will also have to deal with a larger number of data types to analyze. In fact, in most cases, the maintenance data currently analyzable consists of discrete data sets, also known as "value type" monitoring data in the maintenance literature (Jardine, Lin, and Banjevic, 2006), which essentially limits the monitorable quantities to temperature, pressure, etc. Discontinuous value-type data are significantly more difficult to handle than continuous data, for which a large number of well-tested and sophisticated tools (based, for instance, on Fourier analysis, wavelet analysis, time-series analysis, etc.) is available, and present some limits to the applicability of automated anomaly-detection mechanisms for real-time monitoring systems (Chandola, Banerjee, and Kumar, 2009).

In addition, many models of automated machines also record a list of events, including events related to the normal operating conditions of the machines, such as commands issued by the machine operators, completion of manufacturing process, etc., and transfer them to the central management station. This information, also known as "event data" in the maintenance literature, is used to realize higher-order detection of malfunctions or incorrect machine usage through complex event-correlation techniques (Martin-Flatin, Jakobson, and Lewis, 2007).

However, let us point out that the processing of these types of maintenance data, which requires the correlation between value type, continuous time-series, and event data streams, might require significant advances of data analysis techniques, thus calling for both theoretical and applied research.

12.7.3 Leveraging on e-Maintenance Data to Optimize Maintenance Processes and Machine Design

The data collected by e-maintenance platforms will surely become a strategic asset for automated machine manufacturing enterprises, one that can be leveraged to guide and improve both the maintenance processes as well as the design of future machines and components.

In fact, the data collected from e-maintenance platforms represents an invaluable source of information for the restructuring of maintenance processes toward more efficient practices. For instance, the analysis of e-maintenance data could foster the adoption of efficient management practices for expert technical support personnel, which represents a critical issue when dealing with large installations of automated machines. In fact, despite the adoption of sophisticated practices, such as predictive maintenance (Mobley, 2002) and CBM (Jardine, Lin, and Banjevic, 2006), the domain expertise of human personnel has always proved to be a critical asset to enable effective maintenance decisions (Prakash, 2006; Veldman, Klingenberg, and Wortmann, 2011). Unlike large machinery and factory production shop floors, household and similar appliances operate without the presence of technical support personnel, and their maintenance interventions typically require relatively long waiting times between malfunction notifications and the arrival of technicians onsite. It is therefore essential for e-maintenance platforms to provide functions that allow the exploitation of, at best, the expertise of the most skilled technicians, for instance, by enabling the performing of assistance interventions remotely and the building and sharing of knowledge within maintenance teams (Ribeiro, Barata, and Silvério, 2008). These considerations also suggest the opportunity to exploit groupware tools to facilitate the collaboration between technicians in dealing with maintenance operations (Hedjazi and Zidani, 2011).

In addition, the provisioning of e-maintenance data to customers opens up the way to new business models. In fact, this could foster the adoption of performance-based or "power by the hour" contracting, which is becoming more and more common for capital-intensive products in the defense and air transport industries (Kim, Cohen, and Netessine, 2007) and also in the household and similar appliances market. In performance-based contracting, customers do not purchase a product but instead subscribe to a service that warrants the availability of the above-mentioned product (or an equivalent one) in fully operating conditions for a prearranged time interval. The service-oriented business models, of which performance-based contracting represents an important example, has proven to be a very effective way to align a company's offer to its customers' needs (Oliva and Kallenberg, 2003). In this context, e-maintenance solutions, which enable the efficient monitoring and servicing of machines for both management and production, might increase the attractiveness of performance-based contracting and help companies currently adopting a strictly product-based business model to proceed toward a more service-oriented offer.

Finally, the data collected by e-maintenance platforms is also very important to help the evolutionary design of machine components. In fact, by analyzing monitoring data collected from machines, engineers can extrapolate invaluable information. This knowledge could significantly help during the design phase by enabling engineers to perform accurate assumptions about the operating conditions that components will have to face once deployed as well as to avoid over- or underspecifications of the components that could lead to higher manufacturing or maintenance costs.

12.7.4 Deployment Considerations

Along with the trend that is pushing the intelligent management substrate in the machines, one could envision future household and similar appliances that operate (semi-)autonomously, possibly evolving into groups that automatically deal with failures and anomalies. For instance, upon failure, machines could automatically re-dispatch suspended manufacturing processes to other similar machines within the same premises. Or, in case of anomalous readings, co-located machines could exchange maintenance data to assess if the anomaly is due to an environmental factor.

To support the realization of self-management functions at the machine group level, it is possible to think of deploying local monitoring stations that could perform the analysis of maintenance data from all the machines in the premises. This could facilitate maintenance data-analysis operations, which could comparatively evaluate data coming from machines operating in the same environmental conditions and make the e-maintenance platform more resilient to network disconnections, thus effectively implementing self-managed systems that can withstand disconnections from the central remote operations management facility of their e-maintenance platform.

12.8 Conclusions

The recent ICT advances support and foster the realization of large-scale e-maintenance solutions for the remote monitoring and assistance of household and similar appliances. The decrease in costs will soon enable the realization of e-maintenance platforms, which are now affordable (at least) for higher-end appliances, such as food-processing machines, also on lower-end appliances.

The experience with Teorema demonstrates that large-scale e-maintenance platforms are already mature for the adoption in mass-produced equipment. We expect that, in the future, e-maintenance solutions, such as Teorema, will be investigated and put in place for other household and similar appliance markets.

References

Ahrens, J. P., Hendrickson, B., Long, G., Miller, S., Ross, R., and Williamson, D. (2011), "Data-Intensive Science in the US DOE: Case Studies and Future Challenges," *Computing in Science & Engineering*, Vol. 13, No. 6, pp. 14–24.

Alexander, F., Hoisie, A., and Szalay, A. (2011), "Big Data," *Computing in Science & Engineering*, Vol. 13, No. 6, pp. 10–12.

Belqasmi, F., Glitho, R., and Fu, C. (2011), "RESTful Web Services for Service Provisioning in Next-Generation Networks: A Survey," *IEEE Communications Magazine*, Vol. 49, No. 12, pp. 66–73.

Brunello, G. (2003), "Microprocessor-Based Relays – an Enabler to SCADA Integration," *Electricity Today*, Vol. 2003, No. 4, pp. 10–11.

Bryant, R. E. (2011), "Data-Intensive Scalable Computing for Scientific Applications," *Computing in Science & Engineering*, Vol. 13, No. 6, pp. 25–33.

Cai, N., Wang, J., and Yu, X. (2008), "SCADA System Security: Complexity, History and New Developments," in *Proceedings of 6th IEEE International Conference on Industrial Informatics (INDIN 2008)* in Daejeon, Korea, 13–16 July 2008, IEEE, New York, pp. 569–574.

Campos, J. (2009), "Development in the Application of ICT in Condition Monitoring and Maintenance," *Computers in Industry*, Vol. 60, No. 1, pp. 1–20.

Chandola, V., Banerjee, A., and Kumar, V. (2009), "Anomaly Detection: A Survey," *ACM Computing Surveys*, Vol. 41, No. 3, pp. 15:1–15:48.

Chryssolouris, G., Mavrikios, D., Papakostas, N., Mourtzis, D., Michalos, G., and Georgoulias, K. (2009), "Digital Manufacturing: History, Perspectives, and Outlook," *Proceedings of the Institution of Mechanical Engineers, Part B: Journal of Engineering Manufacture*, Vol. 223, No. 5, pp. 451–462.

Crespo-Màrquez, A. (2007), *The Maintenance Management Framework*, Springer-Verlag, London.

Foschini, L., Taleb, T., Corradi, A., and Bottazzi, D. (2011), "M2M-Based Metropolitan Platform for IMS-Enabled Road Traffic Management in IoT," *IEEE Communications Magazine*, Vol. 49, No. 11, pp. 50–57.

Freitas, A., Curry, E., Oliveira, J. G., and O'Riain, S. (2012), "Querying Heterogeneous Datasets on the Linked Data Web – Challenges, Approaches, and Trends," *Internet Computing*, Vol. 16, No. 1, pp. 24–33.

Ganti, R. K., Fan, Y., and Hui, L. (2011), "Mobile Crowdsensing: Current State and Future Challenges," *IEEE Communications Magazine*, Vol. 49, No. 11, pp. 32–39.

Garcia, M. C., Sanz-Bobi, M. A., and del Pico, J. (2006), "SIMAP: Intelligent System for Predictive Maintenance: Application to the Health Condition Monitoring of a Windturbine Gearbox," *Computers in Industry*, Vol. 57, No. 6, pp. 552–568.

Hedjazi, D., and Zidani, A. (2011), "Development of an Industrial e-Maintenance System Integrating Groupware Techniques," *International Journal of Industrial and Systems Engineering*, Vol. 9, No. 2, pp. 227–247.

Holmberg, K., Adgar, A., Arnaiz, A., Jantunen, E., Mascolo, J., and Mekid, S. (Eds.) (2010), *E-maintenance*, Springer-Verlag, London.

Jantunen, E., Emmanouilidis, C., Arnaiz, A., and Gilabert, E. (2011), "E-Maintenance: Trends, Challenges and Opportunities for Modern Industry," in *Proceedings of the 18th IFAC World Congress*, Elsevier, Amsterdam, Holland, pp. 453–458.

Jardine, A., Lin, D., and Banjevic, D. (2006), "A Review on Machinery Diagnostics and Prognostics Implementing Condition-Based Maintenance," *Mechanical Systems and Signal Processing*, Vol. 20, No. 7, pp. 1483–1510.

Katusic, D., Weber, M., Bojic, I., Jezic, G., and Kusek, M. (2012), "Market, Standardization, and Regulation Development in Machine-to-Machine Communications," in *Proceedings of 20th International Conference on Software, Telecommunications and Computer Networks (SoftCOM 2012)*, IEEE, New York, USA, pp. 1–7.

Kim, S.-H., Cohen, M., and Netessine, S. (2007), "Performance Contracting in After-Sales Service Supply Chains," *Management Science*, Vol. 53, No. 12, pp. 1843–1858.

Kunze, U. (2003), "Condition Telemonitoring and Diagnosis of Power Plants Using Web Technology," *Progress in Nuclear Energy*, Vol. 43, No. 1–4, pp. 129–136.

Lazzarini, R., Stefanelli, C., Tortonesi, M., and Virgilli, G. (2013), "E-Maintenance for Household and Similar Appliances," *International Journal of Productivity and Quality Management*, Vol. 12, No. 2, pp. 141–160.

Lebold, M., and Thurston, M. (2001), "Open Standards for Condition-Based Maintenance and Prognostic Systems," in *Proceedings of 5th Annual Maintenance and Reliability Conference (MARCON 2001)*, Gatlinburg, USA.

Lee, J., Ni, J., Djurdjanovic, D., Qiu, H., and Liao, H. (2006), "Intelligent Prognostics Tools and e-Maintenance," *Computers in Industry*, Vol. 57, No. 6, pp. 476–489.

Levrat, E., Iung, B., and Crespo-Màrquez, A. (2008), "E-Maintenance: Review and Conceptual Framework," *Production Planning & Control*, Vol. 9, No. 4, pp. 408–429.

Martin-Flatin, J. P., Jakobson, G., and Lewis, L. (2007), "Event Correlation in Integrated Management: Lessons Learned and Outlook," *Journal of Network and Systems Management*, Vol. 15, No. 4, pp. 481–502.

Mitchell, H. B. (2012), *Data Fusion: Concepts and Ideas*, 2nd Edition, Springer-Verlag, Berlin, Germany.

Mobley, R. K. (2002), *An Introduction to Predictive Maintenance*, 2nd Edition, Butterworth-Heinemann, Boston, USA.

Moubray, J. (1997), *Reliability-Centered Maintenance*, 2nd Edition, Industrial Press Inc., New York, USA.

Muller, A., Crespo-Màrquez, A., and Iung, B. (2008), "On the Concept of e-Maintenance: Review and Current Research," *Reliability Engineering and System Safety*, Vol. 93, No. 8, pp. 1165–1187.

Nakajima, S. (1988), *Introduction to TPM: Total Productive Maintenance*, Productivity Press, Cambridge, MA, USA.

Oliva, R., and Kallenberg, R. (2003), "Managing the Transition from Products to Services," *International Journal of Service Industry Management*, Vol. 14, No. 2, pp. 160–172.

Pfisterer, D., Romer, K., Bimschas, D., Kleine, O., Mietz, R., Truong, C., Hasemann, H., Kröller, A., Pagel, M., Hauswirth, M., Karnstedt, M., Leggieri, M., Passant, A., and Richardson, R. (2011), "SPITFIRE: Toward a Semantic Web of Things," *IEEE Communications Magazine*, Vol. 49, No. 11, pp. 40–48.

Prakash, O. (2006), "Asset Management through Condition Monitoring — How It May Go Wrong: A Case Study," in *Proceedings of 1st World Congress on Engineering Asset Management (WCEAM 2006)*, Springer, New York, USA.

Reshef, D., Reshef, Y., Finucane, H., Grossman, S., McVean, G., Turnbaugh, P., Lander, E., Mitzenmacher, M., and Sabeti, P. (2011), "Detecting Novel Associations in Large Data Sets," *Science*, Vol. 334, No. 6062, pp. 1518–1524.

Ribeiro, L., Barata, J., Silvério, N. (2008), "A High Level e-Maintenance Architecture to Support On-Site Teams," *Enterprise and Work Innovation Studies*, Vol. 4, IET, pp. 129–138.

Sakr, S., Liu, A., Batista, D. M., and Alomari, M. (2011), "A Survey of Large Scale Data Management Approaches in Cloud Environments," *IEEE Communications Surveys & Tutorials*, Vol. 13, No. 3, pp. 311–336.

Shamsuzzoha, A. H. M., and Helo, P. T. (2009), "Reconfiguring product development process in auto industries for mass customization," *International Journal of Productivity and Quality Management*, Vol. 4, No. 4, pp. 400–417.

Sum, C.-S., Harada, H., Kojima, F., Lan, Z., and Funada, R. (2011), "Smart Utility Networks in TV White Space," *IEEE Communications Magazine*, Vol. 49, No. 7, pp. 132–139.

Veldman, J., Klingenberg, W., and Wortmann, H. (2011), "Managing Condition-Based Maintenance Technology," *Journal of Quality in Maintenance Engineering*, Vol. 17, No. 1, pp. 40–62.

Voisin, A., Levrat, E., Cocheteux, P., and Iung, B. (2010), "Generic Prognosis Model for Proactive Maintenance Decision Support: Application to Pre-industrial e-Maintenance Test Bed," *Journal of Intelligent Manufacturing*, Vol. 21, No. 2, pp. 177–193.

Wan Mahmood, W. H., Ab Rahman, M. N. A., Deros, B. M., and Mazli, H. (2011), "Maintenance Management System for Upstream Operations in Oil and Gas Industry: A Case Study," *International Journal of Industrial and Systems Engineering*, Vol. 9, No. 3, pp. 317–329.

Wu, G., Talwar, S., Johnsson, K., Himayat, N., and Johnson, K. (2011), "M2M: From Mobile to Embedded Internet," *IEEE Communications Magazine*, Vol. 49, No. 4, pp. 36–43.

Yang, Q., Barria, J. A., and Green, T. C. (2011), "Communication Infrastructures for Distributed Control of Power Distribution Networks," *IEEE Transactions on Industrial Informatics*, Vol. 7, No. 2, pp. 316–327.

Zhu, J. (2003), "Web Services Provide the Power to Integrate," *IEEE Power and Energy Magazine*, Vol. 1, No. 6, pp. 40–49.

ORTHOGONALLY ENABLING CAPABILITIES

Chapter 13

Using Planning to Adapt to Dynamic Environments

Austin Tate

Contents

13.1 Introduction

Planning is about much more than solving specifically stated problems whereby some goal state is reached from some initial state as efficiently as possible. The real world is a messy place. The current state of the world may be only partially known or observable. The goal, objective, or mission itself may be imprecisely stated, and the agents available to carry out the activities involved may be only partially specified. The model we have of the state, objectives, and agent capabilities may be imperfect. People and systems often should work in harmony as a team to solve problems and accommodate the roles, capabilities, and preferences of the various agents. The real world is also dynamic and changing—the state of the environment, the objectives, and the agents or their capabilities can be in a dynamic state of flux.

Artificial Intelligence (AI) planning and knowledge-rich plan representation techniques have been developed to generate, refine, and adapt plans in highly dynamic situations to provide resilience. They seek to address some of the real messy problems in the world.

Realistic planning systems must allow users and computer systems to cooperate and work together using a "mixed initiative" style. Black box or fully automated solutions are not acceptable in many situations. Studies of expert human problem solvers in stressful or critical situations (Klein, 1998) show that they share many of the problem-solving methods employed by some of the methods studied in AI planning to address these issues (Tate, 2000b).

There is also a need to model domains in which planning takes place, understanding the roles and capabilities of the various human and system agents involved in the planning process and in the domain in which plans are executed, and allow for communication of information about tasks, plans, intentions, and effects between those agents.

This paper argues that a hierarchical task network (HTN) least commitment planning and plan refinement approach—as used for many years in practical planning systems, such as NOAH (Sacerdoti, 1977), Nonlin (Tate, 1977), SIPE (Wilkins, 1988), O-Plan (Currie and Tate, 1991; Tate et al., 1994b) and SHOP (Nau et al., 2005)—provides an intelligible framework for mixed-initiative multi-agent human/system planning environments. When joined with a strong underlying constraint-based plan representation it can provide a framework in which powerful problem solvers, based on search and constraint reasoning methods, can be employed to work in highly dynamic situations and still retain human intelligibility.

I-Plan and its underlying <I-N-C-A> (Issues–Nodes–Constraints–Annotations) ontology is a planner created in the I-X intelligent systems framework, which follows these principles.

13.2 Development of a Flexible AI Planning Approach

Realistic planning systems must allow users and computer systems to cooperate and work together using a "mixed initiative" style. Black box or fully automated solutions are not acceptable in many situations, in which human responsibility is paramount. Highly dynamic environments demand adaptable solutions. Studies of expert human problem solvers in stressful or critical situations show that they share many of the problem-solving methods employed by hierarchical planning methods studied in AI. But powerful solvers and constraint reasoners can also be of great help in parts of the planning process. A more intelligible approach to using AI planning is needed that can use the best "open" styles of planning based on shared plan representations and HTNs and which still allow the use of powerful constraint representations and solvers.

The field of AI planning—that is, reasoning about the activity necessary to achieve stated goals—has a long and distinguished history (Allen et al., 1990; Tate, 1996a). Notwithstanding its successes, most work is based on simplifications and unrealistic general assumptions, which restrict the application of planning algorithms to specific problems under specific conditions. These unrealistic assumptions can be summarized as follows: (a) the presence of an omniscient agent able to formulate centralized, all-encompassing plans; (b) action schemata that capture the totality of conditions under which they are applicable and of effects they bring about; and (c) an environment that is unaffected by external agency, being changed only by the projected actions in a plan.

While research into specialized algorithms has continued, often leading to notable improvements, a shift of emphasis is needed to support planning in dynamic environments and in cooperation with human planners addressing real tasks. One of the key insights is to recognize the value of AI work in the representation of plans rather than in any particular algorithm and that real planning is as much a social activity as a computational task.

This insight guided the development of the open planning architecture (O-Plan) (Currie and Tate, 1991) and its development into one of the first web-based task-support applications (Tate, 1996b; Tate et al., 2003; Tate and Dalton, 2003) and the gradual distillation and refinement of previous plan representations into the <I-N-C-A> ontology (Tate, 1998; Tate, 2003). This model can be used to describe not only plans, but also the planning process itself and, hence, to communicate aspects of this task, raising it to the level of a collaborative social activity in an approach we term intelligible planning (Tate, 2000b).

To encourage and support this shift from automated reasoning to distributed collaboration, a generic set of software tools and documentation, collectively called the I-X intelligent systems suite, has been developed (Tate et al., 2002). I-Plan is a planning system based on these principles. It is part of the I-X suite of intelligent tools. I-Plan is modular and can be extended via plug-ins of various types. It is intended to be a "lightweight" planning system that can be embedded in other applications. In its simplest form, it can provide a small personal planning aid that can be deployed in portable devices and other user-orientated systems to add planning facilities into them. In its more developed forms, it can have the power of longer-established generative HTN AI planners, such as O-Plan.

13.3 I-X: Intelligent Systems Architecture

The I-X approach has five aspects:

1. Systems integration: A broad vision of an open architecture for the creation of intelligent systems that support the "process" for the synthesis of a result or "product." It is based on a "two-cycle" approach, which uses plug-in components to "handle issues" and to "maintain the domain model."
2. <I-N-C-A> ontology: A core notion of the representation of a process or plan as a set of nodes making up the components of the process or plan model along with constraints on the relationship between those nodes. It includes a set of outstanding issues and can maintain annotations for various purposes, including rationale capture.
3. Reasoning: The provision of plug-in reasoning capabilities in the form of "issue handlers" and "constraint managers."
4. Viewers and user interfaces: To support various roles of users performing activities and to provide modules that present the state of the process they are engaged in and the status of the products they are working with.
5. Applications: Work in various application sectors that seeks to create generic approaches (I-Tools) for the various types of tasks in which users may engage. One important application is I-Plan for planning tasks (see Figure 13.1).

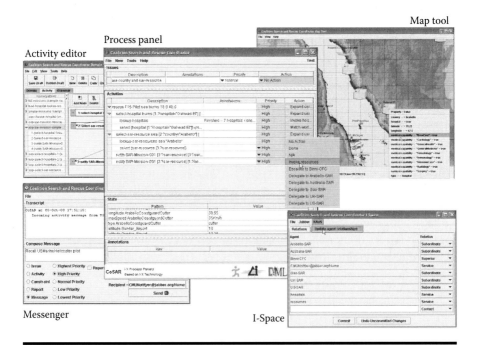

Figure 13.1   I-X task support tools.

13.4 Features of the "Intelligible Planning" Approach

There are a number of features that can encourage an approach to planning that is intelligible to the people responsible for the process and involved in planning and execution:

- Expansion of a high-level abstract plan into greater detail when necessary.
- High-level "chunks" of procedural knowledge (standard operating procedures, best practice processes, tactics, techniques, and procedures, etc.) on a human scale—typically five to eight actions—can be manipulated within the system.
- Ability to establish that a feasible plan exists, perhaps for a range of assumptions about the situation, while retaining a high-level overview.
- Analysis of potential interactions as plans are expanded or developed.
- Identification of problems, flaws, and issues with the plan.
- Deliberative establishment of a space of alternative options perhaps based on different assumptions about the situation involved, of special use ahead of time, in training and rehearsal, and to those unfamiliar with the situation or utilizing novel equipment.
- Monitoring of the execution of events as they are expected to happen within the plan, watching for deviations that indicate a necessity to re-plan (often ahead of this becoming a serious problem).
- Represent the dynamic state of the world at points in the plan and use this for "mental simulation" of the execution of the plan.
- Pruning of choices according to given requirements or constraints.
- Situation-dependent option filtering (sometimes reducing the choices normally open to one "obvious" one).
- Satisficing search to find the first suitable plan that meets the essential criteria.
- Heuristic evaluation and prioritization of multiple possible choices within the constraint search space.
- Uniform use of a common plan representation with embedded rationale to improve plan quality, shared understanding, etc.

The previously described features describe many aspects of problem-solving behavior observed in expert humans working in unusual or crisis situations (Klein, 1998). But they also describe the hierarchical and mixed-initiative approach to planning in AI developed over the last four decades.

13.5 A More Intelligible Framework for AI Planning: The I-X Approach

The I-X approach involves the use of shared models for task-directed communication between human and computer agents who are jointly exploring (via some "process") a range of alternative options for the synthesis of an artifact, such as a design or a plan (termed a "product"). It allows for two levels:

■ Outer level: human-relatable plan representations and HTN planning style for outer level.
■ Inner level: detailed search, constraint solvers, analyzers, and simulations act in this framework to provide feasibility checks, detailed constraints, and guidance.

It also provides for

■ Sharing of issues, processes, and process products between humans and systems described via <I-N-C-A>
■ Secure policy-managed communications, reporting, logging
■ Context, environment, and agent capability–sensitive option generation
■ Links between informal/unstructured outline planning and more structured, detailed planning

The I-X system or agent has two processing cycles (see Figure 13.2):

■ Handle issues
■ Respect domain constraints

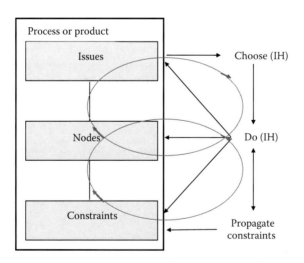

Figure 13.2   I-X approach two cycles: handle issues, propagate constraints.

An I-X system or agent carries out a (perhaps dynamically determined) process, which leads to the production of (one or more alternative options for) a synthesized artifact.

An I-X system or agent views the synthesized artifact as being represented by a set of constraints on the space of all possible artifacts in the domain.

13.6 I-Plan

The I-Plan design provides an extensible framework for adding detailed constraint representations and reasoners into planners. These can be based on powerful automated methods. But this can be done in a context that provides overall human intelligibility.

The I-Plan design is based on two cycles of processing. The first addresses one or more "issues," and the second ensures that constraints in the domain in which processing takes place are checked and respected. So the processing cycles can be characterized as "handle issues, respect constraints."

The emerging partial plan is analyzed to produce a further list of issues and added constraints. A choice of the issues to address is used to drive a workflow-style processing cycle of choosing "issue handlers" (IH) and then executing them to modify the emerging plan state. Checks are then made on the sets of constraints available to check their validity, to add further deduced constraints via propagation, and to signal any indicated or potential constraint violations. In some cases, sophisticated constraint managers can give "maybe" answers when constraints are added, giving vital information on possible fixes or alternatives for adding constraints such that the set of constraints can be made consistent again and problem solving can continue (Dalton et al., 1993; Tate et al., 1994a).

This approach is taken in systems like O-Plan, OPIS (Smith, 1994), DIPART (Pollack, 1994), TOSCA (Beck, 1993), etc. The approach fits well with the concept of treating plans as a set of constraints that can be refined as planning progresses.

Some such systems can also act in a nonmonotonic fashion by relaxing constraints in certain ways. Having the implied constraints or an "agenda" as a formal part of the plan provides an ability to separate the plan that is being generated or manipulated from the planning system and process itself, and this is used as a core part of the I-Plan design.

Mixed initiative planning approaches, for example in O-Plan (Tate, 1994), improve the coordination of planning with user interaction by employing a clearer shared model of the plan as a set of constraints at various levels that can be jointly and explicitly discussed between and manipulated by the user or system in a cooperative fashion. I-Plan adopts this approach.

13.7 <I-N-C-A>

The <I-N-C-A> model is a means to represent plans and activity as a set of constraints. By having a clear description of the different components within a plan, the model allows for plans to be manipulated and used separately from the environments in which they are generated. The underlying thesis is that plans can be represented by a set of constraints on the behaviors possible in the domain being modeled and that plan communication can take place through the interchange of such constraint information.

The <I-N-C-A> representation is intended to utilize a synergy of practical and formal approaches that are stretching the formal methods to cover realistic representations as needed for real problem solving and can improve the analysis that is possible for practical planning systems.

The <I-N-C-A> constraint model provides support for a number of different uses:

■ For automatic and mixed-initiative generation and manipulation of plans and other synthesized artifacts and to act as an ontology to underpin such use
■ As a common basis for human and system communication about plans and other synthesized artifacts
■ As a target for principled and reliable acquisition of plans, process models, and process product information
■ To support formal reasoning about plans and other synthesized artifacts

These cover both formal and practical requirements and encompass the requirements for use by both human and computer-based planning and design systems.

When first designed (Tate, 1996b), <I-N-C-A> was intended to act as a bridge to improve dialogue between a number of communities working on formal planning theories, practical planning systems, and systems engineering process management methodologies. It was intended to support new work then emerging on automatic manipulation of plans, human communication about plans, principled and reliable acquisition of plan information, and formal reasoning about plans. It has since been utilized as the basis for a number of research efforts, practical applications, and emerging international standards for plan and process representations. For some of the history and relationships between earlier work in AI on plan representations, work from the process and design communities and the standards bodies, and the part that <I-N-C-A> played in this see Tate (1998).

In Tate (2000a), the <I-N-C-A> model is used to characterize the plan representation used within O-Plan and is related to the plan refinement planning method used in O-Plan. The <I-N-C-A> work is related to emerging formal analyses of plans and planning. This synergy of practical and formal approaches can stretch the formal methods to cover realistic plan representations as needed for real problem solving and can improve the analysis that is possible for practical planning systems.

We believe the <I-N-C-A> approach is valid in design and synthesis tasks more generally; we consider planning to be a limited type of design activity.

13.8 I-X Approach and I-X Process Panels

I-X and <I-N-C-A> provide a shared intelligible, easily communicated, and extendible conceptual model for objectives, processes, standard operating procedures and plans:

- I   Issues
- N   Nodes/activities
- C   Constraints
- A   Annotations

Intelligent activity planning, execution, monitoring, re-planning, and plan repair is supported via I-Plan and I-P2 (I-X Process Panels); I-P2's aim is a workflow and messaging "catchall" and can take *any* requirement to

- Handle an issue
- Perform an activity
- Respect a constraint
- Note an annotation

I-X deals with these via

- Manual activity
- Internal capabilities
- External capabilities
- Reroute or delegate to other panels or agents
- Plan and execute a composite of these capabilities

It receives reports and interprets them to

- Understand the current status of issues, activities, and constraints
- Understand the current world state, especially the status of process products
- Help the user control the situation

It maintains the current status, models and knowledge.

It copes with partial knowledge of processes and organizations.

It uses representation and reasoning together with the state to seek to present current, context-sensitive, options for action.

It supports a mixed-initiative collaboration model of "mutually constraining things."

13.9 Applications of I-X

- Disaster planning, evacuation operations, military operations in urban terrain, search and rescue
- Rapidly-deployed coalition operations support
- Help desk support
- Computer and systems configuration
- (Multilingual) maintenance procedures aid
- Unusual and emergency procedures assistant

13.10 Comparing the Intelligible Planning Approach to Studies of Expert Human Planners in Dynamic Situations

The features of the "Intelligible Planning" approaches used in O-Plan and I-Plan and described in Section 13.4 are similar to the approaches observed in expert human problem solvers performing in dynamic, stressful, or unusual situations. These observations were made in studies over many years by Klein (1998), and he contrasts these with some automated "black box" AI and algorithmic techniques.

There are different types of planning technology available from the AI community. This is not restricted to a simple kind of search from some known initial state to some final desired state seeking the best solution according to some predefined criteria. Gary Klein's book (1998) on how people make decisions in situations such as military operations, fire fighting, or other life-threatening environments provides a rich set of case studies to show that, in relatively few situations, were deliberative planning techniques in obvious use. People just seemed to be making the "right" choices—or a choice that worked, which was all that was required. They attributed their rapid selection of a suitable course of action to training, experience, or even ESP!

They stated, "I don't plan, I just know what to do." When options were deliberated over and evaluated, the situation for those involved was novel or unusual to their previous experience.

Klein's studies show how people in stressful environments select a course of action and adapt it as circumstances alter. Many of the decisions made by the subjects relate to issues that AI planning researchers are addressing. However, they are far removed from the traditional search style of deliberative plan generation. The AI techniques we describe to address planning in dynamic environments are of much wider utility than simple, fully automated search methods. The planning requirements seen in human problem solving in dynamic situations can be mapped to some of the AI representation and reasoning methods we are bringing to bear on practical planning problems.

- Overall management of the command, planning, and control process steps to improve coordination.
- Expansion of a high-level abstract plan into greater detail when necessary.
- High-level "chunks" of procedural knowledge (standard operating procedures, best practice processes, tactics, techniques, and procedures, etc.) at a human scale—typically five to eight actions—can be manipulated within the system (Klein, p. 52 and p. 58).
- Ability to establish that a feasible plan exists, perhaps for a range of assumptions about the situation, while retaining a high-level overview (Klein, p. 227, "Include only the detail necessary to establish a plan is possible—do not fall into the trap of choreographing each of their movements").
- Analysis of potential interactions as plans are expanded or developed (Klein, p. 53).
- Identification of problems, flaws, and issues with the plan (Klein, p. 63 and p. 71).
- Deliberative establishment of a space of alternative options perhaps based on different assumptions about the situation involved, of special use ahead of time, in training and rehearsal, and to those unfamiliar with the situation or utilizing novel equipment (Klein, p. 23).
- Monitoring of the execution of events as they are expected to happen within the plan, watching for deviations that indicate a necessity to re-plan (often ahead of this becoming a serious problem) (Klein, pp. 32–33).
- AI planning techniques represent the dynamic state of the world at points in the plan and can be used for "mental simulation" of the execution of the plan (Klein, p. 45).
- Pruning of choices according to given requirements or constraints (Klein, p. 94, "singular strategy").
- Situation-dependent option filtering (sometimes reducing the choices normally open to one "obvious" one (Klein, pp. 17–18).
- Satisficing search to find the first suitable plan that meets the essential criteria (Klein, p. 20).
- Anytime algorithms that seek to improve on the best previous solution if time permits.
- Heuristic evaluation and prioritization of multiple possible choices within the constraint search space (Klein, p. 94).
- Repair of plans while respecting plan structure and intentions.
- Uniform use of a common plan representation with embedded rationale to improve plan quality, shared understanding, etc. (Klein, p. 275, seven types of information in a plan).

Gary Klein was asked to comment upon the "Intelligible Planning" approach in Tate (2000b, Appendix) and described in Section 13.4 as compared to his observations of natural problem solving and decision making in humans operating in stressful situations and dynamic environments. He observed the following in this edited personal communication (also from Tate, 2000b, Appendix):

1. I felt a strong kinship with what you are attempting. The effort to use satisficing criteria, the use of anytime algorithms to permit continual improvement, the shift from abstract to detailed plan when necessary, the analysis of interactions in a plan, the identification of flaws in a plan, the monitoring of execution, the use of mental simulation, the representation of a singular strategy, heuristic evaluation, plan repair, and so forth are all consistent with what I think needs to be done.

2. My primary concern is how you are going to do these things ... The discipline of AI can provide constraints that will help you understand any of these strategies in richer detail. But those constraints may also prevent you from harnessing these sources of power.

3. Your slogan, "Search and you're dead," seems right. Unconstrained search is a mark of intellectual cowardice. And it is also not a useful strategy.

Personal communication reply from Austin Tate to Gary Klein (from Tate, 2000b, Appendix):

I want to clarify my use of the slogan, "Search and you're dead," over the last 20 years. This is the headline, but I then clarify what I mean as "(Unconstrained) search and you're dead."

I have found this to be a useful slogan to express my general approach, and it makes for good knockabout fun on panels at conferences. The idea should be to richly describe the constraints known using whatever knowledge is available about the problem and then to seek solutions in that constrained space. We seek to use knowledge of the domain to constrain the use of blind search or "black box" automated methods in ways that are intelligent and intelligible (to humans).

In reality, all planning systems we build have sophisticated search and constraint management components, and it is an aim of our research to be able to utilize the best available in an appropriate context. Search can be a useful tactic in situations in which you are under-constrained and stuck. AI has made enormous advances in constraint management using search and other methods—so much so that some of its proponents argue that we do not need to bother with domain expertise or being knowledge-based about many of the problems we are addressing. It's this latter overenthusiasm for one approach that I seek to counter. Even very powerful search can be made more useful if put into a sensible knowledge-based context. This is, of course, more relevant when humans are involved in the decisions as then a more naturalistic style of mutually progressing toward a solution becomes a key to successful use of the technology.

13.11 Summary

HTN planning could be a useful paradigm to allow for agent operations that can dynamically adapt to a specific context and allow the following:

- Composition of workflows from requirements and component/template libraries
- Coverage of simple through to very complex (preplanned) components
- Execution support, reactive repair, recovery, etc.
- Mixed initiative (people and systems) planning and execution
- Provision of a framework within which more detailed specialized solvers, optimizers, and simulators work

I-X technology and its underlying <I-N-C-A> ontology to represent processes and plans that can act as a flexible, extendable, and intelligible framework to deploy such an approach.

Acknowledgments

The O-Plan and I-X projects were sponsored by the Defense Advanced Research Projects Agency (DARPA) and Air Force Research Laboratory Command and Control Directorate under various programs, including the Planning Initiative and Agent-Based Computing Programs, by the UK Defence Evaluation Research Agency (DERA), and by others. The University of Edinburgh and research sponsors are authorized to reproduce and distribute reprints for their purposes notwithstanding any copyright annotation hereon. The views and conclusions contained herein are those of the authors and should not be interpreted as necessarily representing official policies or endorsements, either expressed or implied, of other parties.

References

Allen, J., Hendler, J. and Tate, A. (1990) *Readings in Planning*, Morgan Kaufmann, San Mateo, CA.

Beck, H. (1993) TOSCA: A Novel Approach to the Management of Job-shop Scheduling Constraints, *Realising CIM's Industrial Potential: Proceedings of the Ninth CIM-Europe Annual Conference*, pp. 138–149, (eds., Kooij, C., MacConaill, P.A. and Bastos, J.).

Currie, K.W. and Tate, A. (1991) O-Plan: The Open Planning Architecture, *Artificial Intelligence* 52(1), Autumn 1991, North-Holland.

Dalton, J., Drabble, B. and Tate, A. (1993) The O-Plan Constraint Associator, *Proceedings of the 13th UK Planning and Scheduling SIG*, Strathclyde University, UK, 14–15 September 1994.

Klein, G. (1998) *Sources of Power—How People Make Decisions*, 1st Edition, MIT Press, 1998.

Nau, D.S., Au, T.-C., Ilghami, O., Kuter, I., Muñoz-Avila, H., Murdock, J.W., Wu, D. and Yaman, F. (2005) Applications of SHOP and SHOP2, *IEEE Intelligent Systems* 20(2) pp. 34–41, Mar.–Apr. 2005.

Pollack, M. (1994) *DIPART Architecture*, Technical Report, Department of Computer Science, University of Pittsburgh, PA 15213, USA.

Sacerdoti, E. (1977) A Structure for Plans and Behaviours. *Artificial Intelligence* series, publ. North Holland.

Smith, S. (1994) OPIS: A Methodology and Architecture for Reactive Scheduling, in *Intelligent Scheduling*, (eds., Zweben, M. and Fox, M.S.), Morgan Kaufmann, Palo Alto, CA., USA.

Tate, A. (1977) Generating Project Networks, *Proceedings of the Fifth International Joint Conference on Artificial Intelligence (IJCAI-77)*, pp. 888–893, Boston, USA, August 1977. Reprinted in "Readings in Planning" (eds., Allen, J., Hendler, J. and Tate, A.), Morgan-Kaufmann, 1990.

Tate, A. (1994) Mixed Initiative Planning in O-Plan2, *Proceedings of the ARPA/Rome Laboratory Planning Initiative Workshop*, (ed., Burstein, M.), Tucson, Arizona, USA, Morgan Kaufmann, Palo Alto.

Tate, A. (ed.) (1996a) *Advanced Planning Technology - Technological Achievements of the ARPA/ Rome Laboratory Planning Initiative (ARPI)*, AAAI Press.

Tate, A. (1996b) Representing Plans as a Set of Constraints—the <I-N-OVA> Model, *Proceedings of the Third International Conference on Artificial Intelligence Planning Systems (AIPS-96)*, pp. 221–228, (ed., Drabble, B.) Edinburgh, Scotland, AAAI Press.

Tate, A. (1998) Roots of SPAR—Shared Planning and Activity Representation, *Knowledge Engineering Review*, Vol. 13, No. 1, March 1998. See also http://www.aiai.ed.ac.uk/ project/spar/.

Tate, A. (2000a) *<I-N-OVA> and <I-N-CA>—Representing Plans and other Synthesized Artifacts as a Set of Constraints*, AAAI-2000 Workshop on Representational Issues for Real-World Planning Systems, at the National Conference of the American Association of Artificial Intelligence (AAAI-2000), Austin, Texas, USA, August 2000.

Tate, A. (2000b) Intelligible AI Planning, in Research and Development in Intelligent Systems XVII, *Proceedings of ES2000, The Twentieth British Computer Society Special Group on Expert Systems International Conference on Knowledge Based Systems and Applied Artificial Intelligence*, pp. 3–16, Cambridge, UK, December 2000, Springer.

Tate, A. (2003) <I-N-C-A>: A Shared Model for Mixed-initiative Synthesis Tasks, *Proceedings of the Workshop on Mixed-Initiative Intelligent Systems (MIIS) at the International Joint Conference on Artificial Intelligence (IJCAI-03)*, pp. 125–130, Acapulco, Mexico, August 2003.

Tate, A. and Dalton, J. (2003) O-Plan: A Common Lisp Planning Web Service, *Proceedings of the International Lisp Conference 2003*, October 12–25, 2003, New York, USA, October 12–15, 2003.

Tate, A., Dalton, J. and Stader, J. (2002) I P2—Intelligent Process Panels to Support Coalition Operations, *Proceedings of the Second International Conference on Knowledge Systems for Coalition Operations (KSCO-2002)*, pp. 184–190, Toulouse, France, April 23–24, 2002.

Tate, A., Drabble, B. and Dalton, J. (1994a) Reasoning with Constraints within O-Plan2, *Proceedings of the ARPA/Rome Laboratory Planning Initiative Workshop*, (ed. Burstein, M.), Tucson, AZ, USA, Morgan Kaufmann, Palo Alto, CA, USA.

Tate, A., Drabble, B. and Kirby, R. (1994b) O-Plan2: An Open Architecture for Command, Planning and Control, in *Intelligent Scheduling*, (eds., Zweben, M. and Fox, M.S.), Morgan Kaufmann, Palo Alto, CA, USA.

Tate, A., Levine, J., Dalton, J. and Nixon, A. (2003) Task Achieving Agents on the World Wide Web, in *Spinning the Semantic Web*, pp. 431–458, (eds., Fensel, D., Hendler, J., Liebermann, H. and Wahlster, W.), Chapter 15, MIT Press, 2003.

Wilkins, D. (1988) *Practical Planning*, Morgan Kaufmann, Palo Alto, 1988.

Chapter 14

Policy-Based Governance of Complex Distributed Systems: What Past Trends Can Teach Us about Future Requirements

Jeffrey M. Bradshaw, Rebecca Montanari, and Andrzej Uszok

Contents

The resilience of an overall distributed system depends on how quickly the system is able to adapt to failures and other dynamic events in order to continue operating as desired. For any distributed system with even moderate complexity, the adaptation must be a largely automated process. Although human operators may not be involved in every decision made by the system, the human operators can establish the guidelines for the overall operation and adaptations performed by the system. In this chapter, Jeff Bradshaw, Andrzej Uszok, and Rebecca Montanari explore policy-based approaches to governing the runtime behavior of complex distributed systems.

14.1 Introduction

Policy-based management has been the subject of extensive research in recent decades (Robinson et al. 1988; Sloman 1994; Chada et al. 2002; Calo et al. 1994; Boutaba 2006). The idea of adopting machine-enforceable policies as means of specifying and governing the behavior of distributed systems emerged as a result of increasing system and network management complexity (Moffet et al. 1993). In order to cater to changing requirements, large distributed systems must be capable of changing and adapting their behavior while they are still running.

Policies are a means to dynamically regulate the behavior of system components without changing code and without requiring the consent or cooperation of the components being governed (Bradshaw et al. 2005; Chada et al. 2002). By changing policies, a system can be continuously adjusted to accommodate variations in externally imposed constraints and environmental conditions.

Policy research was originally focused on the problems of large-scale, enterprise-wide or Internet-wide systems (Boutaba 2006). In these applications, policies have been exploited either to automate access control or to handle network administration tasks, such as configuration, security, recovery, or quality-of-service (QoS).

Considerable effort has been applied to develop expressive representations and sophisticated management systems for specifying, managing, analyzing, and enforcing policies.

Since these early years, the scope of policy management has gone far beyond traditional distributed system and network concerns. New application fields for policy management include, among others, multi-agent systems, pervasive and mobile devices, and autonomic computing systems (Bradshaw et al. 2004, 2012; Bunch et al. 2012; Montanari et al. 2004; Keeney et al. 2003; Lupu et al. 2007; Xu et al. 2011). The novel requirements of these additional application areas, in addition to problems that have arisen in large-scale deployments within more traditional applications, have posed a never-ending stream of challenges to policy researchers.

In this paper, we look at selected trends in research and development of policy management systems for complex distributed systems. We believe that a careful glance in the rearview mirror will help researchers anticipate some of the crucial policy management requirements of the future—a future that portends ever-increasing scale, heterogeneity, decentralization, and complexity.

Rather than organizing this paper by application area, we have approached the subject topically, with a focus on a selection of trends that highlight general common considerations *across* applications that will be driving the direction of policy research and development into the future (Sections 14.2 and 14.3). Section 14.4 will discuss additional implications of these trends and will outline important challenges yet to be seriously addressed.

14.2 Progress and Prospects in Policy Specification Approaches

In this section, we outline the most common policy representation approaches in widely known policy languages. In particular, we reflect on the research results and application demands that have motivated progress for that area. Then, we summarize our views on the most promising directions and needs.

14.2.1 Historical Overview

In network management, policies are applied to automate network administration tasks, such as security, configuration, recovery, or QoS with the promise of reduced maintenance costs and improved flexibility, verifiability, and runtime adaptability. The recognition of policy as a management approach can be traced back to the research of (Robinson et al. 1988; Moffet et al. 1993), in which management policies were described as sets of rules for achieving the scalable management of a distributed computing system and for achieving objectives of optimizing resource usage, cost, revenue, and performance (Boutaba 2006). Policy-based systems for system and network management typically distinguish two different kinds of policies (Chada et al. 2002). *Authorization policies* describe the actions that are allowed (positive authorization) or forbidden (negative authorization) to actors in a specified context and are typically used for security purposes. *Obligation policies* define either the actions subjects must perform when certain conditions are triggered (positive obligation) or else exceptions to such requirements for subjects in a specific context (negative obligation or waiver). Obligation policies are typically exploited for configuration and QoS management.

There are a number of approaches to the definition of policies, and accompanying policy languages, which represent a number of different levels of policy expressiveness and policy enactment semantics. Multiple approaches for policy representation have been proposed, ranging from logic-based languages to special-purpose policy languages that can be processed by machines to generic rule-based (if-then-else) formats (Barnes et al. 1999; Hoagland 2000; Damianou et al. 2000) to more recent ontology-based approaches (Bradshaw et al. 2003a, 2013; Uszok et al. 2003, 2008, 2011; Kagal et al. 2003; Tonti et al. 2003). Policy-based management suffers from fragmentation of approach, partly due to differences in semantics between access control policy languages and resource management policy languages. As a consequence, there is no commonly accepted policy language and no common approach to the engineering of policy-based systems with many languages being proprietary in nature or tied to particular system management products or application scenarios.

Different schema can be used to describe proposed state-of-the-art policy languages. In the following, we focus on the description of the most widely adopted and cited policy languages by analyzing them from two different viewpoints: the policy representation model and the scope of the applicability.

14.2.2 Logic-Based Languages

Logic-based languages are attractive for the specification of security policy because they have a well-understood formalism that is amenable to machine inference (Damianou et al. 2000). Examples of first-order logic-based security policy languages include the logical notation introduced in Chen et al. (1995); the role definition language (RDL) presented in Hayton et al. (1998) and RSL99 (Ahn et al. 1999), and the authorization specification language (ASL) (Jajodia et al. 1997). There are also proposals that exploit deontic logic for security policy representation. One early example had the aim of developing confidentiality policies that incorporated conditional norms (Glasgow et al. 1992). In Cholvy et al. (1997), deontic logic is used to represent security policies with the aim of detecting conflicts. In Ortalo (1998), deontic logic is exploited to express security policies in information systems. Despite the advantages of logic-based languages in terms of expressiveness and analyzability, they assume a strong mathematical background. This can make them difficult to use and understand. Moreover, they are not always directly translatable into efficient implementations.

14.2.3 Rule-Based Approaches

Ponder is the broadest, most influential, and most widely deployed rule-based policy language (Damianou et al. 2001; Sloman et al. 2002). Ponder views policies as rules that define choices in system behavior that will reflect the objectives of the system managers. At a more formal level, a Ponder policy defines a relationship between objects (subjects and targets) of a managed system. Ponder2, the latest incarnation of this language, is a declarative, object-oriented language that supports the specification of several types of management policies for distributed object systems and provides structuring techniques for policies to cater to the complexity of policy administration in large enterprise information systems (Twidle et al. 2009).

A set of tools and services were developed for the specification, analysis, and enforcement of Ponder policies. Thus, the name Ponder became associated not only with the language, but with the entire toolkit. Ponder2 combines a distributed object management system with a domain service, obligation policy interpreter, command interpreter, and authorization enforcement capable of specifying and enforcing both authorization and obligation policies (Twidle et al. 2009). The domain service provides a hierarchical structure for managing objects. The obligation policy interpreter handles event, condition, action rules (ECA). The command interpreter accepts a set of commands, compiled from a high-level language called PonderTalk, via a number of communications interfaces that may perform invocations on a managed object registered in the domain service. The authorization enforcement caters to both positive and negative authorization policies, provides the ability to specify fine-grained authorizations for every object, and implements domain nesting algorithms for conflict resolution.

Other rule-based policy systems adopt an ECA rule paradigm. A popular example is the policy description language from Bell Labs (Lobo et al. 1999) in which a policy is a function that maps a series of events into a set of actions. The language can be described as a real-time specialized production rule system to define policies. It consists of a policy rule corresponding to an obligation policy and of a rule for triggering other events. Policy rules map series of events into sets of actions. The language has clearly defined semantics and has been deliberately limited to a policy specification that is succinct and can be efficiently implemented.

A more recent example is IETF's common policy language (CPL), a standard for representation of both authorizations (with related obligations) as condition-action rules, with an encoding in XML (Verma 2000). We note that recent work on CPL seems to have been focused exclusively on geographic location privacy for the web. RuleML is a generic rule language implemented with the structure of Horn clauses (head <- body) but is evolving toward ECA rule formats. It is sponsored by an international nonprofit organization. DMTF's CIM-SPL (simplified policy language) is a very simple rule language to represent obligation policies in "if-then" form. There is no support for authorization policies (Verma 2000). The TMForum information framework (SID) allows the semantic description of a complex set of interrelated instances, supporting ECA rules that could be used to support obligation policies. The Object Management Group's semantics of business vocabulary and rules (SBVR) provides semantics for business rules. It specifies an XMI encoding of UML instances with some semantic technology support via mapping to ISO Common Logic. Version 1.0 was released in January 2008 with no further updates and no known implementation (The Object Management Group's Semantics of Business Vocabulary and Rules, http://www.omg.org/spec/SBVR/1.0/).

14.2.4 General-Purpose versus Limited-Purpose Policy Languages

The languages described in Section 14.2.3 represent general-purpose policy approaches. Ponder2 can be considered the most significant example of general-purpose policy language because it is designed as an extensible framework that can be used at widely different levels of scale—from small, embedded devices to complex services and virtual organizations.

In addition to general-purpose policy languages, there are a variety of limited-purpose languages that have been developed. Here, our focus will not be on product-specific policy languages (e.g., the impact policy language that is part of IBM Tivoli Business Service Manager) or on simple approaches based on table look-up (e.g., firewall or router policies), but rather will highlight two examples of more broadly conceived languages that were designed to work within a given family of operating environments (e.g., web services) or to serve specific requirements of particular kinds of applications (e.g., access control, digital rights management).

One of the most important of these languages is the web services policy language (WS-Security Policy) specification, an important standard that was developed by a wide consortium from the computing and security industry and has been adopted by the OASIS group (Web Services Policy 1.5 – Primer, http://www.w3.org/TR/2006/WD-ws-policy-primer-20061018). They define a set of XML-based security policy assertions (e.g., protection assertions requiring signing and encryption, token assertions specifying allowed token formats, and supporting token assertions that add functions such as user sign-on with a username token) that apply to SOAP message security, WS-Trust, and WS-Secure Conversation. Policies either can be used at development time to generate code with specified properties, or they can be used during runtime negotiation of security requirements for web service communication. Policies may be attached to WSDL elements. The use of XML for encoding allows great flexibility and ongoing evolution with respect to details in message-based communication with services.

One of the most widely used policy languages for access control is XACML (eXtensible Access Control Markup Language), the subject of a very active OASIS standards group (OASIS eXtensible Access Control Markup Language (XACML) https://www.oasis-open.org/committees/tc_home.php?wg_abbrev=xacml). A number of implementations for XACML are documented on the website, with varying licensing terms, including some that are available for public download. Although XACML was originally conceived as an approach for access control (expressed as condition-effect permit or deny rules), it is now expanding its reach to additional problems. In some cases, obligations related to access control rules can also be specified, but the range of semantics is limited. Like WS-Security Policy, XACML uses an XML encoding for policies.

A variety of rights expression languages are used to provide machine interpretable encodings for digital rights management. The expressions themselves are typically embedded within metadata associated with the media. For example, ODRL (open digital rights language) is an open standard for an XML-based rights expression language (http://www.w3.org/community/odrl/). ODRL is implemented in three serializations: XML, RDF/OWL Ontology, and JSON.

14.2.5 Recent Directions in Policy Languages: OWL-Based Approaches

One of the major advantages of using XML as a policy representation is its straightforward extensibility. A problem with using XML (and many other policy representations) is that its semantics are mostly implicit—meaning is conveyed based on a shared understanding derived from human consensus. Implicit semantics based on convention have potential for ambiguity, often suffer fragmentation into incompatible representations, and require extra effort that could be obviated by a richer representation. For this and other reasons, semantic technologies are replacing XML as a representation of choice in many demanding application domains (Garcia et al. 2005). Semantic technologies, such as the W3C standards RDF and web ontology language (OWL), are built using XML as a foundation, thus sharing its extensibility. However, in contrast to XML, they are able to represent and reason over rich information in great efficiency.

The OWL is a family of knowledge representation languages based on description logic (DL) with a representation in RDF (OWL Web Ontology Language, http://www.w3.org/TR/2004/REC-owl-features-20040210/). OWL supports the specification and use of ontologies that consist of terms representing individuals, classes of individuals, properties, and axioms that assert constraints over them. The axioms can be realized as simple assertions (e.g., Woman is a subclass of Person, hasMother is a property from Person that is inherited by Woman, Woman and Man are disjoint classes) and also as simple rules. OWL contains the necessary constructors for formal description of most policies and information management definitions (OWL Web Ontology Language, http://www.w3.org/TR/2004/REC-owl-features-20040210/; López de Vergara et al. 2004).

There are several advantages to OWL for policy representation. Ontologies simplify the task of governing the behavior of complex environments. The possibility of representing entities and changes in the behavior at multiple levels of abstraction improves the global *expressiveness*. The use of ontologies permits the policy framework to be easily extended by simply adding new concepts to the ontology. In fact, any policy element (e.g., system components, actions, and context) can be described by appropriate concepts and relationships at the desired level of abstraction. In addition, because the semantics of such representations typically are a superset of the semantics of specialized "niche" policy languages, it is possible to convert ontology-based representations into the more specific languages. In traditional languages, this task is usually much trickier. For example, in Ponder, the specification of a communication policy example requires the nontrivial effort to extend the policy language or to convert the policy into a resource control policy. In addition, the possibility of modeling policies at a high-level of abstraction allows users to focus their attention more on high-level management requirements than on implementation details.

An ontology-based description of the policy enables the system to use concepts to describe the environments and the entities being controlled, thus simplifying their description and improving the *analyzability* of the system. As a result, policy frameworks can take advantage of powerful features, such as policy conflict detection and harmonization (Guerrero et al. 2006). In addition, ontology-based approaches simplify access to policy information with the possibility of dynamically calculating relationships between policies and the environment, entities, or other policies, based on ontology relationships rather than fixing them in advance. As with databases, it is possible to access the information provided by querying the ontology according to the ontology schema. This is an advantage in comparison to traditional languages that provide only predefined queries to access information and static representations of policy. How to design the ontology is an application-dependent problem. As in database design, ontologies should be designed and extended consistent with the application context and optimized for the most common queries. Finally, ontologies can also simplify the sharing of policy knowledge among different organizations and applications, thus increasing the possibility for entities to negotiate policies and to agree on a common set of policies. Table 14.1 compares ontology-based approaches to policy to Ponder as a mature representative of the non-ontology-based general-purpose policy language approach.

All this being said, the adoption of ontologies for policy specification requires addressing some technical difficulties. OWL representations present a complex syntax, long declarative descriptions, hyperlinks, and references to external resources that can make it difficult to read (e.g., compare the readability of a Ponder policy with an OWL policy). To improve the *ease of use* of these languages, graphical interfaces or other tools for policy specification are necessary. In addition, *enforceability* is a critical aspect for the OWL approach. Ontology-based policy specification can be difficult to implement in comparison to other policy specifications, such as Ponder, because the high-level specification of ontology-based policies can be far removed from the concrete implementation of the policy enforcement on the systems. Usually the gap between the specification and the implementation of policies cannot be completely overcome in an

Table 14.1   Comparison between OWL-Based and Ponder Approaches

	Semantic Web Languages for Policy Specification	*Ponder* [a]
Expressiveness	Capable of representing concepts and behavior of any complex environment	Capable of controlling specific sorts of behavior within object-oriented systems
	Multiple levels of abstraction	Low level of abstraction: object level
	Easy to extend policy ontology at runtime with new concepts	Extensibility supported by object-oriented inheritance at compile-time
Analyzability	Ontology representation simplifies and directly supports policy reasoning, conflict detection and harmonization	Conflict detection requires transforming policy specification into an event calculus representation
	Simplified access to policy information by querying the ontology	Access to single policy object by API – Access to policy repository to be designed
Ease-of-use	Need of specialized tools to assist unskilled users with policy specification and interpretation	Language specifically designed for simple policy specification and direct readability
Enforceability	High-level specification requires skilled programmers or sophisticated policy automation mechanisms for enforcement	Detailed specifications can be directly mapped into policy enforcement mechanisms
	Policy sharing among heterogenous systems requires an agreement on a common ontology	Policy sharing among heterogeneous systems requires agreement on interfaces

Source: With kind permission from Springer Science+Business Media: "Semantic Web languages for policy representation and reasoning: A comparison of KAoS, Rei, and Ponder," *Proc. of the Second International Semantic Web Conference (ISWC2003)*, LNCS 2870, 2003, Tonti, G. et al.

[a] Used as example of nonsemantic web language.

automated manner but has to be resolved to a greater or lesser degree by human programmers, consistent with the capabilities and features of each platform. Enforcement code generation facilities and libraries of enforcement mechanisms adapted to specific platforms are among the most important features for OWL-based policy management frameworks to provide the enabling of their widespread implementation.

In the following, we describe the key characteristics of some widely known semantic-based policy representations, KAoS, Rei, and AIR, so that we can evaluate the advantages and disadvantages of adopting semantic technologies for policy representation in practice.

14.2.5.1 KAoS

IHMC's KAoS policy services framework is a mature general-purpose policy management system that relies on OWL in the specification, analysis, and enforcement of policy constraints across a wide variety of distributed computing platforms (Bradshaw et al. 2003a, 2013; Uszok et al. 2003, 2008, 2011). KAoS supports both authorization and obligation policies and supports specialized policies and mechanisms of other kinds of policies (e.g., delegation management, policies about space and time). KAoS enables the specification, management, analysis, and enforcement of policies. It provides the KAoS Policy Administration Tool (KPAT) as a graphical interface to assist users in policy specification, revision, and application. In addition, KPAT can be used to browse and load ontologies and to analyze and de-conflict newly defined policies. It supports static (DL-based) and dynamic conflict resolution (collective obligation and planning/resource allocation mechanisms) and the enforcement of policies through a separate software element called the Guard. The breadth of its semantics has allowed its use across multiple application domains and operating environments, including intermittent tactical networks and standalone sensors.

KAoS is implemented in Java. It has well-defined interfaces for programmatic access of functionality and the ability to import and export OWL ontologies. Over the last decade, several government agencies and private organizations have sponsored research and development efforts to mature KAoS. KAoS has been enhanced for scalability, more powerful and flexible reasoning, and for use within distributed enterprises. KAoS has been integrated into IHMC's Luna agent framework (Bunch et al. 2012) as well as several other agent platforms and traditional service-oriented architectures. IHMC and its collaborators have undertaken to translate ontologies to third-party approaches in previous efforts and are currently working to extend these efforts into new arenas (e.g., translation from KAoS ontologies to XACML, translation from natural language documents to KAoS ontologies).

14.2.5.2 Rei

Rei is a policy framework that integrates support for policy specification, analysis, and reasoning in pervasive computing applications (Kagal et al. 2003). Rei has been used in conjunction with the Vigil security framework, Fujitsu's Task Computing project, and with the Groove workspace [http://www.groove.net] within the DARPA Genoa II program [http://www.darpa.mil/iao/GenoaII.htm]. The Rei deontic concept-based policy language allows users to express and represent the concepts of rights, prohibitions, obligations, and dispensations. In addition, Rei permits users to specify policies that are defined as rules associating an entity of a managed domain with its set of rights, prohibitions, obligations, and dispensations. Rei relies on an application-independent ontology to represent the concepts of rights, prohibitions, obligations, dispensations, and policy rules. This allows different elements of a pervasive environment to understand and interpret Rei policies in the correct way. In particular, Rei adopts OWL-Lite to specify policies and can reason over any domain knowledge expressed in either RDF or OWL. A policy basically consists of a list of rules expressed as OWL properties of the policy and a context represented in terms of ontologies that is used to restrict the policy's applicability. Though represented in OWL-Lite, Rei still allows the definition of variables that are used as placeholders as in Prolog. In this way, Rei overcomes one of the major limitations of the OWL language and, more generally, of DLs, that is, the inability to define variables. On the other hand, the choice of expressing Rei rules similarly to declarative logic programs prevents it from exploiting the full potential of the OWL language. In particular, the Rei engine is able to reason about domain-specific knowledge but not about policy specification. There seems to have been no further development activity on Rei since 2005.

14.2.5.3 AIR

AIR (accountability in RDF) is a semantic web-based rule language that supports reasoning on the open web (particularly trust and privacy issues) while focusing on generating and tracking explanations for its inferences and actions and conforming to linked data principles (see Air Policy Language). It relies on web standards although its implementation is proprietary. AIR uses Turtle (Terse RDF Triple Language) and N3 (Notation 3) rule encodings. AIR is not a general-purpose policy language but focuses specifically on trust and privacy issues.

14.3 Selection of Research Issues Relating to Policies and Adaptation

14.3.1 Policy Enforcement and Adaptation

Policies tend to be seen as prescriptive and externally imposed rules whose enforcement ensures a predictable system behavior. Policy-based enforcement approaches exhibit typical features:

1. They work involuntarily with respect to the system components that are governed by policies, that is, without requiring the system components to consent or even be aware of the policies being enforced, thus aiming to guarantee that the system complies with policy.
2. Wherever possible, they are enforced preemptively, preventing in advance buggy, poorly designed, unsophisticated, or malicious system components from doing harm.

These characteristics are well suited for traditional distributed system/network management in which it is crucial to ensure that a system behaves within well-defined boundaries. Considering that, in recent years, the scope of policy management has enlarged, some adjustments to policy enforcement approaches are required to address the specific features of the novel application fields.

14.3.1.1 Self-Regulation of Compliance

In traditional distributed systems, guarantees of policy conformance are typically required. These guarantees are assured through the use of an independent policy enforcement component, independent of the reasoning mechanisms in the software components being regulated (i.e., the subjects of policy).

Some research efforts, especially of those who are studying the development of norms and the influences that determine their degree of adoption in artificial social communities, have looked at issues of self-regulation. In such efforts, the subject of policy relies to a greater or lesser degree on their own reasoning to determine whether or not to adopt a given policy or norm and may engage in negotiation with other subjects about norm compliance (e.g., Grossi et al. 2006; Sen and Airiau 2007). The need for this kind of self-regulation arises especially in multi-agent systems in which the design and development of policy-based frameworks have to deal with a crucial agent dimension, that is, agent autonomy, intended as the capability of an agent to take care of itself and the quality of freedom from outside control (Bradshaw et al. 2003b).

Research on adjustable autonomy in agent systems proposes a hybrid between models of self-regulation and external enforcement. The coupling of autonomy with policy mechanisms allows agent designers to achieve adjustable autonomy with the primary purpose to maintain the system being governed at a sweet spot between convenience (i.e., being able to delegate every bit of an agent's work to the system) and comfort (i.e., the desire not to delegate to the system what it cannot be trusted to perform adequately). Adjustable autonomy gives the agent maximum freedom for local adaptation to unforeseen problems and opportunities while assuring humans that agent behavior will be kept within desired bounds.

Adjustable autonomy requires general engines for reasoning about how to adapt policies and resources to improve system performance and effectiveness for end-users (on whose behalf agents act in the system) while respecting absolute security constraints set by administrators. Along this direction, some initial work has been already performed focusing specifically on trading off operational necessity and security risk. In Bradshaw et al. (2005) formalisms and mechanisms are presented that have been developed for adjustment of policies and resources that will facilitate effective coordination across distributed systems. Kaa (KAoS adjustable autonomy) is a component of KAoS that permits it to perform such adjustments semi-automatically. Assistance from Kaa in making autonomy adjustments might typically be required when it is anticipated that the performance of the current configuration has led to or is likely to lead to failure or unacceptable performance or when there is no set of competent and authorized humans available to make such adjustments themselves. Ultimately, the value of performing an adjustment in a given context is a matter of expected utility: the utility of making the change versus the utility of the status quo. The current implementation of Kaa uses influence-diagram-based decision-theoretic algorithms to determine what, if any, changes should be made in agent autonomy. However, Kaa is designed to allow other kinds of decision-making components to be plugged in if an alternative approach is preferable. When invoked, Kaa first compares the utility of various adjustment options (e.g., increases or decreases in permissions and obligations, acquisition of capabilities, proactive changes to the situation to allow new possibilities), and then—if a change in the status quo is warranted—takes action to implement the recommended alternative.

14.3.1.2 Context-Based Policy Adaptation

In several application fields, for example, multi-agent systems or mobile and autonomic systems, another crucial requirement is the possibility to dynamically change policies to adapt system behavior to unpredictable contexts of operation. Changing policies means, for example, to prolong the validity of acquired permissions even in presence of changes in the conditions that have made the policy applicable or to find alternative permitted/obliged actions in the case the permitted/obliged actions as determined by the current state of the world cannot be performed because of constraints inherent in the current situation. Addressing policy adaptation requires the identification of both appropriate policy representation and enforcement models.

It has been recently recognized that progress in developing adaptive policy systems is represented by the adoption of a policy model that takes context into account. Whereas traditional systems rely on a relatively static characterization of the operating conditions in which changes in the set of both clients (users/devices) and accessible resources are relatively small, rare, or predictable, new application scenarios are characterized by frequent changes in physical user location, in accessible resources, and in the visibility and availability of collaborating entities. The conditions defined at design time to control and govern resource operation and sharing can be unpredictably different from the ones that actually hold at execution time when entities attempt to access some resources. Novel context-aware policy models or systems should be conceived to take into account the high unpredictability, heterogeneity, and dynamicity of the new application fields. Whereas, in traditional policy models, context is an optional element of policy definition that is simply used to restrict the applicability scope of policies, in context-centric solutions, context is the first-class principle that should explicitly guide both policy specification and enforcement process. It should not be possible to define a policy without the explicit specification of the context that makes that policy valid. Context-aware policy models enable policy adaptation in which, with the term "adaptation," we mean the ability of the policy management system to adjust context and policy specifications in order to enable policy enforcement in different, possibly unforeseen, situations. For example, policy adaptation may allow the identification of an alternative context in which permitted/obliged actions can be performed. Context adaptation can be useful to handle the case of dynamic policy conflicts, such as when an entity obliged to perform an action cannot perform it because it is not allowed to.

Considering context as a first-class design principle is a very recent research direction with some context-driven policy model proposals, mainly in the field of access control. The importance of taking context into account for securing pervasive applications is particularly evident in Covington et al. (2001), which allows policy designers to represent contexts through a new type of role called the environment role. Environment roles capture relevant environmental conditions that are used for restricting and regulating user privileges. Permissions are assigned both to roles (both traditional and environmental ones) and role activation/deactivation mechanisms to regulate the access to resources. Environmental roles are similar to our contexts in that they act as intermediaries between users and permissions.

Proteus is a context-aware policy model that is centered around the concept of context and that exploits a semantic-based representation of context and policies (Toninelli et al. 2007). Proteus contexts act as intermediaries between entities and the set of operations that they have to and/or can perform on resources. Proteus policies define, for each context, how to operate on resources. In particular, policies can be viewed as one-to-one associations between contexts and allowed/obliged actions. Entities should and/or can perform only those actions that are associated with the contexts currently in effect (active context), that is, the contexts whose defining conditions match the operating conditions of the requesting entity and of the environment as measured by specific sensors embedded in the system. Ko et al. (2006) proposes an approach that allows the overcoming of the semantic gap between contexts specified in the policy at design time and contexts collected from dynamic context sources in pervasive environments: An access request is allowed if the query context is semantically equivalent to the context specified in the policy rule. The policy model in Ko et al. (2006) also exploits semantic technologies. In particular, contexts and policies are defined by adopting an OWL-based representation, and OWL inference rules are exploited to derive relationships among contexts.

Context-based policy models require, however, taking into account the quality of context (QoC) information used to drive policy decisions. QoC has, in fact, a profound impact on the correct behavior of any context-aware policy framework. Depending, for instance, on the quality of used context data, granting access to a resource might be associated to a variable risk level: The less reliable context information is (i.e., the lower its quality), the higher risk is associated with any access action allowed based on that context information. Using context information with insufficient quality might increase the risk of incorrect access control decisions, thus leading to dangerous security breaches in resource sharing. The importance of considering QoC in designing and managing context-aware systems has recently started to be recognized with few proposed solutions (Buchholz et al. 2003; van Sinderen et al. 2006; Toninelli et al. 2007).

14.3.2 Adaptation Considerations at Runtime

The work on policies for system/network management so far has mainly focused on large, high-bandwidth, fixed networks, for example, enterprise networks, content provider networks, Internet service provider (ISP) networks, etc. Therefore, policy deployment is often based on centralized provisioning and decision-making. The ongoing standardization efforts toward common policy information models and frameworks witness the adoption of centralized architectures for policy enforcement (Verma 2000). IETF and DMTF have jointly produced a set of standards on policy information models and policy management architectures. The two main elements in their model of a policy management system are the policy decision point (PDP), a logical entity that makes policy decisions for itself or for other network elements that request such decisions, and the policy enforcement point (PEP), a logical entity that enforces policy decisions. The PDP is likely to store its policies in a repository, such as a lightweight directory access protocol (LDAP) directory service. The basic interaction between the components begins with the PEP. The PEP will receive a notification or a message that requires a policy decision. Given such an event, the PEP then formulates a request for a policy decision and sends it to the PDP. The PDP returns the policy decision, and the PEP then enforces the policy decision by appropriately accepting or denying the request.

The choice of the architectural model for a policy management system is one of the key factors in the success of such a system and impacts on the overall runtime system functioning efficiency. An architecture advantageous in managing a particular network environment (e.g., a high-bandwidth enterprise network) may not be an appropriate choice for managing another network (e.g., a low-bandwidth mobile wireless network). Hence, it is important to study the performance tradeoffs involved and choose an architecture (or combination of architectures) that suits the requirements of the deployment scenarios. A plethora of policy management architectures have been proposed for as many environments as differentiated services, enterprise networks, utility computing, data centers, wireless networks, optical networks, and middleware systems. However, there is still no real large-scale deployment of policy management as of today.

A taxonomy of policy architectures has been proposed in Phanse et al. (2006) based on various characteristics: (a) the locus of control that represents one or more policy servers or PDPs in a network, capable of making policy-based decisions; (b) the locus of information that refers to the location of the policy information storage module or repository used to store policies and accessed by a policy server to make decisions; (c) the policy distribution model that defines the transfer of policy information between different points in the system; (d) the tiers of control defined as the levels in a network at which policy decisions are made. The work in Phanse et al. (2006) distinguishes three main types of architectures: outsourced, provisioned, and hybrid. In outsourced architectures, all policy decisions are made at a single control tier, controlling the underlying nodes or PEPs. Unlike the outsourced approach, in the provisioned architecture, at least two control tiers are involved, that is, the locus of control and the locus of information are distributed across two or more tiers. The hybrid approach combines features of centralized and distributed architectures. From the analysis study conducted in Phanse et al. (2006) on the various policy architecture types, it emerges that, in order to extend the policy-based approach to newer application fields and networking environments, it is crucial to address the fundamental challenge of adapting the conceptually centralized idea of policy management to a decentralized paradigm applicable to the novel deployment environments. Especially in ad hoc networks, there is the need to build a management framework that is as automated as possible, requiring minimal human intervention, and is intelligent, meaning it is able to learn about changes in networking conditions. This would lead us to a self-organizing or adaptive control structure that automatically reacts to network dynamics.

When deploying a policy-management system, there are additional runtime considerations to address especially when self-regulating and/or context-aware policy enforcement approaches are adopted. For instance, in the case of self-regulating policy enforcement, evaluating options for reallocating tasks and change permissions/obligations among autonomous entities comes at a cost and may introduce performance overhead due to the need of entities to communicate, coordinate, and reallocate responsibilities (Bradshaw et al. 2005). When adopting context-aware policy approaches, additional issues have to be addressed that may increase performance overhead. One significant performance penalty factor derives from the need to integrate context management services within policy frameworks. When, for instance, a resource access request is performed, the context associated with the request needs to be acquired and matched within the context data, defining and activating the policy. In addition, context-aware policy models introduce additional complexity due to the need of evaluating the quality of context information associated to the policy request.

14.3.3 Adaptation as Part of Policy Conflict Resolution, Policy Refinement, and Polycentric Governance

Policy-based management still raises several research issues that have been partially solved. One concern when exploiting policies for ruling system/network behavior is the possibility of conflicts among policies, especially in situations requiring run-time resolution. After exploring current research in policy conflict resolution, we discuss adaptation through policy refinement and through polycentric governance.

14.3.3.1 Policy Conflict Resolution

There are a number of different conflicts that can arise from policies. Conflicts can arise in the set of policies. Conflicts can be modal conflicts, for instance, when a positive and negative authorization apply to the same objects, or application-specific conflicts related to the semantics of the resources and roles in the target and subject domains of policies, such as when two policies permit the same manager to sign checks and approve payments, may conflict with an external principle of separation of duties (Lupu et al. 1999). Conflicts may also arise during the refinement process between the high-level goals and the implementable policies. Some progress has been made in dealing with policy conflicts even though significant challenges remain to be addressed, such as detecting conflicts when arbitrary conditions restrict the applicability of the policies. Chomicki and Lobo present in Chomicki et al. (2003) a formal logic-based framework for detecting and resolving action conflicts in ECA policies. Bandara et al. (2004) present a tool for policy analysis. The tool supports querying a set of policies for validation and review. Validation queries are supported in order to determine the feasibility of a policy. Review queries are used in order to help the administrator analyze the managed system specification and extract specific types of information. In addition, Bandara suggests the use of abducting reasoning and the tool developed for event calculus–based goal elaboration in order to query potential conflicts between policies.

Like many other systems, KAoS originally relied exclusively on numeric policy priority assignments by users to determine how policies should be ranked and de-conflicted. A disadvantage of this approach has been that people may have difficulty assigning meaningful priorities and tracking how a given policy's priority relates to the priorities of other policies, especially when integrating large numbers of policies from different sources. For this reason, KAoS has been extended to use a logical precedence mechanism in addition to numeric priorities (Bradshaw et al. 2013). This allows administrators to specify an almost-infinite variety of precedence relationships among policies, mirroring the kinds of rationale that people use when deciding which policy will trump another (e.g., policies defined by person A take precedence over anyone else's policies; policies of the domain administrator (a role) take precedence over user (another role) policies; more recent policies take precedence over older policies; policies about writing to a specific directory take precedence over policies about writing to the volume; negative authorizations take precedence over positive authorizations).

A final area of ongoing research in policy conflict resolution is for mechanisms to cover situations when specified policies may require more resources than are available in the environment and that need to have, when possible, a fair method of allocation under constraints specified as part of the policy (resource de-confliction). Preliminary work on this problem has been undertaken in the context of QoS policies for network operations (Loyall et al. 2011).

14.3.3.2 Adaptation through Policy Refinement

Policy refinement is the process of deriving a more concrete specification from a higher-level objective. Although the goal of automating refinement of management and security policies from higher-level objectives remains a worthy long-term goal, it is currently not practical except in simple scenarios. A better near-term objective is for tools that provide partial automation of this process through assisting human managers to refine high-level abstract policies into more concrete ones. In Javier et al. (2006), a methodological approach toward the policy refinement problem is provided. In particular, a generic procedure to define policy hierarchies is described, which is essential to achieving systematic policy refinement as well as a policy refinement framework that formalizes the requirements to refine high-level guidelines into executable policies. Initial steps toward a framework for automated distributed policy refinement for both obligation and authorization policies are presented in Craven et al. (2010) in which the process of policy refinement is described as comprising three aspects: decomposition, operationalization, and distribution. In policy decomposition, which is the focus of the work in Craven et al. (2010), policies expressed at higher levels of abstraction are mapped into lower-level policies. The mapping is achieved using policy-independent refinement rules, defined within the scope of an application-specific system model. KAoS provides a "scenario" feature enabling the manual creation of high-level policies to guide the automatic generation of low-level policies. Further research is also needed on defining interfaces for the exchange of policies between the application and the hardware levels in order to effectively enforce policies defined at higher levels.

Typical approaches for addressing policy refinement in real-world systems, such as rule-based action policies, begin to suffer as systems become increasingly distributed, complex, and dynamic in nature. To support policy refinement at runtime, static mapping approaches for policy refinement will need to be replaced by advanced dynamic methods, supported by a planner. Goal policies and real-time utility-based evaluation of policy effectiveness have recently started to be recognized as an attractive basis for representing and managing high-level objectives at runtime (Kephart et al. 2007). Rather than specifying exactly what to do in a specific situation, goal policies specify either a single desired state or one or more criteria that characterize an entire set of desired states. The policy author specifies desired states as strategic goal policies (e.g., commander's intent) and specific tactical policies that best satisfy the needs of the current context would be automatically generated at runtime. The selection of tactics would be based on dynamic measures of utility.

14.3.3.3 Adaptation through Polycentric Governance

Within the framework of resilient systems engineering, Branlat and Woods have discussed important patterns that lead to failure in complex systems (Woods and Branlat 2008). "Bottom-up" approaches to policy refinement can be used to provide support for adaptive performance in the face of stressors and surprise through the principles of polycentric governance (Ostrom 2008). A related notion of organic resilience (Carvalho et al. 2010) was inspired by the concept of "organic computing" proposed in Müller-Scholer (2004). Organic resilience relies heavily on biologically inspired analogs and self-organizing strategies for the management and defense of distributed complex systems. The concept focuses on the design of emergent coordination mechanisms through local gradients and implicit signaling.

The use of collective obligations (van Diggelen et al. 2009a,b) is critical for practical applications of polycentric governance. Whereas an individual obligation is a policy constraint that describes what must be done by a particular individual, collective obligations are used to explicitly represent a given agent's responsibilities within a group to which it belongs without specifying in advance who must do what. In other words, in a collective obligation, it is the group as a whole that becomes responsible with individual members of the group sharing the obligation at an abstract level.

The execution and enforcement of collective obligations requires different mechanisms for different contexts. For some applications, a top-down policy refinement approach, implemented by a specialized planning system and spanning a group of agents, may be the best approach. However, in many cases a biologically inspired "bottom-up" approach might prove more workable. Such an approach requires that the agents themselves, rather than some centralized capability, organize the work. This approach works best when the agents themselves are in the best position to detect local triggers for collective obligations (e.g., potential threats or opportunities) to determine what support they can offer through their own resources and individual capabilities and what information should be shared among peers and with agents elsewhere in the system. The self-organizing nature of the system enables the agents to revisit responsibilities and resource allocations themselves, as needed, on an ongoing basis.

Applied in a manner consistent with polycentric governance, we believe that policy-based collective obligations could provide the regulatory mechanisms to enable effective and coactive coordination algorithms for agents. Moreover, we envision the implementation of policy-learning mechanisms that could rapidly propagate lessons learned about productive and unproductive actions to whole classes of actors.

14.4 Looking Ahead

In the future, we expect the trend of policy to continue to evolve beyond single-purpose approaches that only deal with specific application niches. Instead, there is a desire to be able to formulate, manage, and enforce policies across an entire enterprise, elevating the specialized representations now used for access control, network security, device configuration, and so forth to a rich, general-purpose semantic specification (Uszok et al. 2011). Only in this way can conflicts across diverse components and aspects of the system can be found and resolved and can QoS across the entire distributed system be assured. Some of the current challenges and anticipated solutions by using an enterprise-wide ontology-based approach include the following:

■ Sharing and integrating policies across multiple levels of an enterprise: Ontologies can represent and harmonize policy from different perspectives and at multiple levels of abstraction by semantics, not by convention as must be done using XML.

■ Maintaining high-level strategic policy intent constant while adjusting tactical policies as the situation demands: Ontology-based approaches for policy refinement, goal policies, and collective obligations.

■ Difficulty identifying and resolving static and dynamic policy conflicts: Efficient syntactic and semantic de-confliction and dynamic resource management algorithms.

■ Offline and online reasoning about policy configuration management: If you make a change, the effects are not visible. Use of rich semantics for real-time policy negotiation (e.g., risk-adaptive access control), what if analysis, dealing with rich context (e.g., history and state combinations).

■ Growth in range of uses for policy leads to proliferation of different languages for different application niches: Expressive ontology semantics can represent everything that is in the application niche–specific languages and more and can resolve inconsistencies within them and gaps between them.

■ Current implementations are not integrated, do not provide interfaces with other parts of an enterprise, nor have the ability to reason across domains (spectrum, cognitive radio configuration, network, authentication, logging, etc.): Can provide end-to-end system integration across multiple policy domains.

■ Need to specify policy by people who do not have specialized training: Automatic tools can straightforwardly translate from natural language; other template-based tools allow point-and-click construction of ontology-based policy constraints.

■ Need to deal with legacy systems: Automatic translation to application niche–specific languages.

■ Difficulty in managing large numbers of modular rule-based policies, implications of policies, gaps: Graphical visualization tools for understanding policy at design time; ontologies present relationships among policy concepts as an integrated whole.

∎ Standards driven by pragmatics of application needs, not overarching principles: Collaborate with researchers, users, and vendors in developing a broad, principled approach.

The foremost example of collaboration to develop a broad, principled approach is the NSA-sponsored Federal Digital Policy Management (DPM) initiative in the United States. DPM has chosen an ontology-based approach to policy with sufficient semantics to subsume the more specific approaches to policy specification across all government agencies specifically for this reason (NSA 2012). DPM has adopted the KAoS core ontologies as the basis for future standards efforts (Digital Policy Management 2012).

Although there may be reasons to translate the ontology-based policies into the specialized languages for the purpose of supporting legacy applications and specific enforcement components and devices (e.g., XACML, firewall policies), the focus is on specification and analysis across all application areas using a common ontology-based representation in OWL. Policy automation is a key concept in DPM. According to the organizers of the initiative: "Digital policies are an implementation where operating paradigms and access rules are created and maintained in executable formats that can be processed, downloaded to and enforced by IA devices. Digital Policy Management enables authorized operators to generate, adjudicate, validate, disseminate, and monitor policies."

14.5 Conclusions

Even though each application area has stimulated the research on specific aspects of policy management and proposed specific solutions, all application areas share common considerations. As far as policy languages are concerned, policies can be specified in many different ways, and multiple approaches have been proposed in different application domains. There are, however, some general requirements that any policy representation should satisfy regardless of its field of applicability: *expressiveness* to handle the wide range of policy requirements arising in the system being managed, *simplicity* to ease the policy definition tasks for administrators with different degrees of expertise, *enforceability* to ensure a mapping of policy specifications into implementable policies for various platforms, *scalabilty* to ensure adequate performance, and *analyzability* to allow reasoning about policies. The challenge is to achieve a suitable balance among the objectives of expressiveness, computational tractability, and ease of use for a given application. Other difficulties for the deployment of policy-based management on a large-scale derive from the technical challenges of providing efficient and adaptable policy run-time enforcement models and mechanisms.

References

Ahn, G.-J. et al., "The RSL99 language for role-based separation of duty constraints," *Proc. of the Fourth ACM Workshop on Role-Based Access Control*, ACM Press, 1999.

Air Policy Language, http://dig.csail.mit.edu/TAMI/2007/AIR/.

Bandara, A. K. et al., "A goal-based approach to policy refinement," *Proc. of the 5th IEEE International Workshop on Policies (Policy 2004)*, IEEE Computer Society, 2004.

Barnes, J. F. et al., "CacheL: Language support for customizable caching policies," *Proc. of 4th International Web Caching Workshop*, San Diego, CA, 1999.

Boutaba, R., "Policy-based management: A historical perspective," *Journal of Network System Management*, Springer-Verlag, Vol. 15, No. 4, 2006.

Bradshaw, J. M. et al., "Representation and reasoning for DAML-based policy and domain services in KAoS and Nomads," *Proc. of the Autonomous Agents and Multi-Agent Systems Conference (AAMAS 2003)*, ACM Press, 2003a.

Bradshaw, J. M. et al., "Adjustable autonomy and human-agent teamwork in practice: An interim report on space applications," *Agent Autonomy*, Kluwer, 2003b.

Bradshaw, J. M. et al., "Kaa: Policy-based explorations of a richer model for adjustable autonomy," *Proc. of the 4th International Conference on Autonomous Agents and Multiagent Systems, (AAMAS '05)*, ACM Press, 2005.

Bradshaw, J. M., P. Beautement, M. R. Breedy, L. Bunch, S. V. Drakunov, P. J. Feltovich, R. R. Hoffman, et al. "Making agents acceptable to people," In *Intelligent Technologies for Information Analysis: Advances in Agents, Data Mining, and Statistical Learning*, edited by Ning Zhong and Jiming Liu, pp. 361–400. Berlin, Germany: Springer-Verlag, 2004.

Bradshaw, J. M., M. Carvalho, L. Bunch, T. Eskridge, P. Feltovich, M. Johnson, and D. Kidwell. "Sol: An agent-based framework for cyber situation awareness," *Künstliche Intelligenz*: Vol. 26, No. 2, pp. 127–140, 2012.

Bradshaw, J. M., A. Uszok, M. Breedy, L. Bunch, T. C. Eskridge, P. J. Feltovich, M. Johnson, J. Lott, and M. Vignati. "The KAoS policy services framework," Poster and paper for presentation at the *Eighth Cyber Security and Information Intelligence Research Workshop (CSIIRW 2013)*. Oak Ridge, TN: Oak Ridge National Labs, January 2013.

Buchholz, T. et al., "Quality of context: What it is and why we need it," *Proc. of 10th International Workshop of the HP OpenView University Association (HPOVUA '03)*, 2003.

Bunch, L., J. M. Bradshaw, M. Carvalho, T. Eskridge, P. Feltovich, J. Lott, and A. Uszok. "Human-agent teamwork in cyber operations: Supporting co-evolution of tasks and artifacts with Luna," Invited paper in I. J. Timm and C. Guttmann (eds.), *Multiagent System Technologies: Proceedings of the Tenth German Conference on Multiagent System Technologies (MATES 2012)*, Trier, Germany, 10–12 October 2012. Berlin, Germany: Springer, LNAI 7598, pp. 53–67.

Calo, S. B. et al., "Policy-based management of networks and services," *Journal of Network System Management*, Springer-Verlag, Vol. 11, No. 3, 1994.

Carvalho, M., T. Lamkin, C. Perez. "Organic resilience for tactical environments," In *5th International ICST Conference on Bio-Inspired Models of Network, Information, and Computing Systems (Bionetics)*, Boston, December 2010.

Chada, R. et al., "Policy-based networking," Special Issue, *IEEE Network*, Vol. 16, No. 2, 2002.

Chen, F. et al., "Constraints for role-based access control," *Proc. of the First ACM/NIST Role Based Access Control Workshop*, USA, ACM Press, 1995.

Cholvy, L. et al., "Analyzing consistency of security policies," *Proc. of IEEE Symposium on Security and Privacy*, IEEE Computer Society, 1997.

Chomicki, J. et al., "Conflict resolution using logic programming," *IEEE Transactions Knowledge Data Engineering*, Vol. 15, No. 1, 2003.

Covington, M. J. et al., "Securing context-aware applications using environmental roles," *Proc. of the 6th ACM Symposium on Access Control Models and Technologies (SACMAT 2001)*, ACM, 2001.

Craven, R. et al., "Decomposition techniques for policy refinement," *Proc. of the 6th International Conference on Network and Service Management*, IEEE Computer Society, 2010.

Damianou, N. et al., "Ponder: A language for specifying security and management policies for distributed systems," Imperial College, UK, Research Report, 2000.

Damianou, N. et al., "The Ponder policy specification language," *Proc. of Workshop on Policies for Distributed Systems and Networks (POLICY 2001)*, Springer-Verlag, LNCS 1995, 2001.

"Digital policy management: Be part of the solution, not the problem," 2012. In *VanDyke Technology Group*. http://www.vdtg.com/blog/blog_post_policy_planning.html (accessed May 29, 2013).

García, F. J. et al., "Representing security policies in web information systems," *Proc. of 14th International WWW Conference*, ACM Press, 2005.

Glasgow, J. et al., "A logic for reasoning about security," *ACM Transactions on Computer Systems*, Vol. 10, No. 3, 1992.

Grossi, D., H. Aldewereld, and F. Dignum. "Ubi lex, ibi poena: Designing norm enforcement in e-institutions," In *Coordination, Organizations, Institutions, and Norms in Multi-Agent Systems II*. LNCS 4386, pp. 107–120. Berlin, Germany: Springer, 2006.

Guerrero, A. et al., "Ontology-based policy refinement using SWRL rules for management information definitions in OWL," *Proc. of 17th IFIP/IEEE International Workshop on Distributed Systems: Operations and Management (DSOM 2006)*, Springer-Verlag, LNCS 4269, 2006.

Hayton, R. J. et al., "Access control in an open distributed environment," *Proc. of IEEE Symposium on Security and Privacy*, Oakland, California, USA, 1998.

Hoagland, J., "Specifying and implementing security policies using LaSCO, the language for security constraints on objects," Ph.D. dissertation, UC Davis, 2000.

Jajodia, S. et al., "A logical language for expressing authorisations," *Proc. of the IEEE Symposium on Security and Privacy*, IEEE Computer Society, 1997.

Javier, R. et al., "A methodological approach toward the refinement problem in policy-based management systems," *IEEE Communications Magazine*, Vol. 44, No. 10, 2006.

Kagal, L. et al., "A policy language for pervasive computing environment," *Proc. of IEEE Fourth International Workshop on Policy (Policy 2003)*, IEEE Computer Society, 2003.

Keeney, J. et al., "Chisel: A policy-driven, context-aware, dynamic adaptation framework," *Proc. of the 4th IEEE International Workshop on Policies for Distributed Systems and Networks (Policy 2003)*, IEEE Computer Society, 2003.

Kephart, J. O. et al., "Achieving self-management via utility functions," *IEEE Internet Computing*, Vol. 11, No. 1, 2007.

Ko, H. J. et al., "A semantic context-aware access control in pervasive environments," *Proc. of the 6th International Conference on Computational Science and Applications (ICCSA 2006)*, Springer-Verlag, LNCS 3981, 2006.

Lobo, J. et al., "A policy description language," *Proc. of the AAAI Conference*, 1999.

López de Vergara, J. E. et al., "Applying the web ontology language to management information definitions," *IEEE Communications Magazine*, Vol. 42, No. 7, 2004.

Loyall, J. P., M. Gillen, A. Paulos, L. Bunch, M. Carvalho, J. Edmondson, D. C. Schmidt, A. Martignoni III, and A. Sinclair. "Dynamic policy-driven quality of service in service-oriented information management systems," *Software: Practice and Experience*, Vol. 41, No. 12, pp. 1459–1489, 2011.

Lupu, E. et al., "Autonomous pervasive systems and the policy challenges of a small world!" *IEEE Computer Society*, 2007.

Lupu, E. C. et al., "Conflicts in policy-based distributed systems management," *IEEE Transactions on Software Engineering*, Vol. 25, No. 6, 1999.

Moffet, J. et al., "Policy hierarchies for distributed systems management," *IEEE Journal on Selected Areas in Communications*, Vol. 11, No. 9, ACM Press, 1993.

Montanari, R. et al., "Policy-based dynamic reconfiguration of mobile code applications," *IEEE Computer*, Vol. 37, No. 7, 2004.

Müller-Scholer, C. "Organic computing on the feasibility of controlled emergence," In CODES+ISSS '04: *Proceedings of the International Conference on Hardware/Software Codesign and System Synthesis* (Washington, DC, USA, 2004), IEEE Computer Society, pp. 2–5.

National Security Agency (NSA), Enterprise Services Division, Identity and Access Management Branch "Digital policy management: A foundation for tomorrow," In RSA Conference. https://ae.rsaconference.com/US12/scheduler/eventcatalog/eventCatalog.do (accessed November 29, 2012).

OASIS eXtensible Access Control Markup Language (XACML), https://www.oasis-open.org/committees/tc_home.php?wg_abbrev=xacml.

Ortalo, R., "A flexible method for information system security policy specification," *Proc. of the 5th European Symposium on Research in Computer Security (ESORICS 98)*, Springer-Verlag, LNCS 1485, 1998.

Ostrom, E. "Polycentric systems as one approach for solving collective-action problems," Social Science Research Network, SSRN-id130469, 2008. http://ssrn.com/abstract=1304697 (accessed September 18, 2012).

OWL Web Ontology Language, http://www.w3.org/TR/2004/REC-owl-features-20040210/.

Phanse, K. S. et al., "Modeling and evaluation of a policy provisioning architecture for mobile ad-hoc networks," *Journal of Network System Management*, Springer-Verlag, Vol. 14, No. 2, 2006.

Robinson, D. C. et al., "Domains: A new approach to distributed system management," *Proc. of the Workshop on the Future Trends of Distributed Computing Systems*, IEEE Computer Society, 1988.

Sen, S., and S. Airiau. "Emergence of norms through social learning," Presented at the *International Joint Conference on Artificial Intelligence (IJCAI-07)* 2007, pp. 1507–1512.

Sloman, M. "Policy driven management for distributed systems," *Journal of Network System Management*, Springer-Verlag, Vol. 2, No. 4, 1994.

Sloman, M. et al., "Security and management policy specification," *IEEE Network*, Vol. 16, No. 2, 2002.

The Object Management Group's Semantics of Business Vocabulary and Rules, http://www.omg.org/spec/SBVR/1.0/.

Toninelli, A. et al., "Proteus: A semantic context-aware adaptive policy model," *Proc. of the IEEE Conference on Policy (Policy 2007)*, IEEE Computer Society, 2007.

Tonti, G. et al., "Semantic web languages for policy representation and reasoning: A comparison of KAoS, Rei, and Ponder," *Proc. of the Second International Semantic Web Conference (ISWC 2003)*, Springer-Verlag, LNCS 2870, 2003.

Twidle, K. et al., "Ponder2: A policy system for autonomous pervasive environments," *Proc. of the Fifth International Conference on Autonomic and Autonomous Systems*, IEEE Computer Society, 2009.

Uszok, A. et al., "KAoS policy and domain services: Toward a description-logic approach to policy representation, deconfliction, and enforcement," *Proc. of IEEE Fourth International Workshop on Policy (Policy 2003),* IEEE Computer Society, 2003.

Uszok, A. et al., "New developments in ontology-based policy management: Increasing the practicality and comprehensiveness of KAoS," *Proc. of the IEEE Conference on Policy (Policy 2008)*, IEEE Computer Society, 2008.

Uszok, A., J. M. Bradshaw, J. Lott, M. Johnson, M. Breedy, M. Vignati, K. Whittaker, K. Jakubowski, and J. Bowcock. "Toward a flexible ontology-based policy approach for network operations using the KAoS framework," Presented at the *2011 Military Communications Conference (MILCOM 2011) 2011*, pp. 1108–1114.

van Diggelen, J., Johnson, M., Bradshaw, J. M., Neerincx, M., and Grant, T. Policy-based design of human-machine collaboration in manned space missions. *Proceedings of the Third IEEE International Conference on Space Mission Challenges for Information Technology (SMC-IT)*, Pasadena, CA, 19–23 July 2009a.

van Diggelen, J., Bradshaw, J. M., Johnson, M., Uszok, A., and Feltovich, P. Implementing collective obligations in human-agent teams using KAoS policies. *Proceedings of Workshop on Coordination, Organization, Institutions and Norms (COIN), IEEE/ACM Conference on Autonomous Agents and Multi-Agent Systems*, Budapest, Hungary, 12 May 2009b.

van Sinderen, M. et al., "Supporting context-aware mobile applications: An infrastructure approach," *IEEE Communications Magazine*, Vol. 44, No. 9, 2006.

Verma, D. "Policy based networking: Architecture and algorithms," New-Riders, 2000.

Web Services Policy 1.5 – Primer, http://www.w3.org/TR/2006/WD-ws-policy-primer-20061018.

Woods, D. D., and Branlat, M., "Basic patterns in how complex systems fail," In E. Hollnagel, J. Paries, D. D. Woods, and J. Wreathall (Eds.), *Resilience Engineering in Practice* (pp. 127–143). Burlington, VT, Ashgate Publishing, 2008.

Xu, G. et al., "A policy enforcing mechanism for trusted ad hoc networks," *IEEE Transactions on Dependable and Secure Computing*, Vol. 8, No. 3, 2011.

Chapter 15

Markets and Clouds: Adaptive and Resilient Computational Resource Allocation Inspired by Economics

Peter R. Lewis, Funmilade Faniyi,
Rami Bahsoon, and Xin Yao

Contents

The allocation of computational resources in any complex system is a challenge. In fact, as better defined by the authors of this chapter, resource allocation is a family of problems that is tightly connected with adaptation and resilience. Adaptive systems must be able to change the allocation of their resources in a dynamic way; at the same time, the failure or the lack of some resources are likely to require a re-allocation of other resources in order to continue providing the same functionalities. In this chapter, Peter R. Lewis et al. propose an economics-inspired approach to resource allocation, which not only aims to achieve efficiency, but also takes into consideration the adaptation and resilience aspects.

15.1 Introduction

Since the earliest days of computers, people have sought to apply them to the solving of large and complex problems. Indeed, computers' ability to solve large problems has brought benefits to humanity in fields as wide ranging as chess playing [1] and protein folding [2] among many others. However, key to the continued ability to apply computers to these kinds of problems is finding ways to enable them to scale massively while remaining accessible to those who might use them. For example, it could be argued that the requirement either to own a supercomputer, such as Deep Blue, or have the funds and specialist knowledge to build a distributed platform, such as that used by the Folding@Home project, reduces accessibility.

Grid computing is one technology that attempts to address this. By providing a standard way to access computing power *on tap*, a grid platform allows users to run very large, generic programs, distributed over many computational nodes [3]. Related technologies, such as cloud computing [4], enable a similar standard means of access to potentially unbounded scalable computing, and service oriented architectures [5] provide a framework for distributed computational resources to be componentized and packaged up, such that distributed applications may be constructed from loosely coupled components. As platforms grow, localized failures become more likely, and systems can often no longer be assumed to be of a static nature as their component nodes can be added and removed during runtime. Therefore, key characteristics for resource allocation mechanisms to possess are that they are resilient to failures and also able to adapt in order to obtain high performance, taking advantage of system changes during runtime.

Given such a range of approaches to scaling up computational capabilities, it is not surprising that computational resource allocation in such systems is not a single well-defined problem. Instead, it is perhaps best described as a family of problems, each specific to the particular embodiment but with much in common. At its heart, however, the problem of computational resource allocation can be stated as follows: How should computational resources be made available to users, such as to achieve the objectives of the resource providers, the users, and the system overall? It is important to note here that the term *user* does not apply solely to an end-user of a computer system or its processes, but also to any *user node* that requires the use of a resource from a *provider node*.

In order to answer this question for a particular system, it is, of course, necessary to possess some further understanding of what is required. Does the *how* in the question refer to a particular outcome or endpoint or perhaps, instead, a governing process, a set of rules or parameters to which the allocation must conform? Many approaches [6] focus on fairness and efficiency as global objectives. Furthermore, what are the objectives of the resource provider and user nodes? Are the providing nodes' objectives aligned, and do they align with the objective for the behavior of the system as a whole, if one exists? If there is a conflict or tradeoff in achieving the objectives, how are these to be resolved?

In order to gain some perspective on these issues, it is interesting to consider Foster and Kesselman's [3] characterization of computer systems as they scale. They note that in simple single *end machine* systems, resource allocation is typically dealt with at the operating system level by a kernel or similar program that has absolute control over the resources in the machine. This enables it to achieve a tightly integrated system, but it also provides a bottleneck as resource requests must be fed through the kernel in order to be assigned. In *clusters*, many individual machines can communicate through message passing and file systems. Here increased scale is obtained at the expense of integration as homogeneous nodes are controlled by a single machine responsible for job allocation. Larger still, *intranets* are characterized more by heterogeneity of nodes, which may be under the administrative control of separate entities. Nodes may have different policies for use of their resources, different external demands, and different capabilities. Here issues exist with regard to the availability of global knowledge. Nodes may attempt to map out the computing environment in order to plan the best use of resources although the size and dynamic nature of such networks means that any one node is unlikely to have an accurate view of the system's current state [3]. The final category considered by Foster and Kesselman is perhaps the most interesting, that of *internets*. These forms of network span many organizations, locations, and platforms and are large and heterogeneous. Here there is no central control and often no global objective with regard to resource allocation.

In this chapter, we consider the ability of economics-inspired techniques to achieve efficient allocations while also providing adaptivity and resilience. It has been shown that markets can be used to produce efficient and adaptive allocations in a range of resource allocation scenarios [7]. However, the type of market mechanism used, and how it is deployed, can have a large impact upon resilience. Many mechanisms require a centralized price-fixing process, such as an auctioneer or specialist, introducing a single point of failure. Other approaches use regional supernodes within a network, creating bottlenecks and unnecessary weak points. There are also fully decentralized approaches that may be used although these can introduce additional computational overhead. This chapter reviews these and argues that, of the family of economics-inspired approaches, the retail-inspired posted-offer market mechanism is a promising technique for efficient, adaptive, and resilient computational resource allocation in the presence of increasing scalability.

15.2 Computational Resource Allocation

One prominent way of acquiring the large pool of resources required by modern software systems, such as social networks, is to rent them from the cloud [4]. The cloud makes it possible for infrastructure, platform, and software providers to publicly offer their resources on demand on a pay-per-use or subscription basis. The huge cost savings and rapid elasticity of resources, that is, scaling out and scaling in, have made the cloud a hot topic among academics and in industry. Beyond its appealing business model, cloud computing has raised interesting challenges in terms of how resources may be allocated to satisfy stakeholders' objectives.

Cloud-based systems are continuously faced with the challenge of coping with dynamics and uncertainties at runtime. For example, the mode of use of cloud resources cannot be fully anticipated; hence workload patterns vary frequently. Furthermore, the cloud environment is highly volatile as resources fail and network connections fluctuate in unexpected ways. To be successful, a cloud resource management solution must cater to these uncertainties instead of avoiding them. For these reasons, software agents are often relied upon to act on behalf of users and providers to reach their respective objectives. To clarify our understanding of what an *agent* is, we adhere to the following definition:

> Agents are computer systems [or components of such systems] that are capable of independent, autonomous action in order to satisfy their design objectives. [...] As agents have control over their own behaviour, they must cooperate and negotiate with others in order to satisfy their goals [8].

An agent's autonomy empowers it to decide whether to cooperate with other agents or not, depending on its objective. As opposed to the definition above, cooperation in our work isn't a mandatory requirement; instead, we view agents as self-interested and *fully* autonomous in their ability to make decisions.

The interaction between users and providers in the cloud environment may therefore be modeled using concepts in multi-agent systems (MAS). As advocated by Jennings [9], the MAS analogy is well suited to complex application domains, of which the cloud is an example, characterized by the following:

■ A large number of components
■ Flexible (dis)connection between components
■ Complex component interconnections

Next, we describe the cloud federation model, the objectives of its stakeholders, and the resource allocation problems it presents.

15.2.1 Cloud Federation Model

The emergence of many cloud providers offering various services has propelled the vision of the cloud federation [10,11]. The proponents of the cloud federation model advocate that the next generation of cloud providers will have the capacity to seamlessly interact among themselves, thereby taking advantage of economies of scale [12]. This would afford providers the possibility of outsourcing resources at runtime in the event of failure of any cloud provider in the federation. Opencirrus* is an example of a test bed that is designed for cloud federation research. The cloud federation model is shown in Figure 15.1.

Cloud users interact with the federation via a middleware by submitting their job requests and the associated service level agreements (SLAs). The quality of service (e.g., availability, reliability, and performance) expectations of the cloud users are specified in the SLAs. The middleware layer coordinates interaction with cloud users and interaction among cloud providers in the federation. Each cloud provider is equipped with a cloud manager component, which interfaces with the middleware layer and coordinates the resources of its cloud. All interaction with a cloud provider is via its cloud manager component.

To fully realize the cloud federation model, there are a number of open research problems that must be tackled. They include formalism of a language to inform negotiation among cloud providers at runtime [14], interoperability of data formats and interfaces (APIs) to facilitate inter-cloud communication [15], and middleware layer design for coordinating cloud federation resources [16]. In this chapter, we focus on the design of the cloud federation middleware. According to [10], the two most important tasks of the middleware (referred to as the service manager in their work) are the following:

■ Deploying and provisioning users' jobs based on specified configurations
■ Monitoring and enforcing SLA compliance by throttling the capacity of users' jobs

---

* http://opencirrus.org/.

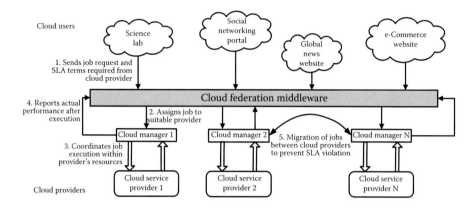

Figure 15.1   Cloud federation model. (From F. Faniyi et al., *Procedia Computer Science*, 9, 0, 1167–1176, 2012.)

These tasks are necessarily geared toward allocating cloud resources to users. Rochwerger et al. [10] further identified two approaches for reaching resource allocation objectives: explicit and implicit approaches. The former involves precise definition of resource-allocation tactics, such as scalability thresholds and the number of instances to launch or suspend. The elastic load balancer in AWS* is an example of a load balancer that follows the explicit approach. Implicit allocation, which is the approach taken in this work, relies on real-time monitoring and adjustment based on high-level service level objectives. Here there is no explicit definition of resource-allocation tactics; instead, the interaction of agents in a *market for resources* yields the allocation for each job request.

The importance of the middleware coordination layer in the cloud federation is well acknowledged [10,11]. This middleware layer (sometimes referred to as the service manager) is the highest level of abstraction responsible for coordination of cloud providers and cloud users in the federation [10]. Importantly, it ensures that cloud users' jobs are allocated to one or more cloud providers who are capable of executing those jobs without violating SLA constraints.

SLA management in the cloud is an active research area. An autonomic resource provision technique was employed by [16] to manage SLAs in federated clouds. While their work considers all phases of the SLA lifecycle, there is no explicit provision for post-negotiation causes of SLA violations, such as a variation in workload. Brandic et al. [17] presented a proposal for SLA management in a single cloud infrastructure. Their work provides a method for mapping low-level resource metrics to high-level cloud-user SLA specifications and deducing the likelihood of SLA violations from this mapping. Another interesting approach is the autonomic resource

---

* http://aws.amazon.com/elasticloadbalancing/.

allocator proposed by Ardagna et al. [18] for managing SLAs of multiple applications running on a single cloud. The authors considered SLA violation from the dimension of workload variation with the objective of maximizing cloud providers' revenue. In reality, a broader set of events may cause these violations, namely, heterogeneous user requests, workload variations, and unavailability of cloud providers in the federation.

15.2.2 Centralization and Decentralization

Classically, resource allocation objectives are achieved in a centralized manner, often relying on a single node responsible for, say, load balancing [19]. As an example, the AWS elastic load balancer functionality is often dedicated to a single virtualized instance, or, in larger sites, to multiple instances, in a centralized fashion. A balanced load, although by no means the only interesting outcome, can be used as an example of a desired resource allocation, an objective against which a particular approach to resource allocation may be tested. Load balancing is additionally, in itself, interesting because it is useful in numerous real-world scenarios, including telecommunications networks, road networks, and electricity and water distribution networks. In many of these domains, even in very large-scale systems, centralization is the usual approach taken [19].

In addition to the explicit-implicit distinction described in Section 15.2.1, resource allocation techniques can be divided into those that are stateless and those that are state-based [20]. Perhaps the most widely known and easily understood stateless approach, used to balance the load on web servers, is round-robin DNS. A more complex example is proportional share scheduling [6], in which resources are allocated to jobs according to a set of predetermined weights. However, stateless approaches such as this are unable to take account of current server load or availability, leading to no guarantee that the desired outcome is achieved. Simple state-based extensions permit the usage of information about the resources being managed and enable the proximity to the desired allocation to be measured. Examples of state-based resource allocation approaches include those that make use of geographical information and previous usage levels in order to determine an appropriate allocation of resource. A useful review and comparison of these approaches in the web server domain may be found in [20].

Centralized resource allocation methods do, however, have a number of drawbacks [21]. These include requirements

- That the environment remain static while the central coordinator is calculating the optimal resource allocation
- That the coordinator has global knowledge of the system and all nodes within it
- That all coordination messages must route through the central point, counteracting the benefit from having resources distributed about the network, reducing scalability [22] and creating a fundamentally brittle system [23]

The Internet, in particular, is a dynamic network, in which the first two requirements are highly unlikely to be met [3]. Brittleness may be mitigated against to a certain degree by introducing backup coordinator nodes; however, even in these cases, the wider system is reliant upon the existence and performance of a small number of key nodes. Failure at these key points in the network may well cripple wider functionality at best [24].

These drawbacks lead to the need for a truly decentralized approach to the allocation of resources that does not rely on a central coordinator [21]. In the field of grid computing, examples include Cao et al.'s [25] hierarchical approach and TURBO [22]. In the latter, allocations are achieved through the reliance on altruistic behavior between cooperating peers, who collaborate in order to reach a global objective. Balanced overlay networks [26] are another effective and generic technique for balancing a load across a decentralized network. In this approach, resource-providing nodes present an estimation of their availability to other local nodes to which they are connected. Newly arriving jobs take a random walk through the network and select the providing node with the highest availability. Upon accepting and completing a job, a provider node updates its availability estimate. In decentralized peer-to-peer storage systems, Surana et al.'s [27] approach may also be used. Here the case is considered when moving loads around the network also uses bandwidth. Their objective is therefore a balance between achieving an even load and minimizing the amount of load moved. Their fully decentralized approach is, in effect, tantamount to performing a centralized calculation at each node, periodically requiring cooperative reassignment of a load, based on global knowledge of the system.

15.2.3 Cooperation, Noncooperation, and Self-Interest

Critically, however, many decentralized approaches either rely on nodes having complete global knowledge or else cooperating to some extent in order to reach a shared objective [28]. As an example of this, in balanced overlay networks [26], resource users are self-interested within the bounds of the providers observed within their random walk although the providers themselves are relied upon both to provide an honest and accurate account of their availability and to facilitate the random walk by exposing their local connections. In the case in which such cooperation may not be relied upon, it is likely that the system's performance would deteriorate significantly. Similarly, Surana et al.'s [27] approach assumes both cooperation between nodes and global knowledge of the system. A noncooperative, decentralized approach to resource allocation does exist in the domain of downloading replicated files. Dynamic parallel access schemes [29,30] make use of self-interested smart clients to increase the speed of file downloads. It is not yet clear, however, how this approach might be generalized to other service-based systems.

Buyya et al. [28] argue that we may not always be able to rely on cooperation between nodes for several reasons. Among these are the possibility that a node behaves erroneously, perhaps due to a software or hardware error, such as a virus; unforeseen circumstances; or an external fault. Large systems are also likely to be noisy systems as data is lost or corrupted in transit and the likelihood of measurements being inaccurate or misreported increases. Finally, limits on and delays in information transmission mean that nodes' actions may be misguided or insufficient. Crucially, Khan and Ahmad [31] show that in any decentralized cooperative approach, global optima can only be achieved when all the nodes cooperate. It is for these reasons that, in seeking high resilience, we look toward approaches that do not rely on the assumed cooperation of nodes.

Some confusion does exist within the literature, however, in the treatment of the terms *noncooperative* and *self-interested*. It is important to note that non-cooperation does not imply self-interest. Indeed, in Khan and Ahmad's [31] study of various games-based resource-allocation methods, they describe a model in which noncooperative agents bid for jobs based on an honest estimation of the estimated time to complete a job. Their agents, although not cooperating, act without consideration of the benefit they expect to derive from their actions. Clearly, such a consideration is a prerequisite for self-interested behavior, and hence the behavior they describe is not self-interested.

Indeed, it is the assumption that an agent will behave either cooperatively or noncooperatively, regardless of its predicament, that is at odds with self-interest. A self-interested agent may behave either cooperatively or noncooperatively at certain times. The key factor is that this decision will be made by the agent, based on whether it is in its own perceived interest to do so. In making this decision, the agent must therefore consider the benefit it expects to gain from the options with which it is faced. If it does not, it cannot be said to be truly self-interested.

Therefore, when considering systems in which nodes are owned or administered by separate parties, such as the very large distributed systems discussed by Foster and Kesselman [3], rather than consider agents on a cooperative/noncooperative spectrum, it may instead be more useful to know whether or not an agent is self-interested. If it is possible to assume this of nodes, then, as will be discussed in the following sections, the models and tools of economics allow for a great deal of progress to be made.

15.2.4 Inspiration from Economics

When selecting components with which to compose an application in a cloud system, appropriate resources may be available from a number of provider nodes. Similarly, large numbers of users may find themselves competing for access to the best resources or a resource at a time more suited to their needs. If individual users and providers are acting in a self-interested manner in these types of computational systems, then the resulting interactions may be thought of as being an economy [7].

Indeed, large computer networks, such as the Internet, made up of heterogeneous individuals with independent objectives, can quite rightly be viewed as social networks as well as purely digital ones. It is perhaps of little surprise then that a social science, such as economics might be useful in solving a problem such as decentralized computational resource allocation because economics itself is concerned with the allocation of resources between individuals with different objectives in human societies. Therefore, in computational networks that are social, to what extent can economic theory be called upon in order to predict and, we hope, design the resource-allocation behavior of complex computational systems when individual nodes are self-interested?

It is perhaps useful at this stage to present some relevant terminology. First, according to Begg et al. [32], economics is "how [a] society resolves the problem of scarcity" (p. 3). Furthermore, they state that "a resource is scarce if the demand at a zero price would exceed the available supply" (p. 5). This is exactly the scenario with which we are faced in the computational resource-allocation problem. There have, of course, been a number of different approaches to this problem in human history, but one that is particularly dominant is the use of markets. Rothbard [33] describes a *free market* as "an array of exchanges that take place in society. Each exchange is undertaken as a voluntary agreement between two people or between groups of people represented by agents." Similarly, Begg et al. [32] define a market as "a set of arrangements by which buyers and sellers are in contact to exchange goods or services" (p. 32). The important factors here are that there is an exchange between two or more individuals and that this exchange is voluntarily entered into by all participants.

In order to facilitate such exchanges, a particular type of good is often agreed-upon to serve as currency, in which case the individual giving away currency in order to obtain another good is termed the *buyer*, and that who receives the currency and gives away the other good is termed the *seller*. It is, of course, not required that this formal delineation be present although it has been argued [34] that an economy will evolve toward common agreement on a particular good to treat as currency, typically that which the individuals find easiest to retain and exchange widely without additional cost. A mechanism through which voluntary exchanges between individuals are facilitated is called an *auction*, and there are many sets of rules for these, leading to a huge range of possible auction types.

A number of auction mechanisms can be found in common use, including the English auction, found among other places on Ebay*; the Dutch auction; the Vickrey auction; and the continuous double auction, often used in stock markets. Cliff [35] gives a useful introduction to and critique of several auction mechanisms, including those listed here, and Friedman and Rust [36] provide a more detailed look at the continuous double auction. Purely electronic markets also make use of a range of auction mechanisms. In designing a mechanism, the aim is typically to achieve an efficient system overall by making use of the self-interested nature of individuals. This is demonstrated by Phelps et al. [37], Byde [38], and David et al. [39]

---

* http://www.ebay.com/.

among others. For many, the ultimate aim of such research is the automation of the mechanism's design, appropriate to individual scenarios [40–43]. Taking Cliff's [43] work as an example of this, a parametrized mechanism design space is specified, which may be searched in order to find high-performing mechanisms for specific scenarios. Results from an evolutionary search demonstrate that classic, human-designed mechanisms are often far from optimal.

15.3 Economics-Inspired Computational Resource Allocation

The application of economic ideas to resource-allocation problems in computational systems is approached in the field of market-based control [7]. Using the terminology of Casavant and Kuhl's [44] taxonomy of scheduling in distributed computing systems, this is a family of distributed mechanisms for dynamic global resource allocation. Such systems work by actions and decisions of resource provider and resource user nodes being automated by the use of software agents interacting in a (possibly artificial) market. The aim of a buyer agent might be to secure the fastest and most reliable resource at the lowest cost for its user. Conversely, a seller agent might aim to maximize the revenue for the resource provider or perhaps generate high levels of business. Whatever the business strategy of the resource provider, the selling agent will be competing with similar agents from other providers for the same resource users. Each agent will therefore have to employ its own strategy for success in the market.

There are several examples of market-based control being used in decentralized computing systems. Brewer [19] proposes the idea of incorporating into a request for web services a notion of its value or cost. It is argued that this, along with the use of smart agents, would allow for responsive adaptation in the presence of changes to the network as well as graceful degradation. Similarly, Gupta et al. [45] argue that, in the provision of virtually zero cost-per-use computational services, a mechanism involving pricing and user self-selection is preferable to the alternative of provider or regulator enforced limits: rationing. More recently, researchers have pursued in-depth study of a market-oriented cloud from various dimensions, including price modeling [46], resource sharing among service providers [47–49], and resource allocation at the hardware layer [50]. A notable example that harnesses a game-theoretic approach is the formulation and study of the service-provisioning problem in cloud systems by Ardagna et al. [51].

Typically, resource-owning or -providing nodes are represented by selling agents, and resource users or tasks are represented by buying agents. Buyers then attempt to purchase sufficient resources to satisfy their task or user's requirements from the set of available sellers. Sellers charge an amount of (either real or artificial) money for the resource, determined by their strategy and dependent on factors, such as the quantity or quality of the resource being provided. Because self-interested buyers can be expected to pay more for resources that they desire more, and self-interested sellers will charge what they can get away with in order to maximize their pay-off, resources will tend to go to those who value them the most. Fundamentally, these approaches attempt to harness the rational behavior of self-interested agents, which interact in some market environment in order to achieve resource allocation without reference to a central authority. Relying upon the theories of economics, through such repeated exchanges between utility-maximizing individuals, efficient resource allocations may be achieved.

15.3.1 Centralized Market Mechanisms

As in human economies, agents in a market-based computational system may inter-act through any of a number of different mechanisms [35]. Common examples include English, Dutch, and Vickrey auctions, in which an auctioneer facilitates the bidding and determines the allocation of resources. When scarcity exists on both the seller and buyer sides, double auctions, such as the continuous double auction and clearing house, provide an alternative approach [36]. Research in the field of automated mechanism design also suggests that other, less obvious, auction mechanisms may lead to more efficient outcomes in certain circumstances [43,52,53].

However, both Cliff and Bruten [54] and Eymann et al. [21] note that, due to the mechanisms employed, a large proportion of market-based control systems are not decentralized because they rely on a centralized price-fixing process rather than the participants, between them, determining prices. This is true of Wolski et al.'s [55] G-Commerce model, which relies upon a central market maker. Cliff and Bruten [54] argue that the presence of such a centralized process or component removes the primary advantage of using a market-based system: its robust, decentralized, self-organizing properties.

Examples of the application of centralized market mechanisms in cloud-based systems include [56] and [57]. In [56], the problem of running independent equal-sized tasks on a cloud infrastructure with a limited budget was studied. The authors concluded that a constrained computing resource-allocation scheme should be benefit-aware, that is, the heuristics for task allocation should incorporate the limited resource in supply within the system. Sun et al. [57] proposed a Nash equilibrium-based continuous double auction (NECDA) cloud resource-allocation algorithm for meeting performance and economic QoS objectives. In each round of the system run, each provider agent determines its requested value based on its workload, and each user agent determines its bid value based on the remaining time and resources [57]. A CDA is then used to decide the outcome resource allocation and the existence of a Nash equilibrium evaluated.

15.3.2 Distributed Market Mechanisms

A number of distributed auction mechanisms have also been proposed [58–60], which do not rely on one central coordinating node. These approaches reduce the fragility associated with reliance upon a single point, provide more scalability, and allow for dynamic composition of auctions. Typically, either the central auctioneer is replaced by a number of local ones, which may communicate through some secure means, or else the auctioneer role is fulfilled by a spare, disinterested node. Double auctions for example, although relying on a specialist to match bids and asks [37], may be decentralized by the presence of multiple specialists between which the participants may choose [61,62]. This is the approach taken by [63], in which multiple specialists were used for service composition in a SaaS cloud. These techniques do reduce bottlenecks at certain points within the network, and the removal of a single node cannot lead to system-wide failure. However, similarly to the replicated round-robin DNS approaches discussed in Section 15.2.2 above, the system is still largely reliant on a small subset of its nodes.

However, it may be possible in systems such as this to scale up the number of auctioneers or specialists in order to achieve a suitable degree of redundancy and decentralization. This issue is an active area of research although intuition suggests that, with all else equal, a system that relies upon a set of super-nodes cannot provide the level of resilience of a system without such a need even if the super-nodes were present in abundance. Approaches such as this also raise questions of motivation for those acting as super-nodes as participation fees, for example, are set by auctioneers in most cases [61]. Therefore, if an approach exists without the need for such complexity, it should be preferred.

A further alternative is that individual provider nodes themselves host independent auctions for their resources. This approach is applied to computational resource allocation in Spawn [64]. Here, users' agents bid in sealed-bid auctions hosted by providers' agents for their resources. In order to be effective, this requires a high level of strategic ability on the part of buyers as they must decide in which auctions to participate. Of course, consumers may win multiple auctions, and questions then arise of how to handle these situations. Literature exists that explores the dilemma faced by buying agents bidding in multiple auctions, such as that by Gerding et al. [65,66] although, again, this adds complexity.

15.3.3 Bargaining

Cliff and Bruten [54] conclude from their critique that, rather than depend upon a central node, such as an auctioneer, market mechanisms should instead rely on the ability of intelligent agents to bargain between themselves in order to arrive at acceptable prices. This approach is taken in the AVALANCHE [67] and CATNET [21,23,68,69] systems. These take inspiration from agent-based computational economics (ACE) [70], an agent-based modeling technique that attempts to replicate the dynamics of human markets with complex cognitive agents.

These approaches are those that attempt to replicate human markets the most faithfully because they rely on highly developed strategies as agents negotiate bilaterally in order to determine the provision of a resource. It is likely, in this approach, that the development and operation of such strategies will themselves require significant computational overhead. Although these approaches are indeed effective and widely applicable, if a simpler alternative exists, it should be preferred when possible. An additional point of interest is that, in the mechanism used in CATNET [21], resource-providing nodes are relied upon to forward requests to neighboring hosts. They do this without any consideration of the effect of this on their own interests, which would appear to be at odds with the self-interested nature of the agents. The study of bargaining agents is a topic of ongoing research [71–73] and has a relevance in economics more widely than only for computational resource allocation.

15.3.4 Retail Markets and the Posted-Offer Mechanism

Although they do not discuss them in detail, Cliff and Bruten [54] also briefly mention retail markets as an alternative to auctions and bilateral bargaining. The mechanism used in modern retail markets is usually referred to as the *posted-price* or *posted-offer* model [74,75] although, in online content delivery, it is sometimes referred to as the *quoted-price* model [76]. It is a fully decentralized approach to the determination of price without the need for complex bilateral negotiation and provides a potentially simpler alternative.

Wang [77] provides an interesting comparison of auction-based and posted-offer selling and shows that auctions are more commonly used in human markets in which there is a greater dispersal of valuations of the good among the buyers. When buyer valuations are more similar, however, he favors the posted-offer market mechanism. This can be reconciled with the idea that, according to the most common mechanism-design objectives, there exists no single dominant mechanism [53]. For an example of this in a specific case, the impossibility result due to Myerson and Satterthwaite [78] shows that no double auction can simultaneously be efficient and budget balanced while also ensuring that at least one participant would not be better off using a different mechanism. It is therefore appropriate that research into computational resource allocation continues to consider the impact of a range of mechanisms.

The application of posted-offer markets [74] to computational resource allocation is the most recently proposed technique in the market-based control family [79]. The posted-offer mechanism is a process in which sellers of multiple units of a good each post one price or offer, and buyers subsequently respond by stating the quantity they wish to purchase from each seller. Exchanges then occur between buyers and sellers at these prices and quantity values. Technically, the reverse process in which buyers quote prices and sellers state quantities is also a posted-offer mechanism although it is less commonly encountered. Importantly, price quotations cannot be changed during the exchange period: No further negotiation is permitted, substantially reducing the burden on agents and simplifying the allocation process.

Some prior examples of the use of similar mechanisms in computational resource allocation do exist in the literature although they are not faithful implementations and make additional assumptions. Chavez et al. [80] use an approach of this type in Challenger, in which offers are broadcast to the nodes in a network although, instead of using price, bids contain an honest reporting of a job's priority. This honesty means that there is no competition between nodes, and as discussed in Section 15.2.3, this is not self-interested behavior. Xiao et al. [81] describe their system, GridIS, in which buyers broadcast job requests and sellers reply by posting offers to perform them at a price. However, the behavior of the sellers used requires certain global information in determining their price, both in the form of the latest accepted market price, which, in a posted-offer mechanism, is considered to be private information, and also the level of aggregate supply of all the providers in the network. Again, the assumption of private information forbids this too.

One of the most faithful implementations of the posted-offer mechanism in decentralized computational resource allocation is that by Kuwabara et al. [82] although they do not describe it as such. They propose an approach in which sellers quote prices for their resources, and buyers subsequently decide the quantity (which may be zero) to purchase from each seller. Their analysis determines the quantities provided at the equilibria at which the markets arrive and present this as a stable outcome allocation of resources. This is indeed fully decentralized because no central component, such as an auctioneer or specialist, is used; prices are determined privately by the sellers and then posted via a broadcast mechanism. More recently, we have extended this approach, investigating the behavior of posted-offer markets used for computational resource allocation in a range of homogeneous and heterogeneous contexts [79,83].

15.3.5 Applicability of Economic-Inspired Approaches

Many computational resource providers, for example those in the cloud, bid to attract users by promising "elastic" service provisioning with near-infinite scalability. In reality, clouds, just like data centers, are resource-constrained and prone to failures at both node and network levels. However, in contrast to conventional data centers, cloud providers face the problem of not being able to fully anticipate the workload patterns imposed on their infrastructure in advance. These conditions make it hard to promise high qualities of service without incurring significant cost.

Economics-inspired methods offer a potential solution to alleviate this problem, providing increased resilience to internal unexpected changes (server or network failures) and adaptivity to dynamics caused by external sources (e.g., a spike or dwindle in workload) [84]. Ongoing research in this area investigates the use of market mechanisms to manage interaction of computing nodes in cloud systems. Results from cloud simulation studies (e.g., [83]) indicate that novel resource allocation methods inspired by economics can be more resilient to node failures. Due to the inherent decentralization of many market mechanisms, they offer the capability to manage resources at the scale of cloud federations [13]. In the following section, we will give an introduction to this work, providing an example of how posted-offer markets may be used for resource allocation both in an abstract problem and cloud federations.

15.4 Cloud Resource Allocation Using Posted-Offer Markets

Cloud federations are example of distributed environments in which the posted-offer mechanism may be used to allocate resources. A cloud federation consists of single cloud providers who exchange (or trade) resources in order to improve their SLA compliance levels. Buyya et al. [11] envisioned the federated (or intercloud) model as an environment that could flexibly respond to variations in workload, network, and resource conditions by dynamically coordinating multiple clouds in the federation. Because it is infeasible for a cloud provider to have data centers in every country, the federated cloud environment offers the additional benefit of rapidly scaling to meet the needs of geographically distributed cloud users than any single cloud provider [11]. The RESERVOIR project [10] also sets out a vision similar to [11] for an open federated cloud-computing model to address the limited scalability of single cloud providers and lack of interoperability among them.

15.4.1 Motivating Example: Service Selection Problem

Consider a hypothetical online shopping cart application that dynamically composes its services to meet customer orders. An order typically consists of one or more products, which may be purchased and shipped from a pool of diverse services. For simplicity, we restrict the composition to four abstract services in a sequential workflow pattern (Figure 15.2), and an order contains only one product. For more advanced workflow patterns see Jaeger et al. [85].

The responsibilities of the four services are defined as follows:

- Service A: Renders the company's product catalogue in a browser and provides a means for customers to place orders.
- Service B: Provides selected product(s) in the customer order at a specified cost. Supposing N product supplier services are available, possible options are $B_1, B_2, \ldots, B_N$, of which only one is selected per product.

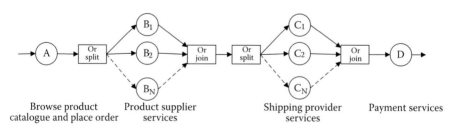

Browse product catalogue and place order   Product supplier services   Shipping provider services   Payment services

Figure 15.2 Online shopping cart service composition. (From F. Faniyi et al., *Procedia Computer Science*, 9, 0, 1167–1176, 2012.)

■ Service C: Offers shipping services for product(s) in the customer order within a specified delivery time and at a cost. For N shipping service providers, possible options are $C_1, C_2, ..., C_N$.

■ Service D: Provides payment service to collect funds from customers on behalf of the company.

Services A and D belong to the company; hence they are static. On the other hand, services B and C are provided by software-as-a-service cloud service providers (CSPs), which are selected dynamically at runtime. This implies that product supplier services ($B_1, ..., B_N$) are substitutable; similarly, supplier services ($C_1, ..., C_N$) are substitutable, subject to the following constraints:

■ Minimize product cost
■ Minimize delivery time and shipping cost

The company's SLA objective is to minimize the cost of meeting orders (i.e., product and shipping cost) without exceeding the promised delivery time.

The setup of the cloud federation is shown in Figure 15.3. The online shopping cart company interfaces with the cloud federation to find concrete CSP instances of the product supplier and shipping services. For each requested service, the order and associated SLA terms are submitted to the cloud federation. Here, an order request i is denoted by O_i. To meet the SLA constraint specified in customer orders, the following simplified SLA models are defined for the dynamic services.

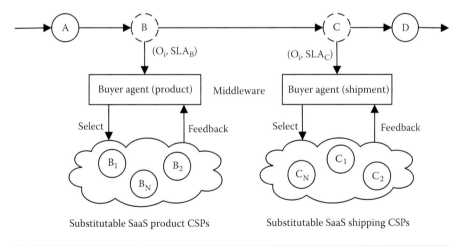

Figure 15.3 Online shopping cart application interfacing with two specialized cloud federations. The buyer agent (product) acts on behalf of the application to select product supplier service, and the shipment buyer agent selects shipment services on behalf of the application. (From F. Faniyi et al., *Procedia Computer Science*, 9, 0, 1167–1176, 2012.)

Each CSP (a seller in the posted-offer market) publishes its cost and delivery time offerings via its cloud manager interface. In terms of the posted-offer mechanism, this represents the posting of an offer. Buyer agents are specialized for their respective objectives and hence select CSPs according to their offers. These CSPs are then instantiated in the workflow.

15.4.2 Bertrand-Based Load Balancing

Bertrand's [86] model of economic competition is one of the simplest to account for the interactions between individual sellers who compete on price to provide homogeneous goods. The posted offer mechanism is qualitatively similar to Bertrand's model in that it also accounts for sellers that compete on price to provide a homogeneous good to a population of buyers. In the types of computational resource-allocation problems investigated here, including the above example, the good is considered homogeneous because the buyers do not care from whom they purchase equivalent resources or services so long as it fulfills the necessary requirements.

Many market-based resource-allocation mechanisms, such as those discussed in Section 15.3, are concerned with achieving system-wide efficient allocations. In this section, however, we consider how to achieve particular outcome resource allocations in a given scenario. The approach, which is described more fully in [79] begins from the starting point of a desired allocation of resources, which the system designer or owner wishes to achieve. An artificial market is then created in order to bring this allocation into effect under the assumptions of decentralization and self-interest.

By means of an abstract problem model, consider a scenario consisting of a set of resource-providing nodes (or CSPs in the above example), S, each member of which provides an equivalent, quantitatively divisible resource π, which may vary only in price. The members of S are assumed to be self-interested. Subsequently, imagine a large population of resource users or buyers, B, each member of which aims to consume some of the resource π (e.g., use the service), at regular intervals.

If s_i is a node in S and b_j is a node in B, q_{ij} is used to denote the quantity of the resource π provided by s_i to b_j. The total quantity of π provided by s_i at a given instant, its *load*, l_{s_i}, is, therefore,

$$l_{s_i} = \sum_{j=1}^{|B|} q_{ij}. \tag{15.1}$$

As an example of a desired outcome resource allocation, we consider the ability of Bertrand competition to bring about a balanced load, such that at any instant, each resource-providing node in S is providing an equal amount of π across the population of resource users. A particular resource allocation such as this, a configuration for the provision of π by the nodes in S at a given instant, may be expressed by the vector $\overrightarrow{L_S} = \langle l_{s_1}, l_{s_2}, \ldots, l_{s_n} \rangle$, where $n = |S|$. For convenience and ease of comparison between scenarios, we often normalize this vector by the total resource being provided. An evenly balanced load may therefore be written as $\left\langle \frac{1}{n}, \frac{1}{n}, \ldots, \frac{1}{n} \right\rangle$.

Although this is a trivial problem when central control or cooperation may be assumed, here the objective is to achieve this using only self-interest in a fully decentralized manner with no central or regional control and with only private information available.

15.4.2.1 Mechanism and Assumptions

A posted offer mechanism is used to decide what quantity of the resource π is provided to which user node and from which provider node. At a given instant, a resource-providing node, $s_i \in S$, advertises π at the price $P_{s_i}^\pi$ per unit via a broadcast mechanism. Each resource user, a buyer in this case, then has the option of purchasing some of the resource π should it be in its interest to do so at the price offered. The system iterates, with sellers able to independently adapt their prices to the market conditions over time.

Each time-step, each buyer, if it chooses to buy, may purchase any amount of π from any number of resource providers in S, subject to the constraint that the total amount purchased per time-step is equal to its total required quantity (here often normalized to one unit). If no offer from any seller in S is acceptable, the buyer may instead purchase nothing. These constraints mean therefore that $\sum_{i=1}^{|S|} q_{ij} \in \{0,1\}$ for all $b_j \in B$.

15.4.2.2 Buyer Behavior

In this model, both buyers and sellers accrue a payoff, or utility gain, from their interactions in the marketplace. For buyers, this is deemed to be the value they associate with the price paid subtracted from the value they associate with the purchased resource. If buyer b_j's unit valuation of π is denoted by $v_{b_j}^\pi$, then its payoff from a unit transaction with s_i will be $v_{b_j}^\pi - p_{s_i}^\pi$. Because any buyer accepting a price above $v_{b_j}^\pi$ would lead to a negative payoff, this is its reserve or limit price. From a buyer's perspective, if a seller's price would not lead to a negative payoff for the buyer, then the price is described as being *acceptable*. S_{b_j} is used to denote the subset of S that contains exactly those sellers in S whose price is *acceptable* to buyer b_j. When buyers are homogeneous insofar as they have the same reserve prices, such that $v_{b_j}^\pi = v^\pi$, $\forall b_j \in B$, a set of sellers acceptable to the buyer population B exists and is denoted as S_B. Of course, $S_B \subseteq S$, or more precisely, $S_B = \left\{ s_i : s_i \in S, p_{s_i}^\pi \leq v^\pi \right\}$.

As with sellers, buyers are assumed to be self-interested and boundedly rational at least insofar as they prefer higher payoffs to lower ones. As with real economic actors, this is manifested through the following of some behavioral strategy. The strategy incorporates a decision function, which, given a situation, describes the quantity (which may be zero) to buy from each seller. A similar approach is taken by Greenwald and Kephart [87], who model buyers as either hyper-rational *bargain hunters*, seeking out the best possible price, or else *time savers* who will purchase from any *acceptable* seller, chosen at random.* In our work [79], we consider these two buyer-behavior models and also a third behavior called *spread buyers*, simple risk-averse buyers, who prefer to spread their purchases across a number of sellers. The possibility of complex and arbitrary buyer-decision functions means that there may not be a straightforward mapping between sellers' prices and buyer valuations and the subsequent outcome allocation. Determining the outcome is therefore nontrivial.

Although buyers may adopt any of a number of behavioral strategies, in this chapter, three representative buyer types are considered. These are Greenwald and Kephart's [87] hyper-rational *bargain hunters* and *time savers* and a further type, a risk-averse *spread buyer* behavior [88]. These are now described.

Bargain hunters always attempt to maximize their instantaneous payoff. In each iteration, they check the prices of all the sellers, selecting the one seller who provides the most attractive offer (i.e., the lowest price). If this price is acceptable, then the buyer purchases its entire unit of π from that seller. In the event that more than one seller provides an equally attractive and acceptable offer, the buyer purchases an even proportion of π from each such seller. This is the basic model of consumers used by Bertrand [86].

Time savers do not check the price of every seller in the system when deciding from whom to buy. Instead, they select a seller at random, and if its price is acceptable, then they purchase the entire unit of π from that seller. If it is not, then they continue selecting previously unchecked random sellers until they find an acceptable price. If no seller has an acceptable price, then they purchase nothing.

Spread buyers are simple risk-averse agents, preferring to spread their purchases across a number of sellers. At each time-step, the buyer looks at all the available offers and purchases a proportion of π from each seller with a price below $v_{b_j}^{\pi}$, relative to the expected utility gain from purchasing from that seller. Specifically, the quantity purchased by buyer b_j from seller s_i is determined according to the following calculation:

$$q_{ij} = \frac{\left(v_{b_j}^{\pi} - p_{s_i}^{\pi}\right)}{\left(nv_{b_j}^{\pi} - \sum_{k=1}^{n} p_{s_k}^{\pi}\right)}. \tag{15.2}$$

Spread buyers only consider those sellers with an acceptable price.

It is worth reinforcing that although three buyer behaviors are considered here, many other potential behaviors will exist and can be analyzed using a game-theoretic methodology.

---

* Greenwald and Kephart [87] refer to *time savers* as *any seller* or *type A* buyers.

15.4.2.3 Seller Behavior

Sellers also receive a payoff, defined by their payoff function. Seller s_i's payoff is denoted as P_{s_i}. In its simplest form, this is its revenue from the sale of π:

$$P_{s_i} = \sum_{j=1}^{|B|} p_{s_i}^{\pi} q_{ij}, \qquad (15.3)$$

or indeed

$$P_{s_i} = p_{s_i}^{\pi} \times l_{s_i}. \qquad (15.4)$$

Clearly, a seller wishing to maximize its revenue would aim to increase both its price and the quantity of its resource sold to the buyers, its market share. However, as we have seen from the buyers' behavior, the market share will depend upon the relationship between its price and those of its competitors; specifically, a higher price is likely to lead to a lower market share.

15.4.2.4 Outcome Behaviors and Allocations

One motivation for employing an artificial market is that competition between self-interested sellers drives the system toward equilibrium. It is at this equilibrium that the system is stable in the long term, and thus we refer to the allocation of resources in this stable state as the *outcome resource allocation*.

The model described here is, in essence, a generalized version of the Bertrand game [86]. The classic Bertrand game consists of two sellers, both of whom offer to sell a certain homogeneous good to a population of buyers. Each seller must decide what price to charge for the good and then supply the quantity subsequently demanded by the buyers. The buyers in the classic Bertrand game behave hyper-rationally, as with the *bargain hunters* studied here, always buying from the seller with the lowest price or half from each seller if the prices are identical.

In this game, either seller can take the entire market by offering a price only fractionally lower than its competitor. However, because this applies to both sellers, the noncooperative Nash equilibrium for the game is for both sellers to charge as little as possible, their zero-profit price. If each seller's costs are equal, then the equilibrium price for each seller will also be equal. This leads to the sellers sharing the market equally at equilibrium, and it is this basic idea that provides us with a balanced load in the simplest case.

However, in the more general case, in which buyers may follow any of a number of strategies, calculating the expected outcome resource allocation may be a more complex task. In [79], we describe and exemplify a game-theoretic methodology for calculating the expected outcome resource allocation by determining the sellers' best response at each iteration. This is done by solving payoff equations constructed from the given buyer behavior. This enables us to identify the Nash equilibrium outcome, in which each and every seller's best response is equal to its previous position.

In the following illustrations, it is assumed that the buyers have an identical reserve price, $v^\pi = 300$, and therefore that we have a single acceptable set of sellers, S_B. Any seller in S but not in S_B will, of course, attract no buyers at all and will hence receive no payoff and have a load of zero. For the sake of clarity, in the remainder of this section, only those sellers in S_B are considered.

Bargain Hunters: Let us first consider a scenario with two identical resource-providing nodes, such that $S = \{s_1, s_2\}$, each with costs of zero. Recalling the sellers' payoff function, given in Equation 15.4, we have that

$$P_{s_1} = p_{s_1}^\pi \times l_{s_1}.$$ (15.5)

and

$$P_{s_2} = p_{s_2}^\pi \times l_{s_2}.$$ (15.6)

As in Bertrand competition, B is a large population of hyper-rational buyers, *bargain hunters*, as described in Section 15.4.2.2. Recalling the decision function for these buyers and the assumption that each buyer wishes to purchase exactly one unit of π, we may therefore say that

$$P_{s_1} = \begin{cases} |B| \times p_{s_1}^\pi & \text{if } p_{s_1} < p_{s_2}; \\ 0.5 \times |B| \times p_{s_1}^\pi & \text{if } p_{s_1} = p_{s_2}; \\ 0 & \text{otherwise.} \end{cases}$$ (15.7)

and the equivalent for s_2, respectively.

From a game-theoretic perspective, given an observed value for their competitor's price, both s_1 and s_2 will wish to respond with the best response. In this case, this will be to undercut the competitor's price, if possible, in order to receive the payoff given by the first case in Equation 15.7. The competing seller will, of course, act similarly, leading to a price war in which each undercuts the other until their zero-payoff price is reached. Assuming that a seller would rather not participate than receive a negative payoff, once $p_{s_1} = p_{s_2} = 0$, the rational course of action is to maintain a price of 0, accepting the second case.

Recalling that the current load on a resource-providing node is given by Equation 15.1 above, we therefore have that, at equilibrium,

$$l_{s_1} = 0.5 \times |B|,$$ (15.8)

and

$$l_{s_2} = 0.5 \times |B|.$$ (15.9)

This is indeed an evenly balanced load, that is,

$$\overrightarrow{L_S} = \left\langle \frac{1}{2}, \frac{1}{2} \right\rangle. \tag{15.10}$$

The theory of Bertrand competition (which is described more fully in [86]) demonstrates that when competing on price alone, two sellers are enough for the perfectly competitive outcome described here. Because the same logic applies to larger number of sellers, this evenly balanced outcome also holds for larger systems under the same assumptions. This idea was first presented in [88] and elaborated upon in [79].

Time Savers: Intuitively, a population of *time savers* will possess less of the *all or nothing* nature of *bargain hunters* as each will prefer potentially any seller whose price is acceptable. Considering the simple two-node example described above, what outcome should we expect with a population of *time savers*? Recalling that only those sellers in S_B are considered at present, the payoff for s_1 and s_2 should be expected to be

$$P_{s_1} = \frac{p_{s_1}^{\pi}}{|S_B|} \tag{15.11}$$

$$P_{s_2} = \frac{p_{s_2}^{\pi}}{|S_B|} \tag{15.12}$$

Here, unlike with *bargain hunters*, there is no advantage for a seller in undercutting the price of a competing seller because this will only serve to reduce its payoff. Instead, the dominant position is to charge the highest possible price while still remaining in S_B; the equilibrium is at $p_{s_1} = p_{s_2} = v^{\pi}$.

Similarly to *bargain hunters*, however, because $p_{s_1} = p_{s_2}$, then $\overrightarrow{L_S} \approx \left\langle \frac{1}{2}, \frac{1}{2} \right\rangle$. Note that due to the probabilistic nature of the buyers' decision function, the allocation will tend toward this as the probabilities average out.

Spread Buyers: For a population of *spread buyers* as described in Section 15.4.2.2, the sellers' payoff functions for the simple two-node case are

$$P_{s_1} = \sum_{j=1}^{|B|} \frac{v^{\pi} - p_{s_1}^{\pi}}{2v^{\pi} - \left(p_{s_1}^{\pi} + p_{s_2}^{\pi}\right)} \times p_{s_1}^{\pi} \tag{15.13}$$

and

$$P_{s_2} = \sum_{j=1}^{|B|} \frac{v^{\pi} - p_{s_2}^{\pi}}{2v^{\pi} - \left(p_{s_2}^{\pi} + p_{s_1}^{\pi}\right)} \times p_{s_2}^{\pi}. \tag{15.14}$$

Sellers s_1 and s_2 will each then attempt to maximize their respective payoff function as before. The outcome resource allocation occurs when the system is at equilibrium. Figure 15.4a illustrates an example payoff function for s_1 when $v^\pi = 300$ and $p_{s_2}^\pi = 250$.

Clearly, the best response price for s_1 is less than $p_{s_2}^\pi$; in fact, in this instance, it is 217.71. However, given this value as $p_{s_1}^\pi$ subsequently, s_2 is then faced with the payoff function illustrated in Figure 15.4b. Of course, s_2 will respond to this value for $p_{s_1}^\pi$. Its best response is, in this case, 204.92. By using the sellers' payoff functions to iteratively calculate each seller's best response, this particular system is found to be at equilibrium when $p_{s_1}^\pi = p_{s_2}^\pi = 200$.

Clearly, at this point, the market share and hence the load of each seller is also equal: $\overrightarrow{L_S} = \left\langle \dfrac{1}{2}, \dfrac{1}{2} \right\rangle$.

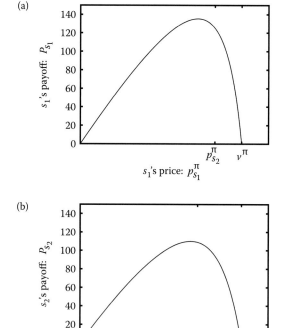

Figure 15.4 (a) Seller s_1's payoff function with one competitor and a population of *spread buyers*, and (b) s_2's subsequent payoff function from s_1's best response.

15.4.3 Deployment of Posted-Offer Markets in the Cloud

Here we present an evaluation of homogeneous buyer and seller populations under two cases of the cloud service selection case study. Two buyer strategies, namely, *time savers* and *bargain hunters* (cf. Section 15.4.2.2) are considered to understand if a tradeoff exists between the timeliness of meeting an order and the selling price. In all experiments, results presented are averaged over 10 independent simulation runs to account for stochasticity.

The two cases considered are shown in Table 15.1. An order request, O_i, has a priority—high, medium, or low—that indicates the urgency of the request. For each order request, a buyer agent is assigned to search the seller agents (CSPs) capable of fulfilling that order.

Given an SLA, the buyer utility function is defined as

$$U_b(O_i) = w_b + (k^*\beta_{price}) \tag{15.15}$$

The value of w_b is initialized based on the order's priority. For results presented here, the tuple is defined as (Priority, w_b) = {(High, w_b = 2), (Medium, w_b = 1), (Low, w_b = 0)}. SLA priorities are randomly assigned to orders, following a normal distribution; k is a sensitivity factor for tuning the valuation of the buyer agent, and $k = 0.1$ for all experiments considered here. The value of β_{price} is derived from the summation of nonfunctional (NF) attributes of the buyer agent, that is, the availability, reliability, and performance. Each NF attribute is randomly initialized to a value in the interval [80, 99.999].

Given an order request, the seller utility is defined as

$$U_s(O_i) = w_s^{\cdot}\theta_{price} \tag{15.16}$$

The value of w_s is initialized based on the rule (Priority, w_s) = {(High, w_s = 0.1), (Medium, w_s = 0.01), (Low, w_s = 0.001)}. Similar to β_{price}, the value of θ_{price} is derived from summation of NF attributes of the seller agent and set to 100 for the three NF attributes.

It is worth noting that this formulation ensures that it is always possible to meet an order although the time spent making the trading decision is non-determinant.

Table 15.1   Scenarios

	No. Orders	Arrival Rate	No. Service B	No. Service C
Case A	50	10 ticks	25	25
Case B	100	10 ticks	25	25

The overhead of using a trading strategy is measured by the number of seller agents inspected before a trading decision is made. From Figure 15.5, it can be observed that, in both cases considered, *time savers* incurred a lower overhead than *bargain hunters*. Therefore, *time savers* strictly dominate *bargain hunters* when timeliness is the critical factor.

However, on the price dimension, Figure 15.6 shows that, in both cases, *bargain hunters* always meet the order at lower prices when compared to *time savers*. This strict dominance indicates that a tradeoff exists between price and timeliness when considering these strategies.

In practice, this indicative result may be used to guide the design of real software agents. That is, specific software agents may be deployed in order to implement the appropriate strategy for the context of the order at hand.

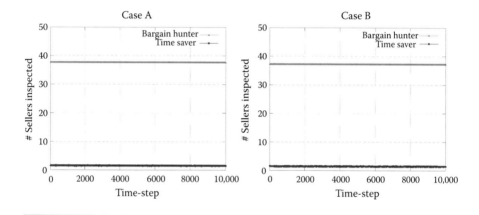

Figure 15.5   Trading overhead.

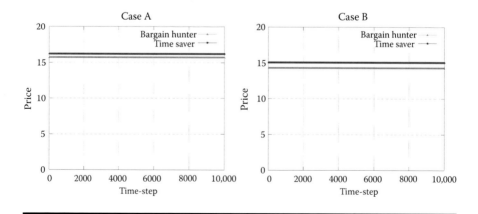

Figure 15.6   Trading price.

15.5 Conclusions

Emerging paradigms for the development and deployment of massively distributed computational systems allow resources to span many locations, organizations and platforms, connected through the Internet. In such systems, both resource-providing and resource-using nodes may arrive, organize, and dissipate as computational capabilities are formed and reformed as needed without reference to a central authority or coordinator.

As these systems mature, it is predicted that the majority of their interactions will be carried out by autonomous software agents on behalf of their owners. In such distributed systems, in which there exists a distribution of work to be done or a resource to be provided about a network of nodes, neither control nor even full knowledge of key resources may be assumed as they may be owned or administered by different organizations or individuals and as such have independent objectives. There is a need to find novel ways to understand and autonomically manage and control these large, decentralized, and dynamic systems. As part of this, there remains the problem of how to allocate distributed resources among the nodes in an adaptive and resilient way.

In this chapter, we have described a range of techniques that take inspiration from economics. These provide methods for modeling such problems and reconciling conflicting nodes' objectives in an adaptive manner. In particular, game theoretic analysis is a useful tool with which to reason about the interactions between self-interested adaptive agents. A number of different approaches to implementing economics-inspired resource allocation have been proposed. However, these approaches vary in terms of the resilience (or lack of resilience) that they provide. Single- and double-sided auctions, typically either require a centralized price-fixing process, such as an auctioneer or specialist, or else regional super-nodes able to perform this function in a distributed manner. Both approaches require information to be channeled through one or more coordinating nodes, introducing weak points in the system and potentially creating bottlenecks. An alternative to this is bilateral bargaining, and this shows a great deal of promise as a fully decentralized and more resilient approach. However it seems likely that this requires highly complex agent capabilities throughout the system, which will come with their own computational overhead. Furthermore, when agents are unable to fulfill this role, they will most likely be disadvantaged. The simpler retail-inspired posted-offer market mechanism provides a further promising alternative. Here agents are not required to possess complex strategic capabilities, reducing computational overhead, and the mechanism does not require global or regional coordination nodes, increasing resilience.

We have shown how posted-offer markets may be applied to an abstract problem model, motivated by the service selection problem in cloud computing in order to achieve a balanced load across multiple resource providers. We outlined a methodology for analyzing outcome resource allocations given arbitrary buyer behavior models. We argued that different buyer behavior types may indeed be relevant for deployment in cloud-based systems because they possess different characteristics representative of users' preferences over quality of service attributes.

In this chapter, we focused on cloud computing as a case study, but economics-inspired techniques for resource allocation can equally be applied to a wide range of computational and engineering problems. Other recent examples include the use of auctions to adaptively allocate object-tracking responsibilities among nodes in smart camera networks [89] and for conflict resolution in multi-user active music systems [90]. In the smart camera network case, analysis of the performance of the system in the presence of node and network failures and node additions during runtime have shown high levels of resilience and advantageous adaptivity [91]. On the subject of computational resource allocation generally, there is much knowledge transfer between research in clouds and in other decentralized systems, and each application area brings with it its own set of challenging assumptions.

From a conceptual perspective, future research into market-based control and economics-inspired computation must therefore consider that behaviors of participating agents cannot be assumed to be theoretically optimal and may adapt in unpredictable ways. Similarly, mechanisms used may need to vary by deployment as different mechanisms themselves possess characteristics suitable for different assumptions and quality of service requirements, most notably resilience. It is therefore important that research into economics-inspired computational resource allocation continues to consider a wide range of behavioral strategies and market mechanisms.

Acknowledgments

This research was supported by the EPiCS project and received funding from the European Union Seventh Framework Programme under grant agreement n° 257906. http://www.epics-project.eu/. The research leading to this chapter was substantially carried out while Dr. Lewis was with CERCIA, University of Birmingham, United Kingdom.

References

1. F. Hsu, *Behind Deep Blue: Building the Computer that Defeated the World Chess Champion.* Princeton University Press, 2002.
2. A. L. Beberg, D. L. Ensign, G. Jayachandran, S. Khaliq, and V. S. Pande, "Folding@home: Lessons from eight years of distributed computing," in *IEEE International Symposium on Parallel and Distributed Processing (IPDPS)*, pp. 1–8, 2009.
3. I. Foster and C. Kesselman, eds., *The Grid: Blueprint for a New Computing Infrastructure.* Morgan-Kaufman, 1999.
4. R. Buyya, C. S. Yeo, and S. Venugopal, "Market-oriented cloud computing: Vision, hype, and reality for delivering IT services as computing utilities," in *2008 10th IEEE International Conference on High Performance Computing and Communications*, pp. 5–13, IEEE, 2008.

5. M. P. Singh and M. N. Huhns, *Service-Oriented Computing: Semantics, Processes, Agents.* Chichester, West Sussex: John Wiley and Sons, 2005.

6. K. Lai, "Markets are dead, long live markets," *ACM SIGecom Exchanges*, vol. 5, no. 4, pp. 1–10, 2005.

7. S. H. Clearwater, ed., *Market-Based Control: A Paradigm for Distributed Resource Allocation.* Singapore: World Scientific, 1996.

8. P. Ciancarini and M. Wooldridge, "Agent-oriented software engineering," in *Software Engineering, 2000. Proceedings of the 2000 International Conference on,* pp. 816–817, 2000.

9. N. Jennings, "Agent-oriented software engineering," in *Multi-Agent System Engineering* (F. Garijo and M. Boman, eds.), vol. 1647 of *Lecture Notes in Computer Science*, pp. 1–7 Springer Berlin Heidelberg, 1999.

10. B. Rochwerger, D. Breitgand, E. Levy, A. Galis, K. Nagin, I. M. Llorente, R. Montero et al., "The reservoir model and architecture for open federated cloud computing," *IBM Journal of Research and Development,* vol. 53, no. 4, pp. 4:1–4:11, 2009.

11. R. Buyya, R. Ranjan, and R. Calheiros, "Intercloud: Utility-oriented federation of cloud computing environments for scaling of application services," in *Algorithms and Architectures for Parallel Processing* (C.-H. Hsu, L. Yang, J. Park, and S.-S. Yeo, eds.), vol. 6081 of Lecture Notes in Computer Science, pp. 13–31, Springer Berlin/ Heidelberg, 2010.

12. A. Celesti, F. Tusa, M. Villari, and A. Puliafito, "How to enhance cloud architectures to enable cross-federation," in *Cloud Computing (CLOUD), 2010 IEEE 3rd International Conference on,* pp. 337–345, 2010.

13. F. Faniyi, R. Bahsoon, and G. Theodoropoulos, "A dynamic data-driven simulation approach for preventing service level agreement violations in cloud federation," *Procedia Computer Science,* vol. 9, no. 0, pp. 1167–1176, 2012.

14. I. Brandic, D. Music, and S. Dustdar, "Service mediation and negotiation boot-strapping as first achievements towards self-adaptable grid and cloud services," in *Proceedings of the 6th International Conference Industry Session on Grids Meets Autonomic Computing*, GMAC '09, (New York), pp. 1–8, ACM, 2009.

15. N. Loutas, E. Kamateri, F. Bosi, and K. Tarabanis, "Cloud computing interoperability: The state of play," in *Cloud Computing Technology and Science (CloudCom), 2011 IEEE Third International Conference on,* pp. 752–757, 2011.

16. P. Rubach and M. Sobolewski, "Autonomic SLA management in federated computing environments," in *Proceedings of the 2009 International Conference on Parallel Processing Workshops*, ICPPW '09, (Washington, DC, USA), pp. 314–321, IEEE Computer Society, 2009.

17. I. Brandic, V. C. Emeakaroha, M. Maurer, S. Dustdar, S. Acs, A. Kertesz, and G. Kecskemeti, "LAYSI: A layered approach for SLA-violation propagation in self-manageable cloud infrastructures," in *Proceedings of the 2010 IEEE 34th Annual Computer Software and Applications Conference Workshops,* COMPSACW '10, (Washington, DC, USA), pp. 365–370, IEEE Computer Society, 2010.

18. D. Ardagna, M. Trubian, and L. Zhang, "SLA based resource allocation policies in autonomic environments," *J. Parallel Distrib. Comput.,* vol. 67, pp. 259–270, 2007.

19. E. A. Brewer, "Lessons from giant-scale services," *IEEE Internet Computing,* vol. 5, no. 4, pp. 46–55, 2001.

20. V. Cardellini, M. Colajanni, and P. S. Yu, "Dynamic load balancing on web server systems," *IEEE Internet Computing*, vol. 3, no. 3, pp. 28–39, 1999.

21. T. Eymann, M. Reinicke, O. Ardaiz, P. Artigas, L. D. de Cerio, F. Freitag, R. Messeguer et al., "Decentralized vs. centralized economic coordination of resource allocation in grids," *Lecture Notes in Computer Science*, pp. 9–16, 2004.

22. R. Alfano and G. D. Caprio, "TURBO: An autonomous execution environment with scalability and load balancing features," in *Proceedings of the IEEE Workshop on Distributed Intelligent Systems: Collective Intelligence and its Applications (DIS06)*, pp. 377–382, 2006.

23. O. Ardaiz, P. Artigas, T. Eymann, F. Freitag, L. Navarro, and M. Reinicke, "The catallaxy approach for decentralized economic-based allocation in grid resource and service markets," *Applied Intelligence*, vol. 25, no. 2, pp. 131–145, 2006.

24. V. Ramasubramanian and E. G. Sirer, "Perils of transitive trust in the domain name system," Tech. Rep. TR2005-1994, Cornell University, Ithaca, NY, USA, 2005.

25. J. Cao, D. P. Spooner, S. A. Jarvis, and G. R. Nudd, "Grid load balancing using intelligent agents," *Future Generation Computer Systems*, vol. 21, no. 1, pp. 135–149, 2005.

26. J. S. A. Bridgewater, P. O. Boykin, and V. P. Roychowdhury, "Balanced overlay networks (BON): An overlay technology for decentralized load balancing," *IEEE Transactions on Parallel and Distributed Systems*, pp. 1122–1133, 2007.

27. S. Surana, B. Godfrey, K. Lakshminarayanan, R. Karp, and I. Stoica, "Load balancing in dynamic structured peer-to-peer systems," *Performance Evaluation*, vol. 63, no. 3, pp. 217–240, 2006.

28. R. Buyya, D. Abramson, J. Giddy, and H. Stockinger, "Economic models for resource management and scheduling in grid computing," *Concurrency and Computation: Practice and Experience*, vol. 14, no. 13–15, pp. 1507–1542, 2002.

29. P. Rodriguez and E. W. Biersack, "Dynamic parallel access to replicated content in the internet," *IEEE/ACM Transactions on Networking*, vol. 10, no. 4, pp. 455–465, 2002.

30. R. Chang, M. Guo, and H. Lin, "A multiple parallel download scheme with server throughput and client bandwidth considerations for data grids," *Future Generation Computer Systems*, vol. 24, no. 8, pp. 798–805, 2008.

31. S. U. Khan and I. Ahmad, "Noncooperative, semi-cooperative, and cooperative games-based grid resource allocation," in *20th International Parallel and Distributed Processing Symposium*, 2006, p. 10, 2006.

32. D. K. Begg, S. Fischer, and R. Dornbusch, *Economics*. London: McGraw-Hill, 7th ed., 2002.

33. M. N. Rothbard, "Free market," in *The Concise Encyclopedia of Economics*. http://www.econlib.org/.

34. R. M. Starr, "Commodity money equilibrium in a Walrasian trading post model: An example," eScholarship Repository, University of California, 2006.

35. D. Cliff, "Minimal-intelligence agents for bargaining behaviors in market-based environments," Tech. Rep. HPL-97-91, HP Labs, Bristol, UK, 1997.

36. D. P. Friedman and J. Rust, *The Double Auction Market: Institutions, Theories, and Evidence*. Boulder, Colorado, USA: Westview Press, 1993.

37. S. Phelps, S. D. Parsons, and P. McBurney, "An evolutionary game-theoretic comparison of two double-auction market designs," in *Proceedings of the 6th Workshop on Agent Mediated Electronic Commerce*, (New York), 2004.

38. A. Byde, "A comparison between mechanisms for sequential compute resource auctions," in *Proceedings of the Fifth International Joint Conference on Autonomous Agents and Multi-agent Systems* (Hakodate, Japan), 2006.

39. E. David, A. Rogers, N. R. Jennings, J. Schiff, S. Kraus, and M. H. Rothkopf, "Optimal design of English auctions with discrete bid levels," *ACM Transactions on Internet Technology*, vol. 7, no. 2, p. 12, 2007.

40. N. R. Jennings, P. Faratin, A. R. Lomuscio, S. D. Parsons, C. Sierra, and M. Wooldridge, "Automated negotiation: Prospects, methods and challenges," *International Journal of Group Decision and Negotiation*, vol. 10, no. 2, pp. 199–215, 2001.

41. A. Byde, "Applying evolutionary game theory to auction mechanism design," in *IEEE International Conference on E-Commerce*, pp. 347–354, 2003.

42. V. Conitzer and T. Sandholm, "Self-interested automated mechanism design and implications for optimal combinatorial auctions," in *Proceedings of the 5th ACM Conference on Electronic Commerce*, (New York), pp. 132–141, ACM Press, 2004.

43. D. Cliff, "Explorations in evolutionary design of online auction market mechanisms," *Journal of Electronic Commerce Research and Applications*, vol. 2, no. 2, pp. 162–175, 2003.

44. T. L. Casavant and J. G. Kuhl, "A taxonomy of scheduling in general-purpose distributed computing systems," *IEEE Transactions on Software Engineering*, vol. 14, no. 2, pp. 141–154, 1988.

45. A. Gupta, D. O. Stahl, and A. B. Whinston, "The economics of network management," *Communications of the ACM*, vol. 42, no. 9, pp. 57–63, 1999.

46. Y. C. Lee, C. Wang, A. Y. Zomaya, and B. B. Zhou, "Profit-driven service request scheduling in clouds," in *Proceedings of the 2010 10th IEEE/ACM International Conference on Cluster, Cloud and Grid Computing*, CCGRID '10, (Washington, DC, USA), pp. 15–24, IEEE Computer Society, 2010.

47. I. Fujiwara, K. Aida, and I. Ono, "Applying double-sided combinational auctions to resource allocation in cloud computing," in *Proceedings of the 2010 10th IEEE/IPSJ International Symposium on Applications and the Internet*, SAINT '10, (Washington, DC, USA), pp. 7–14, IEEE Computer Society, 2010.

48. B. Song, M. M. Hassan, and E.-N. Huh, "A novel heuristic-based task selection and allocation framework in dynamic collaborative cloud service platform," in *Proceedings of the 2010 IEEE Second International Conference on Cloud Computing Technology and Science*, CLOUDCOM '10, (Washington, DC, USA), pp. 360–367, IEEE Computer Society, 2010.

49. S. Shang, J. Jiang, Y. Wu, G. Yang, and W. Zheng, "A knowledge-based continuous double auction model for cloud market," in *Proceedings of the 2010 Sixth International Conference on Semantics, Knowledge and Grids*, SKG '10 (Washington, DC, USA), pp. 129–134, IEEE Computer Society, 2010.

50. X. You, X. Xu, J. Wan, and D. Yu, "RAS-M: Resource allocation strategy based on market mechanism in cloud computing," in *ChinaGrid Annual Conference, 2009. ChinaGrid '09. Fourth*, pp. 256–263, 2009.

51. D. Ardagna, B. Panicucci, and M. Passacantando, "Generalized Nash equilibria for the service provisioning problem in cloud systems," *Services Computing, IEEE Transactions on*, vol. PP, no. 99, p. 1, 2012.

52. V. Walia, A. Byde, and D. Cliff, "Evolving market design in zero-intelligence trader markets," in *Proceedings of the IEEE International Conference on E-Commerce*, pp. 157–163, IEEE Computer Society, 2003.

53. S. Phelps, P. McBurney, and S. Parsons, "Evolutionary mechanism design: A review," *Autonomous Agents and Multi-Agent Systems*, vol. 21, no. 2, pp. 237–264, 2009.

54. D. Cliff and J. Bruten, "Simple bargaining agents for decentralized market-based control," Tech. Rep. HPL-98-17, HP Laboratories, Bristol, UK, 1998.

55. R. Wolski, J. Plank, J. Brevik, and T. Bryan, "Analyzing market-based resource allocation strategies for the computational grid," *International Journal of High Performance Computing Applications*, vol. 15, no. 3, p. 258, 2001.

56. W. Shi and B. Hong, "Resource allocation with a budget constraint for computing independent tasks in the cloud," in *Cloud Computing Technology and Science (CloudCom), 2010 IEEE Second International Conference on*, pp. 327–334, 2010.

57. D. Sun, G. Chang, C. Wang, Y. Xiong, and X. Wang, "Efficient Nash equilibrium based cloud resource allocation by using a continuous double auction," in *Computer Design and Applications (ICCDA), 2010 International Conference on*, vol. 1, pp. V1–94 –V1–99, 2010.

58. M. Esteva and J. Padget, "Auctions without auctioneers: Distributed auction protocols," in *Lecture Notes in Artificial Intelligence*, vol. 1788, ch. II, pp. 20–28, Berlin, Germany: Springer-Verlag, 2000.

59. D. Hausheer and B. Stiller, "Peermart: The technology for a distributed auction-based market for peer-to-peer services," in *Proceedings of the IEEE International Conference on Communications*, vol. 3, pp. 1583–1587, 2005.

60. H. Kikuchi, S. Hotta, K. Abe, and S. Nakanishi, "Distributed auction servers resolving winner and winning bid without revealing privacy of bids," in *Proceedings of the Seventh International Conference on Parallel and Distributed Systems: Workshops* (Washington, DC, USA), p. 307, IEEE Computer Society, 2000.

61. J. Niu, K. Cai, S. Parsons, E. Gerding, P. McBurney, T. Moyaux, S. Phelps, and D. Shield, "JCAT: A platform for the TAC market design competition," in *Proceedings of the 7th International Joint Conference on Autonomous Agents and Multiagent Systems: Demo Papers*, pp. 1649–1650, International Foundation for Autonomous Agents and Multiagent Systems, 2008.

62. E. Robinson, P. McBurney, and X. Yao, "How specialised are specialists? Generalisation properties of entries from the 2008 and 2009 TAC market design competitions," in *Agent-Mediated Electronic Commerce. Designing Trading Strategies and Mechanisms for Electronic Markets* (E. David, E. Gerding, D. Sarne, and O. Shehory, eds.), vol. 59, of Lecture Notes in Business Information Processing, pp. 178–194, Springer Berlin Heidelberg, 2010.

63. V. Nallur and R. Bahsoon, "A decentralized self-adaptation mechanism for service-based applications in the cloud," *Software Engineering, IEEE Transactions on*, 2012. doi:10.1109/TSE.2012.53.

64. C. A. Waldspurger, T. Hogg, B. A. Huberman, J. O. Kephart, and S. Stornetta, "Spawn: A distributed computational economy," *IEEE Transactions on Software Engineering*, vol. 18, no. 2, pp. 103–117, 1992.

65. E. H. Gerding, R. K. Dash, D. C. K. Yuen, and N. R. Jennings, "Optimal bidding strategies for simultaneous Vickrey auctions with perfect substitutes," in *Proceedings of the 8th Workshop on Game Theoretic and Decision Theoretic Agents* (Hakodate, Japan), 2006.

66. E. H. Gerding, R. K. Dash, D. C. K. Yuen, and N. R. Jennings, "Bidding optimally in concurrent second-price auctions of perfectly substitutable goods," in *Sixth International Joint Conference on Autonomous Agents and Multiagent Systems (AAMAS-07)* (Honolulu, Hawaii, USA), pp. 267–274, 2007.

67. T. Eymann, "Co-evolution of bargaining strategies in a decentralized multi-agent system," in *AAAI Fall 2001 Symposium on Negotiation Methods for Autonomous Cooperative Systems*, 2001.

68. T. Eymann, B. Padovan, and D. Schoder, "The catallaxy as a new paradigm for the design of information systems," in *Proceedings of the 16th World Computer Congress of the International Federation for Information Processing*, 2000.

69. T. Eymann, M. Reinickke, O. Ardaiz, P. Artigas, F. Freitag, and L. Navarro, "Self-organizing resource allocation for autonomic networks," in *Database and Expert Systems Applications, 2003. Proceedings. 14th International Workshop on*, pp. 656–660, 2003.

70. L. Tesfatsion and K. Judd, *Handbook of Computational Economics*. Elsevier, 2006.

71. E. H. Gerding and J. A. La Poutr, "Bilateral bargaining with multiple opportunities: Knowing your opponent's bargaining position," *IEEE Transactions on Systems, Man and Cybernetics, Part C: Applications and Reviews*, vol. 36, no. 1, pp. 45–55, 2006.

72. F. Lopes, M. Wooldridge, and A. Q. Novais, "Negotiation among autonomous computational agents: Principles, analysis and challenges," *Artificial Intelligence Review*, vol. 2008, no. 29, pp. 1–44, 2008.

73. A. Chandra, P. S. Oliveto, and X. Yao, "Co-evolution of optimal agents for the alternating offers bargaining game," in *EvoApplications*, vol. 1, pp. 61–70, 2010.

74. C. R. Plott and V. L. Smith, "An experimental examination of two exchange institutions," *The Review of Economic Studies*, pp. 133–153, 1978.

75. J. Ketcham, V. L. Smith, and A. W. Williams, "A comparison of posted-offer and double-auction pricing institutions," *The Review of Economic Studies*, vol. 51, no. 4, pp. 595–614, 1984.

76. S. Jagannathan and K. C. Almeroth, "Price issues in delivering e-content on-demand," *ACM SIGecom Exchanges*, vol. 3, no. 2, pp. 18–27, 2002.

77. R. Wang, "Auctions versus posted-price selling," *The American Economic Review*, vol. 83, no. 4, pp. 838–851, 1993.

78. R. B. Myerson and M. A. Satterthwaite, "Efficient mechanisms for bilateral trading," *Journal of Economic Theory*, vol. 29, no. 2, pp. 265–281, 1983.

79. P. Lewis, P. Marrow, and X. Yao, "Resource allocation in decentralised computational systems: An evolutionary market-based approach," *Autonomous Agents and Multi-Agent Systems*, vol. 21, no. 2, pp. 143–171, 2010.

80. A. Chavez, A. Moukas, and P. Maes, "Challenger: A multi-agent system for distributed resource allocation," in *Proceedings of the First International Conference on Autonomous Agents*, 1997.

81. L. Xiao, Y. Zhu, L. M. Ni, and Z. Xu, "GridIS: An incentive-based grid scheduling," in *Proceedings of the 19th IEEE International Parallel and Distributed Processing Symposium*, p. 65b, 2005.

82. K. Kuwabara, T. Ishida, Y. Nishibe, and T. Suda, "An equilibratory market-based approach for distributed resource allocation and its applications to communication network control," in *Market-Based Control: A Paradigm for Distributed Resource Allocation* (S. H. Clearwater, ed.), pp. 53–73, Singapore: World Scientific, 1996.

83. F. Faniyi and R. Bahsoon, "Self-managing SLA compliance in cloud architectures: A market-based approach," in *Proceedings of the 3rd international ACM SIGSOFT symposium on Architecting Critical Systems, ISARCS '12* (New York), pp. 61–70, ACM, 2012.

84. F. Faniyi and R. Bahsoon, "Engineering proprioception in SLA management for cloud architectures," in *Software Architecture (WICSA), 2011 9th Working IEEE/IFIP Conference on*, pp. 336–340, 2011.

85. M. Jaeger, G. Rojec-Goldmann, and G. Muhl, "QoS aggregation for web service composition using workflow patterns," in *Enterprise Distributed Object Computing Conference, 2004. EDOC 2004. Proceedings. Eighth IEEE International*, pp. 149–159, 2004.

86. A. Mas-Colell, M. D. Whinston, and J. R. Green, *Micro-Economic Theory*. Oxford: Oxford University Press, 1995.

87. A. R. Greenwald and J. O. Kephart, "Shopbots and pricebots," *Proceedings of the Sixteenth International Joint Conference on Artificial Intelligence*, vol. 1, pp. 506–511, 1999.

88. P. R. Lewis, P. Marrow, and X. Yao, "Evolutionary market agents for resource allocation in decentralised systems," in *Parallel Problem Solving From Nature—PPSN X, vol. 5199 of Lecture Notes in Computer Science,* pp. 1071–1080, Springer, 2008.

89. L. Esterle, P. R. Lewis, X. Yao, and B. Rinner, "Socio-economic vision graph generation and handover in distributed smart camera networks," *ACM Transactions on Sensor Networks*, vol. 10, issue 2, no. 20, 2014.

90. A. Chandra, K. Nymoen, A. Voldsund, A. Jensenius, K. Glette, and J. Torresen, "Market-based control in interactive music environments," in *Post-proceedings of 9th Int. Symposium on Computer Music Modeling and Retrieval (CMMR 2012)* (S. Y. et al., ed.), LNCS, Springer, 2013 (To appear).

91. L. Esterle, B. Rinner, P. R. Lewis, and X. Yao, "Improved adaptivity and robustness in decentralised multi-camera networks," in *Proceedings of the Sixth International Conference on Distributed Smart Cameras (ICDSC 2012)*, pp. 1–6, IEEE Press, 2012.

Chapter 16

Instrumentation-Based Resource Control

Alex Villazón and Walter Binder

Contents

Resource management is a precursor to any form of dynamic adaptation. Before a system can make decisions about reallocating resources–computational, storage, or communication–the system must have awareness of the ressource requirements and consumption by existing processes (services or applications) and the resources available within different systems. Once allocation decisions are made, the systems must also be able to limit the consumption of the resources so that processes do not exceed the allocations. In this chapter, Alex Villazón and Walter Binder summarize their long-term efforts on resource management with the Java programming language.

16.1 Introduction

Accessing information about resource consumption (CPU, memory, network) allows applications to better adapt to changing conditions, for example, making scheduling, provisioning, or migration decisions [1,2]. Several techniques have been explored to gather resource-consumption information. For example, *hardware performance counters* [3] are used to collect resource usage information for performance analysis and tuning in grid-based systems [4]. In systems based on virtual environments, such as those of mobile object systems studied in the past or more recent cloud-based systems, resource reification (i.e., making resource information available for manipulation) has been achieved at the operating system (OS) level through special interfaces [5], the virtual machine (VM) level through customizations (extension of the VM) [6–8], or at a higher level through program transformation techniques [9–11]. The latter is particularly interesting because it avoids using nonstandard APIs, thus ensuring portability of applications. Providing resource control facilities on top of standard runtime systems is challenging to achieve.

VM technology has become mature and is widely used, from server and desktop applications to mobile devices. Examples of this success are the Java Virtual Machine (JVM) [12], several JavaScript VMs [13,14], the Dalvik VM of the Android OS [15], and the Common Language Runtime (CLR) VMs for the .Net framework [16,17], just to cite some of them. All these platforms are based on similar principles of abstractions of the underlying hardware and the use of OS-independent intermediate language (bytecode). Furthermore, the use of such intermediate languages has made VMs an interesting target of a plethora of different programming languages. For example, in addition to the Java programming language, the JVM is the target of dozens of programming languages, such as Scala [18], Ceylon [19], Groovy [20], or Clojure [21]. Different compilation and bytecode instrumentation techniques have been developed to optimize, analyze, profile, test, debug [22–25], or even enable interoperability between different languages and VMs [18,21].

Accounting for and controlling the resource consumption of applications and of individual software components is crucial in VM-based environments that run components on behalf of external clients in order to protect the host from malicious or badly programmed code. Resource accounting may also provide valuable feedback about actual usage by end-clients and thus enable precise billing and provisioning policies. Also, resource awareness can help developers to optimize the execution of applications in environments with limited computational or energy capabilities, such as modern mobile applications on smartphones or similar devices. For example, resource awareness can be used to move heavy computations from a mobile device to the cloud when a device battery is too low and gather the results later on. Such a scenario requires special resource management support.

Even though Java was designed with special emphasis on portability, it unfortunately lacks mechanisms for resource management that could, for example, be used to limit the resource consumption of hosted components, to charge the clients for the resource consumption of their deployed components, or to trigger migration of computations. On the one hand, prevailing solutions for resource control in Java either are incomplete or rely on native code, on low-level resource control mechanisms offered by the underlying operating system, or on a modified JVM [7,8,26]. Consequently, these systems are not well suited to being deployed in heterogeneous environments, in which a wide variety of different hardware platforms and operating systems have to be supported. On the other hand, resource control with the aid of program transformations through *portable bytecode instrumentation* offers an important advantage over prevailing approaches because it is independent of any particular JVM and underlying operating system. Such an approach works with standard Java runtime systems and may be integrated into existing environments. Furthermore, it enables resource control within embedded systems based on modern Java processors, which provide a JVM implemented in hardware that cannot be easily modified [27].

This is the essence of the approach described in this chapter, which is completely based on program transformations: The bytecode of existing applications is instrumented in order to make the resource consumption of programs explicit. Such an approach enables resource management in a platform-independent manner by accounting, respectively limiting the number of bytecodes that a thread (or component) may execute.

16.2 General Approach to Portable Resource Management

One of the main reasons for building a portable resource management framework is to make it as easy to deploy as any Java application. There are, of course, technical obstacles because there is a tradeoff between, on the one hand, the low-level information and control required for the resource and especially CPU management and, on the other hand, the difficulty of implementing the JVM and associated runtime libraries across a wide variety of devices and operating systems.

All prevailing approaches somehow rely on the support of native, that is, non-portable, code for their implementation. Some specialized JVMs were implemented with dedicated support for resource control [7,8,26]. Also, a resource control library for Java [9] has been proposed, which uses native code for CPU control in combination with some light bytecode instrumentation to enable proper cooperation with the OS via native code libraries.

Our approach, in contrast, draws its portability from the combination of byte-code instrumentation and runtime libraries implemented in pure Java. Figure 16.1 depicts the situation in which all Java code, including applications; possible additional middleware; and the standard Java class library—also know as Java development Kit (JDK) classes—(except the subset that is implemented in native code) is made CPU-manageable through bytecode instrumentation (a process symbolized by the round arrow flowing through an instrumentation mechanism); a small, but adaptable (pure Java) management library is required at runtime. This offers an important advantage over the other approaches because it is independent of any particular JVM and underlying operating system.

Traditionally, the CPU consumption of a program is measured in seconds. This approach, however, has several drawbacks: It is platform-dependent (for the same program and input, the CPU time differs, depending on hardware, operating system, and virtual machine), and measuring it accurately may require platform-specific features (such as special OS functions), limiting the portability of the CPU management services, and the resulting CPU consumption may not be easily reproducible as it may depend on external factors, such as the system load.

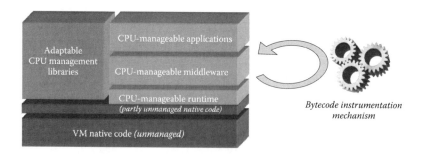

Figure 16.1 **Overview of the portable Java CPU management.**

For these reasons, we use the number of executed JVM bytecode instructions as our CPU consumption metric. Although this metric is not directly translatable into real CPU time, it has many advantages:

■ Platform-independence: The number of executed bytecode instructions is a platform-independent, dynamic metric [28]. It is independent of the hardware and virtual machine implementation (e.g., interpretation versus just-in-time compilation). However, the availability of different versions and implementations of the JDK may limit the platform-independence of this metric.

■ Reproducibility: For deterministic programs, the CPU consumption measured in terms of executed bytecode instructions is exactly reproducible if the same JDK version is used. However, reproducibility cannot be guaranteed for programs with nondeterministic thread scheduling.

■ Comparability: CPU consumption statistics collected in different environments are directly comparable because they are based on the same platform-independent metric.

■ Portability and compatibility: Because counting the number of executed bytecode instructions does not require any hardware or OS-specific support, it can be implemented in a fully portable way. Our CPU management scheme is implemented in pure Java, and it is compatible with any standard JVM.

There are two different techniques for the implementation of the instrumentation mechanism: one based on low-level bytecode instrumentation libraries and the other based on a high-level instrumentation framework. In our work, we explored both approaches, and that will be detailed in Section 16.3.

For effective resource accounting, it is essential that the instrumentation is performed with full method coverage, that is, the instrumentation should cover all methods executing in the JVM (which have a bytecode representation), including methods in dynamically generated or loaded classes as well as in the JDK classes. Running a JVM with an instrumented JDK library can be challenging, notably when structural modifications are made to critical classes and notably because the bootstrapping process of the JVM can be perturbed. We discuss these issues related to the transformations that are necessary to enable CPU accounting followed by our approach in Section 16.2.5.

16.2.1 CPU Accounting Scheme

In the proposed approach, the bytecode of existing applications is instrumented in order to make the resource consumption of programs explicit. This technique is called *resource reification* [29]. Thus, instrumented programs will unknowingly keep track of the number of executed bytecode instructions for CPU accounting. More precisely, each thread permanently accounts for its own CPU consumption, expressing it as the number of executed JVM bytecode instructions. This constitutes a platform-independent measurement unit, which has some practical advantages. Periodically, each thread will aggregate the collected information concerning its own CPU consumption within an account that is shared with a number of other threads. We call this approach *self-accounting*. During these information update routines, the thread will also execute management code, for example, to ensure that a given resource quota is not exceeded (the component may be terminated if there is a hard limit on the total number of bytecode instructions it may execute, or threads may be delayed in order to meet a restriction placed on execution rates). In this way, the proposed CPU management scheme does not rely on a dedicated supervisor thread because the management activity is distributed among all threads in the system, thus effectively implementing a form of *self-control*.

Hence, and this is for us a guarantee of portability and reliability, we do not rely on the underlying scheduling of the JVM, which is left loosely specified in the Java language, to make it easier to implement Java across a wide variety of environments: Although some JVMs seem to provide preemptive scheduling, ensuring that a thread with high priority will execute whenever it is ready to run, other JVMs do not respect thread priorities at all.   .

16.2.2 Bytecode Instrumentation Scheme

Concerning the bytecode instrumentation schemes, our two main design goals are to ensure portability (by following a strict adherence to the specification of the Java language and VM) and performance (introducing minimal overhead due to the additional instructions inserted into the original classes).

The main idea of the instrumentation for CPU accounting is to have a thread-local bytecode counter that is updated in each *basic block of code* (BBC), according to the number of bytecodes in the BBC, and in some strategic locations (beginning of methods, exception handlers, and loops), polling code is inserted to determine whether the bytecode counter has exceeded a given threshold. In this case, a user-defined resource-management policy is invoked by the exceeding thread. Here we use the term BBC as the longest possible sequence of bytecode instructions in which only the first instruction may be the target of a branch and in which the last instruction is one that changes the control flow (i.e., a branch, return, etc., but not an invocation).

Each thread has an associated ThreadCPUAccount, which is shown in Figure 16.2. During normal execution, each thread updates the consumption counter of its ThreadCPUAccount. In order to prevent overflows of the consumption counter and to schedule regular activation of the shared management tasks, the counter is checked against an adjustable granularity limit. More precisely, each time the counter has been incremented by granularity (executed bytecodes), its value will be registered and reset to an initial value by the invocation of a consume() method. In other words, each thread invokes the consume() method of its ThreadCPUAccount when the local consumption counter exceeds a certain limit defined by the granularity

```
public final class ThreadCPUAccount {
  public int consumption;
  private long aggregatedConsumption = 0;
  private int granularity;

  private boolean consumeInvoked = false;
  private volatile CPUManager manager;

  public ThreadCPUAccount(CPUManager m) {
    manager = m;
    granularity = manager.getGranularity();
    consumption = -granularity;
  }

  public ThreadCPUAccount(int g) {
    manager = null;
    granularity = g;
    consumption = -granularity;
  }

  public void setManager(CPUManager m) {
    manager = m;
  }

  public void consume() {
   long amountCons = (long)consumption + granularity;
   if (manager == null) {
     aggregatedConsumption += amountCons;
     consumption = -granularity;
   } else {
     granularity = manager.getGranularity();
     consumption = -granularity;
     if (consumeInvoked) {
       aggregatedConsumption += amountCons;
     } else {
       amountCons += aggregatedConsumption;
       aggregatedConsumption = 0;
       consumeInvoked = true;
       manager.consume(amountCons);
       consumeInvoked = false;
     }
   }
  }
  ...
}
```

Figure 16.2   Excerpt of the ThreadCPUAccount class.

variable. In order to optimize the comparison of whether the consumption counter exceeds the granularity, the counter runs from granularity to zero, and when it equals or exceeds zero, the consume() method is called.

In order to apply this CPU accounting scheme, methods of applications are instrumented in the following way:

■ Insertion of conditionals in order to invoke the consume() method periodically. The rationale behind these rules is to minimize the number of checks of whether consume() has to be invoked for performance reasons and to make sure that malicious code cannot execute an unlimited number of bytecode instructions without invocation of consume(). The conditional "if (cpu.consumption >= 0) cpu.consume();" is inserted in the following locations (the variable cpu refers to the ThreadCPUAccount of the currently executing thread):

a. In the beginning of each method. This ensures that the conditional is present in the execution of recursive methods. For performance reasons, the insertion in the beginning of methods may be omitted if each possible execution path terminates or passes by an already inserted conditional before any method/constructor invocation. In other words, leaf methods may be omitted (especially accessors of abstract data types) as well as entry points of methods inside which the first method invocation happens after another inserted conditional, typically inside or after a loop.

b. In the beginning of each loop.

c. In the beginning of each JVM subroutine. This ensures that the conditional is present in the execution of recursive JVM subroutines.

d. In the beginning of each exception handler.

■ The run() method of each class that implements the Runnable interface is instrumented according to Figure 16.2 in order to invoke consume() before the thread terminates. After the thread has terminated, its ThreadCPUAccount becomes eligible for garbage collection.

■ Finally, the instructions that update the consumption counter are inserted at the beginning of each accounting block. In order to reduce the accounting overhead, the conditionals inserted before are not considered as separate accounting blocks. The number of bytecode instructions required for the evaluation of the conditional is added to the size of the accounting block they precede.

16.2.3 Instrumentation Example

Figure 16.3 illustrates how the sample method (see left portion of Figure 16.3) is transformed using the proposed CPU accounting scheme (see right side of Figure 16.3). For the sake of simplicity this transformation is shown as Java source code even though the actual instrumentation occurs at bytecode level. We assume

```
void f() {                    void f(ThreadCPUAccount cpu) {
                                  cpu.consumption += ...;
                                  if (cpu.consumption >= 0)
                                      cpu.consume();
    X;                            X;
    while (true) {                while (true) {
                                      cpu.consumption += ...;
                                      if (cpu.consumption >= 0)
                                          cpu.consume();
        if (C) {                      if (C) {
                                          cpu.consumption += ...;
            Y;                            Y;
        }                             }
                                      cpu.consumption += ...;
        Z;                            Z;
    }                             }
}                             }
```

Figure 16.3 Original (left) and instrumented (right) sample method for CPU accounting.

that the code block *X* includes a method invocation; hence, the conditional at the beginning of the method cannot be omitted. Here we do not show the concrete values by which the consumption variable is incremented; these values are calculated statically and represent the number of bytecodes that are going to be executed in the next accounting block.

16.2.4 Obtaining Reference to the Associated ThreadCPUAccount

As can be observed in the instrumented method (right) of Figure 16.3, the ThreadCPUAccount reference was passed by argument to the instrumented method, thus changing its profile. In reality, this works in the ideal case in which we know that absolutely all code has been transformed, that is, when both callers and callees refer to the same method profiles, and the account reference can be simply and directly transmitted in a chain. But in order to cope with special cases in which this is not possible, we have to leave in all classes stubs with the original method profiles, which act as wrappers for the methods with the additional ThreadCPUAccount argument, as in Figure 16.4.

Now, how can the getCurrentAccount() method know which account is associated with the current thread? Unfortunately, the JVM provides no means for applications to be alerted each time a thread switch occurs. Therefore, we have to call the standard Thread.currentThread() method each time this information is needed.

```
void f() {
    // Obtain a reference to the account
    // of the current Thread:
    ThreadCPUAccount cpu = ThreadCPUAccount.getCurrentAccount();

    // Invoke the real method:
    f(cpu);
}
```

Figure 16.4 Sample wrapper method for CPU accounting.

Then, using the reference to the current thread, we can obtain the associated CPU account through a hash table, or we can patch the java.lang.Thread class and add the needed reference directly to its set of instance fields. The second solution is more efficient while still respecting the language specifications. The getCurrentAccount() method does exactly this.

16.2.5 Bootstrapping with an Instrumented JDK

The functionality of getCurrentAccount() described before is, however, not valid during the bootstrapping of the JVM. During this short but crucial period, there is an initial phase in which the Thread.currentThread() pretends that no thread is executing and returns the value null (this is, in fact, because the Thread class has not yet been asked by the JVM to create its first instance). As a consequence, in all code susceptible of being executed during bootstrapping, that is, in the JDK, as opposed to application code, we have to make an additional check whether the current thread is undefined; for those cases, we have to provide a dummy, empty ThreadCPUAccount instance, the role of which is to prevent all references to the consumption variable in the instrumented JDK from generating a NullPointerException. This special functionality is provided by the jdkGetCurrentAccount() method, which replaces the normal getCurrentAccount() whenever we instrument JDK classes.

A second issue that arises when bootstrapping the JVM with an instrumented JDK is that we have no means for being informed when the bootstrapping is terminated. This is important because, before that moment, we are not allowed to actually use classes of the JDK inside the implementation of our runtime support, like ThreadCPUAccount. If we did, it would disturb the normal class loading sequence of the JVM and most probably make it crash. But unless we want to implement all our runtime support from scratch and not only ignore the benefits of code reuse for our own base classes, but also impose the same constraints to third-party implementations of CPUManager classes (where it will be), we have to know from what moment we are allowed to use helper classes like java.util.Vector. Our solution is as follows: Application classes, that is, all those that are not part of the JDK, are loaded by the JVM at the end of the bootstrapping. We can thus try to detect when such non-JDK classes are initialized by the JVM because that will signal the end of the bootstrapping. Concretely, our approach exploits the static initializer method, named <clinit>, which may exist inside every class for the sake of initializing class (static) variables. Whenever the JVM initializes a class, it will invoke the <clinit> method of that class if it exists. Our instrumentation scheme inserts an invocation to the runtime system inside every <clinit> method (and, if necessary, create it from scratch) of every non-JDK class.

At that call, the more advanced functionalities of CPU management can be loaded and launched. This latter step is performed through reflection (using the java.lang.reflect API) in order to hide such dependencies away from the JVM; if the dependencies were explicit, the JVM might try to load those classes eagerly and, of course, break the bootstrapping.

A third issue concerns the insertion of wrapper methods in the JDK. At some instances, native code in the JDK assumes that it will find a required piece of information at a statically known number of stack frames below itself. This is unfortunately incompatible with the generation of wrapper methods as described previously because at runtime, it would induce additional stack frames that would break the kind of native methods mentioned above. For this reason, the JDK cannot take advantage of the more efficient wrapper insertion scheme and has to invoke jdkGet-CurrentAccount at the beginning of every method. We can however, at the expense of a comprehensive class hierarchy analysis process, completely duplicate most JDK methods so that there always is one version with the original profile and a second with the additional ThreadCPUAccount argument.

16.2.6 Implementation of the ThreadCPUAccount

Normally, each ThreadCPUAccount refers to an implementation of CPUManager (see Figure 16.5), which is shared between all threads belonging to a component. The first constructor of ThreadCPUAccount requires a reference to a CPUManager. The second constructor, which takes a value for the accounting granularity, is used only during bootstrapping of the JVM (manager == null). If the JDK has been rewritten for CPU accounting, the initial bootstrapping thread requires an associated ThreadCPUAccount object for its proper execution. However, loading complex user-defined classes during the bootstrapping of the JVM is dangerous

```
public interface CPUmanager   {
  public int getGranularity () ;
  public void consume (long c) ;
}
```

Figure 16.5   The (simplified) CPUManager interface.

as it may violate certain dependencies in the classloading sequence. For this reason, a ThreadCPUAccount object can be created without previous allocation of a CPUManager implementation so that only two classes are inserted into the initial classloading sequence of the JVM: ThreadCPUAccount and CPUManager. Both of them only depend on java.lang.Object.

After the bootstrapping, the setManager(CPUManager) method is used to associate ThreadCPUAccount objects that had been allocated during the bootstrapping with a CPUManager. As the variable manager is volatile, the thread associated with the ThreadCPUAccount object will notice the presence of the CPUManager upon the following invocation of consume().

After the bootstrapping, the granularity variable in ThreadCPUAccount is updated during each invocation of the consume() method. Thus, the CPUManager implementation may allow changing the accounting granularity dynamically. However, the new granularity does not become active for a certain thread immediately but only after this thread has called consume().

The consume() method of ThreadCPUAccount passes the locally collected information concerning the number of executed bytecode instructions to the consume(long) method of the CPUManager, which implements custom scheduling policies. As sometimes consume(long) may execute a large number of instructions and the code implementing this method may have been instrumented for CPU accounting as well, it is important to prevent a recursive invocation of consume(long). We use the flag consumeInvoked for this purpose. If a thread invokes the consume() method of its associated ThreadCPUAccount while it is executing the consume(long) method of its CPUManager, it simply accumulates the information on CPU consumption within the aggregatedConsumption variable of its ThreadCPUAccount. After the consume(long) method has returned, the thread will continue normal execution, and upon the subsequent invocation of consume(), the aggregatedConsumption will be taken into account. During bootstrapping, a similar mechanism ensures that information concerning the CPU consumption is aggregated internally within the aggregatedConsumption field as a CPUManager may not yet be available.

16.2.7 Aggregating CPU Consumption

Normally, each ThreadCPUAccount refers to an implementation of CPUManager, which is shared between all threads belonging to a component. The CPUManager implementation is provided by the middleware developer and implements the actual CPU accounting and control strategies, for example, custom scheduling schemes. The methods of the CPUManager interface are invoked by the ThreadCPUAccount implementation as depicted in Figure 16.6. In this figure, it is shown that the increment is performed directly on the *consumption* variable for reasons of efficiency; due to the same concern for performance, the per-thread accounting objects are not user-extensible.

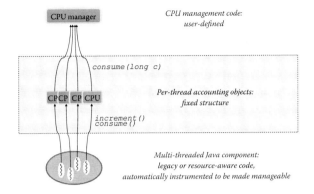

Figure 16.6 **Overview of the CPU management hierarchy.**

For resource-aware applications, the CPUManager implementation may provide application-specific interfaces to access information concerning the CPU consumption of components, to install notification callbacks to be triggered when the resource consumption reaches a certain threshold, or to modify resource control strategies.

Whenever a thread invokes *consume()* on its ThreadCPUAccount, this method will, in turn, report its collected CPU consumption data (stored in the consumption field) to the CPUManager associated with the ThreadCPUAccount (if any) by calling *consume(long)*; *consume()* also resets the consumption field. The *consume(long)* method implements the custom CPU accounting and control policy. It may simply aggregate the reported CPU consumption (and write it to a log file or database), it may enforce absolute limits and terminate components that exceed their CPU limit, or it may limit the execution rate of threads of a component (i.e., putting threads temporarily to sleep if a given execution rate is exceeded). This is possible without breaking security assumptions because the *consume(long)* invocation is synchronous (i.e., blocking) and executed directly by the thread to which the policy applies.

getGranularity() returns the accounting granularity currently defined for a ThreadCPUAccount associated with the given CPUManager. It is an adjustable value, which defines the frequency (and thus, indirectly, the overhead) of the management activities: It has to be adapted to the number of threads under supervision in order to prevent excessive delays between invocations to consume(long).

16.3 Bytecode Instrumentation

In this section we discuss different approaches for the implementation of the proposed instrumentation scheme to enable resource management. First, an overview of the different mechanisms for bytecode instrumentation is depicted. Then we discuss the use of low-level instrumentation techniques and discuss its limitations. Finally, we describe a high-level approach inspired by aspect-oriented programming (AOP) [30] and how the transformations required for resource reification can be concisely expressed hiding low-level instrumentation

16.3.1 Overview of Bytecode Instrumentation

There are several ways to instrument bytecode in Java. In theory, instrumentation can be made using any programming language that is able to understand the structure of a Java class (i.e., can disassemble it) and can write the correctly formed instrumented bytecode. Even though some instrumentation tools were implemented in native code [31], pure Java bytecode engineering libraries have been developed, such as BCEL [32], ASM [33], Javassist [34], or Soot [35], and are used in different areas, such as security [36,37], compilers [38,39], distributed systems [40–42], debugging [25,43], static and dynamic analysis [23,24,44–46], or AOP [39,47], to cite some of them.

Bytecode instrumentation libraries can be used externally to the JVM (i.e., used to perform offline instrumentation through external tools) or allow users to install bytecode transformations that can be integrated with the JVM. This is possible because Java supports dynamic bytecode instrumentation through the java.lang.instrument API as well as using native code agents through the Java Virtual Machine Tool Interface (JVMTI) [48]. The advantage of dynamic bytecode instrumentation is that it enables instrumentation of dynamically loaded and generated classes, which may escape static instrumentation.

Figure 16.7 shows different possible approaches to bytecode instrumentation. In Figure 16.7a, static instrumentation is done by an external tool (which runs in a separate JVM to instrument the code beforehand). The bytecode of application and/or the JDK classes are instrumented offline, and the JVM is later bootstrapped with the resulting instrumented code. In Figure 16.7b, the instrumentation is done dynamically in the same JVM running the instrumented application. This is the typical setting when using the java.lang.instrument API that enables load-time instrumentation. Finally, in Figure 16.7c, the instrumentation is performed

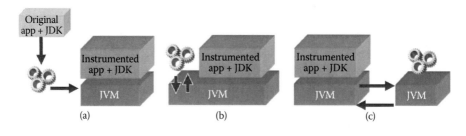

Figure 16.7 Bytecode instrumentation approaches. (a) Instrumentation with external tool (static); (b) Instrumentation in the same JVM (dynamic); (c) Instrumentation in separate JVM (dynamic).

dynamically but in separate JVM. The instrumentation is done by an instrumentation server, which communicates through a given inter-process communication (IPC) mechanism with the client JVM. This setting isolates the instrumentation from the target JVM and reduces issues related to the instrumentation of the JDK. This setting typically uses a tiny native agent on the client JVM (using the JVMTI interface) that communicates with the instrumentation server. Depending on the used IPC, the server can be implemented in pure Java (e.g., when sockets are used) otherwise a native layer may be required.

16.3.2 Low-Level Instrumentation-Based Implementation

The instrumentation scheme for CPU accounting described in Section 16.5 was implemented in the context of the Java Resource Accounting Framework, Second Edition (JRAF-2) [10,11]. It was based on BCEL [32], a low-level bytecode instrumentation library. JRAF-2 followed the approach of Figure 16.7a, that is, an instrumentation tool statically instrumented in both application and JDK classes. The instrumentation scheme to deal with bootstrapping issues explained in Section 16.2.5 was applied by the tool.

Even though pure static instrumentation can be sufficient in some settings, such an approach is rather limited to enable comprehensive instrumentation, notably because generated classes or dynamically loaded classes may escape instrumentation and therefore limit resource management in our case. We also explored how to adapt JRAF-2 to follow the approach Figure 16.7b, that is, enabling dynamic and load-time instrumentation for resource reification. We encountered several issues related to restrictions imposed by the instrumentation API (java.lang.instrument), notably the instrumentation of previously loaded classes. For instance, new fields or methods cannot be added, and method signatures cannot be changed. Therefore, it is not compatible with some of the transformations promoted in our approach. Our research was therefore oriented toward a more general instrumentation framework, which included the forms of transformations required for resource accounting.

In [49], we introduced *Framework for Exhaustive Rewriting and Reification with Advanced Runtime Instrumentation* (FERRARI), a generic bytecode instrumentation framework supporting the comprehensive instrumentation of the whole JDK, including all core classes, as well as dynamically loaded classes. FERRARI was neither designed as an application-specific framework nor a low-level bytecode instrumentation toolkit. Instead, it aimed at generating the necessary program logic to enable general-purpose instrumentation of JDK and application classes while giving advanced support for bootstrapping with an instrumented JDK. FERRARI followed a combination of the approaches in Figure 16.7a and 16.7b, that is, it consisted of a static instrumentation tool and a runtime instrumentation agent. The static instrumentation tool targeted instrumentation of JDK classes whereas application classes were instrumented at load-time by the agent. FERRARI provided a flexible interface, enabling different user-defined instrumentations (UDIs) written in pure Java to control the instrumentation process. Both the static instrumentation of JDK classes and dynamic load-time instrumentation of application classes were controlled by a UDI. UDIs may change method bodies, add new methods (with minor restrictions), and add fields (with some restrictions). The core of FERRARI was implemented with BCEL [32], but UDIs were free to use any bytecode instrumentation library.

FERRARI offered generic mechanisms to ensure complete instrumentation coverage of any code in a system that has a corresponding bytecode representation. To this end, it ensured that UDI-inserted code was not executed before the JVM has completed bootstrapping and provided support for temporarily bypassing the execution of inserted code for each thread during load-time instrumentation. The latter mechanism is essential, for example, to prevent that instrumented code itself invokes instrumented code, resulting in infinite recursion otherwise. Also, FERRARI ensured that the classes used for instrumentation by UDIs (BCEL or other) were not instrumented.

Even though FERRARI solved several issues related to comprehensive instrumentation, the instrumentations themselves were yet based on low-level approaches, that is, the UDIs were mostly implemented by directly manipulating bytecode instructions though BCEL, ASM, or other instrumentation library, thus making the implementation of new instrumentation tools difficult and error-prone, often requiring high development and testing effort. In addition, the resulting tools are often complex and difficult to maintain and to extend.

16.3.3 High-Level Instrumentation-Based Implementation

In our pursuit to simplify the development of tools based on instrumentation, we explored higher-level abstractions based on AOP [30]. AOP is a powerful approach enabling a clean modularization of crosscutting concerns, such as error checking and handling, synchronization, context-sensitive behavior, monitoring and logging, and debugging support. It enables the developer to specify as aspects certain points of interest (called *join points**) in the execution of a program that can be intercepted and to execute use-defined code, for example, before or after those points of interest. Researchers have explored the use of AOP languages to simplify the development of different software-engineering tools, such as profilers, debuggers, or testing tools [43,50,51], which, in many cases, can be specified as aspects in a concise manner. Hence, in a sense, AOP can be regarded as a versatile approach for specifying some program instrumentation at a high level, hiding low-level implementation details from the programmer.

In AOP frameworks, the instrumentation is performed by the *aspect weaver*, which is a utility that takes the instructions described in an aspect and generates

---

* Aspects specify *pointcuts* to intercept certain points in the execution of programs (so-called *join points*), such as method calls, field accesses, etc. *Advices* are executed *before, after,* or *around* the intercepted join points. Advices have access to contextual information of the join points.

the final implementation code by instrumenting the application to which the aspect is woven. In the case of AspectJ [47], which is an AOP language and framework for Java, the aspect weaving is done directly at the bytecode level, thus enabling aspect weaving into compiled Java applications. AspectJ provides AOP capabilities for Java, allowing new functionality to be systematically added to existing programs.

Even though AOP gives a higher-level abstraction to simplify the implementation of tools, prevailing aspect weavers, such as AspectJ [47] or *abc* [39], failed to weave aspects in the Java class library, thus strongly restricting the applicability of AOP. For example, aspects can be used for building profilers in a few lines of code [50,51], but the impossibility of weaving aspects into JDK classes causes incomplete profiles, limiting the practical value of this approach. In [52], we tackled this problem by leveraging FERRARI's UDI approach so as to integrate the AspectJ weaver on top of FERRARI such that aspects can be woven also into the JDK classes, thus enabling aspect weaving with full coverage.

Concerning the applicability of AOP to express the required transformations for the CPU accounting mechanism and for certain dynamic analyses, unfortunately mainstream AOP languages and the corresponding aspect weavers (e.g., AspectJ or *abc*) lack some features that are essential to this purpose. Notably, the absence of pointcuts at the level of individual BBCs and bytecode instructions make the use of existing AOP frameworks impracticable.

To cope with these limitations, we developed Domain-Specific Language (DiSL) (for bytecode instrumentation) [53,54], an AOP-inspired language and weaver specially designed to ease the implementation of efficient bytecode instrumentations for dynamic analysis framework, including the instrumentations required to enable resource management. DiSL borrows several principles from AOP and enhances them with an *open join point model*, that is, DiSL allows *any region of bytecode to be used as a join point*. For example, DiSL allows us to capture not only method execution or field access (as supported by mainstream AOP languages, such as AspectJ), but also the execution of specific bytecodes or basic blocks.

DiSL follows the approach shown in Figure 16.7c, that is, the execution of instrumented code and the instrumentation itself are performed in separate JVMs. A tiny client-side DiSL native agent captures every classloading event using a JVMTI callback hook. The agent receives information about every captured class, including the bytecode as a byte array. The agent uses a socket to communicate with the DiSL server JVM to send the original class. The server JVM runs the DiSL weaver, instruments the class according to the DiSL instrumentation description, and sends back the instrumented code to the client JVM, where it is eventually defined. All the logic to run instrumented code and handle bootstrapping issues is added to the woven code by the server. This approach greatly simplifies the issues of instrumentation on the same JVM, notably those related to infinite recursion, classloading perturbations due to classes used by the instrumentation framework itself, and those of bootstrapping with instrumented code.

16.4 Recasting JRAF-2 Instrumentation in DiSL

In this section, we describe how the instrumentation scheme to enable resource management (JRAF-2) can be expressed in DiSL. First, we give an overview of DiSL and its main features related to the proposed recast. For a complete description of all the features, please refer to [53,54]. Then, we show the actual recast of JRAF-2 expressed as a DiSL instrumentation and compare it with a low-level instrumentation based solution.

16.4.1 DiSL Overview

DiSL is an AOP-inspired *domain-specific embedded language* [55], which has Java as its host language. That is, DiSL programs are implemented in Java, and annotations are used to express where and how programs are to be instrumented. DiSL allows *any region of bytecodes to be used as a join point*, thus following an *open join point model*. DiSL provides an extensible mechanism for marking user-defined bytecode regions, thus enabling the definition of new join points. The marked regions correspond to join point *shadows*, commonly used by aspect weavers [39,56]. Shadows are normally not exposed to the aspect developer but rather used internally by the weaver to insert calls to advice methods. In DiSL, however, the shadows are reified and exposed to the developer.

Advice in DiSL are expressed in the form of *code snippets* that are always inlined. DiSL instrumentation aspects—simply called *instrumentations*—describe where and how snippets are to be applied to the base program under instrumentation.

DiSL provides efficient access to both static and dynamic context information of the base program. All static information (including class, method, and bytecode properties) are exposed to the developer. Static properties at the basic block and bytecode level are exposed, too (e.g., basic block index, number of bytecodes in a basic block, or the opcode of a captured bytecode).

The DiSL weaver uses the bytecode abstraction provided by ASM [33], which is used to navigate through different structures efficiently. DiSL is publicly available to download.*

16.4.1.1 Instrumentations, Snippets, and Markers

DiSL *instrumentations* are standard Java classes with DiSL-specific annotations. An instrumentation can only have *snippets* that are static methods annotated with @Before, @After, @AfterReturning, or @AfterThrowing. Snippets are defined as static methods because their body is used as template that is instantiated and inlined at the matching join points in the base program. Snippets do not return any value and must not throw any exception (that is not caught by a handler in the snippet).

---

* http://disl.ow2.org/.

```
public class SampleInstrumentation {

 @Before(marker = BodyMarker.class)
 static void onMethodEntry() {
   ... // snippet body
 }

 @After(marker = BasicBlockMarker.class)
 static void afterBasicBlock() {
   ... // snippet body
 }
}
```

Figure 16.8   Example DiSL instrumentation.

Because of DiSL's open join point model, pointcuts are not hardcoded in the language but defined by an extensible library of *markers*. Markers are standard Java classes implementing a special interface for join point selection. DiSL provides a rich library of markers, including those for method body, basic block, individual bytecode, and exception handler. In addition, the developer may extend existing markers or implement new markers from scratch.

The marker class is specified in the marker attribute in the snippet annotation. The weaver takes care of instantiating the selected marker, matching the corresponding join points, and weaving the snippets. Figure 16.8 shows a sample instrumentation in which different markers are used. The BodyMarker defines a shadow corresponding to the full method body whereas the BasicBlockMarker defines a shadow corresponding to the basic block of code in the method body. The weaver takes care of inlining the corresponding snippet body before the method body and after every single basic block of code.

16.4.1.2 Thread Local Variables

DiSL supports thread-local variables with the @ThreadLocal and @Inheritable ThreadLocal annotations. This mechanism extends java.lang.Thread by inserting the annotated field. While the inserted fields are instance fields, thread-local variables must be declared as static fields in the instrumentation class. These fields must be initialized to the default value of the field's type.* The DiSL weaver translates all access to thread-local variables in snippets into bytecodes that access the correspond-

---

* During JVM bootstrapping, in general, inserted code cannot be executed because it may introduce class dependencies that can violate JVM assumptions concerning the class initialization order. Hence, threads created during bootstrapping could not initialize inserted thread-local fields in the beginning.

ing field of the currently executing thread. The @InheritableThreadLocal annotation ensures that new threads "inherit" the value of a thread-local variable from the creating thread. Note that the standard Java class library offers classes with similar semantics (java.lang.ThreadLocal and java.lang.InheritableThreadLocal). However, accessing fields directly inserted into java.lang.Thread results in more efficient code.

16.4.1.3 Context Information

Snippets have access to complete static context information (i.e., static reflective join point information). This allows, for example, the gathering of information about the method, basic block, or even bytecode instruction that is executed. To this end, snippets can take an arbitrary number of static context references as arguments. Methods in static context classes return constants: primitive values, strings, or class literals. The reason for this restriction is that DiSL stores the results of the static context method directly in the constant pool of the woven class. DiSL provides an extensible library of static context classes commonly used, such as, MethodStaticContext, BasicBlockStaticContext, and BytecodeStaticContext.

16.4.2 JRAF-2 Instrumentation in DiSL

The instrumentation for CPU reification for resource management as described in Section 16.2.2 can be expressed concisely in few lines of DiSL code. Figure 16.9 shows two classes, one corresponding to the runtime class CPUManager (is not instrumented and is used by the instrumented code only) and the second class, JRAF2Instrumentation, corresponding to the DiSL instrumentation.

The instrumentation class defines a single snippet method beforeBasicBlock(), which is marked with the @Before annotation and the BasicBlockMarker for weaving the snipped code before every single basic block of the instrumented method. The snippet receives as argument the BasicBlockStaticContext, which is used to gather context information of every woven basic block of bytecode. A @ThreadLocal counter variable is defined to hold the actual number of executed bytecodes. The counter is updated on every snippet by gathering the size of the corresponding basic block (counter += bbsc.getBBSize()). This information is provided at weave time by the BasicBlockStaticContext instance passed as argument.

Now, as described Section 16.2.2, the instrumentation scheme adds a conditional to certain locations to invoke resource management code so as to avoid executing an infinite number of bytecode instructions without control. These locations are notably the beginning of each method and the beginning of each loop. This is clearly shown in the sample instrumented method of Figure 16.3 in Section 16.2.2. The snippet has a conditional that verifies if the current basic block is the first in a loop (bbsc.isFirstInLoop()), and thanks to the basic block index (bbsc.getBBIndex()), we can identify if the basic block is the first one on the method body (i.e., index equal to zero). The conditional also contains the check

```
// Runtime class (not used for instrumentation)
public class CPUManager {
    public static volatile int THRESHOLD = ...;
    public static void runResourceManagementPolicy(Thread t, int counter) { ... }
    ...
}



// DiSL instrumentation class
public class JRAF2Instrumentation {
    @ThreadLocal
    static int counter = 0;

    @Before(marker = BasicBlockMarker.class)
    static void beforeBasicBlock(BasicBlockStaticContext bbsc) {

        counter += bbsc.getBBSize();
        if((bbsc.isFirstInLoop() || bbsc.getBBIndex()==0 ) &&
            counter >= CPUManager.THRESHOLD) {
            CPUManager.runResourceManagementPolicy(Thread.currentThread(), counter);
            counter = 0;
        }
    }
}
```

Figure 16.9 **Instrumentation for resource reification in DiSL, JRAF-2 recast.**

of whether the counter value has exceeded a given threshold so as to invoke the resource management policy code. Once the management code is invoked, the thread local counter is reset to zero.

Note that first part of the conditional that checks if the basic block is the first of the loop or if it is the very first block of the method does not actually generate any bytecode but is partially evaluated by the DiSL weaver [54]. Thus, the only code that is inlined (and only if the first part of the conditional is evaluated to true during weaving) is the check of the counter against the threshold, together with the corresponding invocation to the management code. Without partial evaluation, every basic block would potentially contain useless code, which is statically known during instrumentation and can therefore be removed.

Compared to the original implementation of JRAF-2, which has about 10,000 lines of code (dealing also with low-level issues of instrumenting the Java class library), the DiSL version is very compact and generates equivalent bytecode. There are small differences between the DiSL recast and the instrumentation scheme described Section 16.2.2, notably in the way to invoke resource management code. We intentionally show this difference to reinforce the argument that high-level instrumentations enable very flexible adaptations while keeping the core instrumentation scheme consistent. For example, changing the resource management policy code with a pluggable interface can be easily done without affecting the instrumentation itself. In the case of the original JRAF-2 low-level instrumentation version, this had required large amount of development effort.

16.5 Related Work

Altering Java semantics via bytecode instrumentation has been used for many purposes. In the following, we present related work on the two main subjects of this chapter, namely resource management in Java and different approaches to bytecode instrumentation, including those based on AOP principles.

16.5.1 Resource Management in Java

Java does not support resource management. Some research has been conducted to consider resource management as an integral part of the Java language. This effort resulted in the resource consumption API [57] as part of MVM [26] and later proposed as a starting point for the Java specification request (JSR-284) [58]. The framework complemented Java with resource consumption management and control with reservation features. The API encapsulated resource policies in *resource domains*, which were part of the *isolate* API [59]. Isolates were introduced as an abstraction of a single Java task running in a VM. This was a very sound base, but it has unfortunately not been released to the general public.

Prevailing approaches to provide resource control in Java-based platforms rely on a modified JVM, on native code libraries, or on program transformations. For instance, the KaffeOS [7], Aroma VM [8], and the MVM [26] are specialized JVMs supporting resource control. JRes [9] is a resource control library for Java, which uses native code for CPU control and rewrites the bytecode of Java programs for memory control. For CPU control, some light bytecode instrumentation was also applied to enable proper cooperation with the OS via native code libraries. This was not bytecode-level accounting as this seemed prohibitive in terms of performance. In JRes, information was obtained by polling the OS about the CPU consumption of threads and therefore requires a JVM with OS-level threads, which is not necessarily always the case.

Our proposal for CPU management is built on the idea of self-accounting. Such an approach solves one important weakness of all existing solutions based on a polling supervisor thread: the fact that the Java specification does not formally guarantee that the supervisor thread will ever be scheduled, whatever its priority is set to. Any resources consumed will be accounted for by the consuming thread itself (provided that the consuming code is implemented in Java and not in some native language), and if required, the thread will eventually take self-correcting measures.

Another advantage is the use of a portable, hardware-independent unit of measurement for CPU consumption (based on the number of executed bytecodes). In distributed applications, this is a clear advantage because we may then envision platform-independent "execution contracts" between heterogeneous hosts as well as the specification of platform-independent schedulers and scheduling policies.

Concerning limitations, the major hurdle of our approach is that it cannot directly account for the execution of native code. Some solutions can be implemented to attenuate this limitation, such as adding wrapper libraries and running some calibration process to evaluate the actual consumption of native system calls [11]. Part of our techniques to ensure full method coverage also includes the use of native method prefixing, offered by the JVMTI [48] in order to wrap native methods with bytecode versions that are amenable to instrumentation.

16.5.2 Bytecode Instrumentation and AOP

Java supports bytecode instrumentation using native code agents through the JVMTI [48] as well as portable bytecode instrumentation through the java.lang. instrument API. Several bytecode instrumentation libraries have been developed, such as BCEL [32], ASM [33], Javassist [34], BIT [22], or Soot [35]. They provide low-level APIs to manipulate bytecode instructions and are used in a wide range of research areas. While program manipulation at the bytecode level is very flexible, because the possible bytecode transformations are not restricted, tool development is tedious and error-prone, often requiring high development and testing effort. Specifying bytecode instrumentation at a higher level using AOP [30] is a promising alternative in order to reduce tool development time and cost.

AspectJ [47] is the de-facto standard AOP language and framework for Java. The AspectJ weaver is based on bytecode instrumentation to weave aspects into application code [56]. The weaver supports both offline and dynamic weaving through the java.lang.instrument API. Several researchers have explored the use of AspectJ and other AOP languages to simplify the development of profilers [50], debuggers [43], or testing tools [51], thus using AOP as a high-level, flexible, and more general-purpose instrumentation mechanism. In order to allow developers to simplify extensions to AspectJ, the AspectBench Compiler (*abc*) [39] uses extensible compiler techniques. For example, it allows developers to define new pointcuts [60–62] through extensible front-end and back-end.

The use of AOP has been also explored as an approach to enable runtime monitoring [63]. Because AspectJ lacks low-level pointcuts to capture enough details to cover all monitoring needs (e.g., weaving of statements, BBCs, loops, or local variable accesses), the authors present two new pointcuts to intercept basic blocks and loops. Unfortunately, no concrete use case is described. The extension is implemented with *abc* and supports only minimal context information. In contrast, DiSL enables support to capture arbitrary regions of bytecodes, thanks to its open join point model and provides the extensible library of markers to implement monitoring, profiling, and resource management tools.

One of the main limitations of the AspectJ and *abc* weavers is the impossibility of weaving aspects in the Java class library (JDK classes). This strongly restricts the applicability of AOP for building tools that require instrumentation with full method coverage, such as profilers and resource management frameworks. Our work on bringing the standard AspectJ weaver on top of the FERRARI framework

[52] and DiSL's support for full method coverage [53] can be considered a step toward bringing the full potential of AOP to develop a wide range of new tools, which otherwise would require larger development effort.

16.6 Conclusion

In this book chapter, we presented a mechanism for indirectly controlling CPU consumption in Java applications by counting and limiting the number of byte-codes executed by the threads. This approach is fully portable, works with any standard JVM, and allows for pluggable custom CPU management policies. Our system is a valuable, low-level building block for creating resource-aware applications, that is, applications that can adapt their behavior depending on their resource consumption.

We first presented an initial implementation of our approach, called JRAF-2, using low-level bytecode instrumentation. The system is complex as it has to deal with the complications of instrumenting also the JDK. Thanks to our high-level bytecode instrumentation framework DiSL, which is built on top of AOP principles, we were able to recast JRAF-2 in a few lines of code. The DiSL source code corresponding to JRAF-2 as presented here is directly usable in practice and offers the same behavior as the initial version of JRAF-2. The DiSL-based version of JRAF-2 offers a convenient starting point for other researchers to explore resource management on the Java platform.

Acknowledgments

We would like to thank Jarle Hulaas for his contributions on the JRAF-2 framework, to Philippe Moret for his work on the FERRARI framework, to Danilo Ansaloni for his contribution on bringing AspectJ on top of FERRARI, to Lukas Marek and Yudi Zheng for their work on the design and implementation of DiSL, and to all the people that used our tools.

References

1. V. Poladian, J. P. Sousa, D. Garlan, and M. Shaw. Dynamic Configuration of Resource-Aware Services. In *Proceedings of the 26th International Conference on Software Engineering (ICSE '04)*. IEEE Computer Society, Washington, DC, USA, Pages 604–613, 2004.
2. M. Jones, D. Rosu, and M. Rosu. CPU Reservations and Time Constraints: Efficient, Predictable Scheduling of Independent Activities. *Proceedings of 16th ACM Symposium on Operating Systems Principles (SOSP)*, 1997.

3. J. Dongarra, K. London, S. Moore, P. Mucci, D. Terpstra, H. You, and M. Zhou. Experiences and Lessons Learned with a Portable Interface to Hardware Performance Counters, *PADTAD Workshop, IPDPS 2003*, Nice, France, April 26, 2003.

4. T. Fahringer, A. Jugravu, S. Pllana, R. Prodan, C. Seragiotto Jr., and H. L. Truong. ASKALON: A Tool Set for Cluster and Grid Computing. *Concurrency and Computation: Practice and Experience*, Volume 17, Pages 143–169, Wiley InterScience, 2005.

5. S. Browne, J. Dongarra, N. Garner, K. London, and P. Mucci. A Scalable Cross-Platform Infrastructure for Application Performance Tuning Using Hardware Counters, *Proceedings of SuperComputing 2000 (SC '00)*, Dallas, November 2000.

6. M. Johnson, H. McCraw, S. Moore, P. Mucci, J. Nelson, D. Terpstra, V. Weaver, and T. Mohan. PAPI-V: Performance Monitoring for Virtual Machines, *CloudTech-HPC 2012*, Pittsburgh, PA, September 10–13, 2012.

7. G. Back, W. Hsieh, and J. Lepreau. Processes in KaffeOS: Isolation, Resource Management, and Sharing in Java. In *Proceedings of the Fourth Symposium on Operating Systems Design and Implementation (OSDI '2000)*, San Diego, CA, USA, October 2000.

8. N. Suri, J. M. Bradshaw, M. R. Breedy, P. T. Groth, G. A. Hill, R. Jeffers, T. S. Mitrovich, B. R. Pouliot, and D. S. Smith. NOMADS: Toward a Strong and Safe Mobile Agent System. In C. Sierra, G. Maria, and J. S. Rosenschein, editors, *Proceedings of the 4th International Conference on Autonomous Agents (AGENTS-00)*, Pages 163–164, NY, June 3–7, 2000. ACM Press.

9. G. Czajkowski, and T. von Eicken. JRes: A Resource Accounting Interface for Java. In *Proceedings of the 13th Conference on Object-Oriented Programming, Systems, Languages, and Applications (OOPSLA-98)*, Volume 33, 10 of ACM SIGPLAN Notices, Pages 21–35, New York, October 18–22, 1998. ACM Press.

10. W. Binder, J. G. Hulaas, and A. Villazón. Portable Resource Control in Java. *ACM SIGPLAN Notices*, Volume 36, Issue 11, Pages 139–155, November 2001. Proceedings of the 2001 ACM SIGPLAN Conference on Object Oriented Programming, Systems, Languages and Applications (OOPSLA '01).

11. W. Binder, and J. Hulaas. A Portable CPU-Management Framework for Java. *IEEE Internet Computing*, Volume 8, Issue 5, Pages 74–83, September/October 2004.

12. T. Lindholm, and F. Yellin. *The Java Virtual Machine Specification*. Addison-Wesley, Reading, MA, USA, second edition, 1999.

13. The Tamarin JavaScript Engine. Web pages at https://developer.mozilla.org/Tamarin/.

14. The V8 JavaScript Engine. Web pages at http://code.google.com/p/v8/.

15. The Dalvik VM. Web pages at http://code.google.com/p/dalvik/.

16. The .Net Framework. Web pages at http://www.microsoft.com/net/.

17. The Mono Project. Cross-platform, open source, .Net development framework. Web pages at http://www.mono-project.com/.

18. M. Odersky, L. Spoon, and B. Venners. *Programming in Scala*. Artima Press, 2nd edition, 2010.

19. The Ceylon Language. Web pages at http://ceylon-lang.org/.

20. Groovy: A dynamic language for the Java platform. Web pages at http://groovy.codehaus.org/.

21. Clojure. Web pages at http://clojure.org/.

22. H. B. Lee, and B. G. Zorn. BIT: A Tool for Instrumenting Java Bytecodes. In *Proceedings of the USENIX Symposium on Internet Technologies and Systems (ITS-97)*, Pages 73–82, Berkeley, December 8–11, 1997.

23. B. Dufour, L. Hendren, and C. Verbrugge. *J: A Tool for Dynamic Analysis of Java Programs. In *OOPSLA '03: Companion of the 18th Annual ACM SIGPLAN Conference on Object-Oriented Programming, Systems, Languages, and Applications*, Pages 306–307, New York, 2003. ACM Press.

24. S. Artzi, S. Kim, and M. D. Ernst. ReCrash: Making Software Failures Reproducible by Preserving Object States. In J. Vitek, editor, *ECOOP '08: Proceedings of the 22th European Conference on Object-Oriented Programming*, Volume 5142 of *Lecture Notes in Computer Science*, Pages 542–565, Paphos, Cyprus, 2008. Springer-Verlag.

25. R. Lencevicius, U. Hölzle, and A. K. Singh. Dynamic Query-Based Debugging of Object-Oriented Programs. *Automated Software Engineering Journal*, Volume 10, Issue 1, Pages 39–74, 2003.

26. G. Czajkowski, and L. Daynès. Multitasking Without Compromise: A Virtual Machine Evolution. In *ACM Conference on Object-Oriented Programming, Systems, Languages, and Applications (OOPSLA '01)*, Tampa Bay, FL, October 2001.

27. W. Binder, and B. Lichtl. Using a Secure Mobile Object Kernel as Operating System on Embedded Devices to Support the Dynamic Upload of Applications. *Lecture Notes in Computer Science*, Volume 2535, 2002.

28. B. Dufour, K. Driesen, L. Hendren, and C. Verbrugge. Dynamic Metrics for Java. *ACM SIGPLAN Notices*, Volume 38, Issue 11, Pages 149–168, November 2003.

29. A. Villazón, and W. Binder. Portable Resource Reification in Java-Based Mobile Agent Systems. In *Fifth IEEE International Conference on Mobile Agents (MA-2001)*, Atlanta, GA, USA, December 2001.

30. G. Kiczales, J. Lamping, A. Menhdhekar, C. Maeda, C. Lopes, J.-M. Loingtier, and J. Irwin. Aspect-Oriented Programming. In M. Akşit and S. Matsuoka, editors, *Proceedings of European Conference on Object-Oriented Programming*, Volume 1241, Pages 220–242. Springer-Verlag, Berlin, Heidelberg, and New York, 1997.

31. K. O'Hair. BCI (Bytecode Instrumentation). https://weblogs.java.net/blog/kellyohair/archive/2005/05/bytecode_instru.html.

32. M. Dahm. Bytecode engineering. In *Java-Information-Tage 1999 (JIT '99)*, September 1999. http://commons.apache.org/bcel/.

33. The ASM Project. Web pages at http://asm.ow2.org/.

34. S. Chiba. Load-Time Structural Reflection in Java. In *Proceedings of the 14th European Conference on Object-Oriented Programming (ECOOP '2000)*, Volume 1850 of *Lecture Notes in Computer Science*, Pages 313–336. Springer Verlag, Cannes, France, June 2000.

35. R. Vallée-Rai, E. Gagnon, L. J. Hendren, P. Lam, P. Pominville, and V. Sundaresan. Optimizing Java Bytecode Using the Soot Framework: Is It Feasible? In *Compiler Construction, 9th International Conference (CC 2000)*, Pages 18–34, 2000.

36. C. Bernardeschi, N. De Francesco, G. Lettieri, and L. Martini. Checking Secure Information Flow in Java Bytecode by Code Transformation and Standard Bytecode Verification. *Software Practice and Experience*. Volume 34, Issue 13, November, 2004.

37. A. Chander, J. C. Mitchell, and I. Shin. Mobile Code Security by Java Bytecode Instrumentation. In *2001 DARPA Information Survivability Conference & Exposition (DISCEX)*, 2001.

38. M. Batchelder, and L. Hendren. Obfuscating Java: The most pain for the least gain. In *Proceedings of the 16th International Conference on Compiler Construction (CC '07)*, S. Krishnamurthi and M. Odersky (Eds.). Springer-Verlag, Berlin, Heidelberg, Pages 96–110. 2007.

39. P. Avgustinov, A. S. Christensen, L. J. Hendren, S. Kuzins, J. Lhoták, O. Lhoták, O. de Moor et al., abc: An Extensible AspectJ Compiler. In *AOSD '05: Proceedings of the 4th International Conference on Aspect-Oriented Software Development*, pages 87–98, New York, 2005. ACM Press.

40. P. McGachey, A. L. Hosking, J. Eliot, and B. Moss. Pervasive Load-Time Transformation for Transparently Distributed Java. *Electronic Notes in Theoretical Computer Science (ENTCS)*. Volume 253, Issue 5, December, 2009.

41. E. Tilevich, and Y. Smaragdakis. J-Orchestra: Enhancing Java Programs with Distribution Capabilities. *ACM Transactions on Software Engineering Methodology*. Volume 19, Number 1, Article 1, August, 2009.

42. D. Weyns, E. Truyen, and P. Verbaeten. Distributed Threads in Java. In *Proceedings of the International Symposium on Parallel and Distributed Computing (ISPDC 2002)*, 2002.

43. L. D. Benavides, R. Douence, and M. Südholt. Debugging and Testing Middleware with Aspect-Based Control-Flow and Causal Patterns. In *Middleware '08: Proceedings of the 9th ACM/IFIP/USENIX International Conference on Middleware*, Pages 183–202, New York, 2008. Springer-Verlag New York, Inc.

44. M. Dmitriev. Profiling Java Applications Using Code Hotswapping and Dynamic Call Graph Revelation. In *WOSP '04: Proceedings of the Fourth International Workshop on Software and Performance*, pages 139–150. ACM Press, 2004.

45. W. Binder, J. Hulaas, P. Moret, and A. Villazón. Platform-Independent Profiling in a Virtual Execution Environment. *Software: Practice and Experience*, Volume 39, Issue 1, Pages 47–79, 2009. http://dx.doi.org/10.1002/spe.890.

46. F. Chen, T. F. Serbanuta, and G. Rosu. JPredictor: A Predictive Runtime Analysis Tool for Java. In *ICSE '08: Proceedings of the 30th International Conference on Software Engineering*, Pages 221–230, New York, 2008. ACM Press.

47. G. Kiczales, E. Hilsdale, J. Hugunin, M. Kersten, J. Palm, and W. G. Griswold. An Overview of AspectJ. In J. L. Knudsen, editor, *Proceedings of the 15th European Conference on Object-Oriented Programming (ECOOP-2001)*, Volume 2072 of *Lecture Notes in Computer Science*, Pages 327–353, 2001.

48. Oracle Corporation. JVM Tool Interface (JVMTI) version 1.2. Web pages at http://docs.oracle.com/javase/6/docs/platform/jvmti/jvmti.html, 2007.

49. W. Binder, J. Hulaas, and P. Moret. Advanced Java Bytecode Instrumentation. In *PPPJ '07: Proceedings of the 5th International Symposium on Principles and Practice of Programming in Java*, Pages 135–144, New York, 2007. ACM Press.

50. D. J. Pearce, M. Webster, R. Berry, and P. H. J. Kelly. Profiling with AspectJ. *Software: Practice and Experience*, Volume 37, Issue 7, Pages 747– 777, June 2007.

51. A. Villazón, W. Binder, and P. Moret. Flexible Calling Context Reification for Aspect-Oriented Programming. In *AOSD '09: Proceedings of the 8th International Conference on Aspect-oriented Software Development*, Pages 63–74, Charlottesville, VA, USA, March 2009. ACM Press.

52. A. Villazón, W. Binder, and P. Moret. Aspect Weaving in Standard Java Class Libraries. In *PPPJ '08: Proceedings of the 6th International Symposium on Principles and Practice of Programming in Java*, Pages 159–167, New York, September 2008. ACM Press.

53. L. Marek, A. Villazón, Y. Zheng, D. Ansaloni, W. Binder, and Z. Qi. DiSL: A Domain-Specific Language for Bytecode Instrumentation. In *Proceedings of the 11th International Conference on Aspect-Oriented Software Development (AOSD)*, 2012.

54. Y. Zheng, D. Ansaloni, L. Marek, A. Sewe, W. Binder, A. Villazón, P. Tuma, Z. Qi, and M. Mezini. Turbo DiSL: Partial Evaluation for High-level Bytecode Instrumentation. *50th International Conference on Objects, Models, Components, Patterns (TOOLS-2012)*, Prague, Czech Republic, May 2012, *Lecture Notes in Computer Science (LNCS)*, Volume 7304, Pages 353–368.

55. P. Hudak. Modular Domain Specific Languages and Tools. In *Proceedings of Fifth International Conference on Software Reuse*, Pages 134–142. IEEE Computer Society Press, 1998.

56. E. Hilsdale, and J. Hugunin. Advice Weaving in AspectJ. In *AOSD '04: Proceedings of the 3rd International Conference on Aspect-Oriented Software Development*, Pages 26–35. ACM Press, 2004.

57. G. Czajkowski, S. Hahn, G. Skinner, P. Soper, and C. Bryce. *A Resource Management Interface for the Java™ Platform*. Technical Report. Sun Microsystems, Inc., Mountain View, CA, USA, 2003.

58. Java Community Process. JSR-284: Resource Consumption Management API, http://www.jcp.org/en/jsr/detail?id=284.

59. Java Community Process. JSR-121: The Application Isolation API Specification, Java Community Process, http://www.jcp.org/en/jsr/detail?id=121.

60. E. Bodden, and K. Havelund. Racer: Effective Race Detection Using AspectJ. In *International Symposium on Software Testing and Analysis (ISSTA)*, Seattle, WA, July 20–24, 2008, Pages 155–165, New York, July 2008. ACM Press.

61. B. De Fraine, M. Südholt, and V. Jonckers. StrongAspectJ: Flexible and Safe Pointcut/Advice Bindings. In *AOSD '08: Proceedings of the 7th International Conference on Aspect-Oriented Software Development*, Pages 60–71, New York, 2008. ACM Press.

62. S. Akai, S. Chiba, and M. Nishizawa. Region Pointcut for AspectJ. In *ACP4IS '09: Proceedings of the 8th Workshop on Aspects, Components, and Patterns for Infrastructure Software*, Pages 43–48, New York, 2009. ACM Press.

63. J. Cook, and A. Nusayr. Using AOP for Detailed Runtime Monitoring Instrumentation. In *WODA 2008: The Sixth International Workshop on Dynamic Analysis*, New York, July 2009. ACM Press.

Index

Page numbers followed by f and t indicate figures and tables, respectively.